Lecture Notes in Mathematics

Edited by A. Dold, B. Eckmann and F. Takens

1431

J.G. Heywood K. Masuda
R. Rautmann V.A. Solonnikov (Eds.)

The Navier-Stokes Equations Theory and Numerical Methods

Proceedings of a Conference
held at Oberwolfach, FRG, Sept. 18–24, 1988

Springer-Verlag
Berlin Heidelberg New York London
Paris Tokyo Hong Kong Barcelona

Editors

John G. Heywood
Dept. of Mathematics, University of British Columbia
121–1984 Mathematics Rd., Vancouver B.C. V6T 1Y4, Canada

Kyûya Masuda
Dept. of Mathematics, University of Tokyo
Hongo, Tokyo 113, Japan

Reimund Rautmann
Dept. of Mathematics, University of Paderborn
Warburger Str. 100, 4790 Paderborn, Germany

Vsevolod A. Solonnikov
Mathematical Institute of the Academy of Sciences of the USSR
Fontanka 27, Leningrad 191011, USSR

Mathematics Subject Classification (1980): 35Q10, 76D05; 35A05, 35A07, 35A35, 35A40, 31B20, 35B25, 35B30, 35B32, 35B35, 35B40, 35B45, 35B65, 35C15, 35K22, 35R35, 45D05, 45L05, 45L10, 65J15, 65M25, 65M30, 65N30, 65R20, 68J05, 76D05, 76D10, 76E15, 76E30, 76N10, 76N15, 76R10, 76V05, 76Z10, 82A50

ISBN 3-540-52770-2 Springer-Verlag Berlin Heidelberg New York
ISBN 0-387-52770-2 Springer-Verlag New York Berlin Heidelberg

Printing and binding: Druckhaus Beltz, Hemsbach/Bergstr.
2146/3140-543210 – Printed on acid-free paper

This volume is dedicated

to the memory of

Professor Konstantin I. Babenko
(† 1987)

and

Professor Lamberto Cattabriga
(† 1989)

Preface

The organizers of this first conference on the Navier-Stokes equations at Oberwolfach were J.G. Heywood, K. Masuda, R.Rautmann , and V.A. Solonnikov. Forty-four participants (including 7 from Japan and 9 from the USSR) discussed new results which were presented in 43 lectures, additional talks, and movies.

These results concerned many aspects of the Navier-Stokes theory, including problems in unbounded domains, free boundary problems, problems for compressible and visco-elastic fluids, the theory of stability and attractors, and recent advances in numerical methods of flow computation based on new theoretical approaches.

The organizers thank Professor Barner and the members of the Forschungsinstitut Oberwolfach for making this meeting possible. The bringing together of various groups, from different directions in mathematics and its applications, all joined together by a common interest in the Navier-Stokes equations, was extremely interesting and will surely be fruitful in stimulating further progress. The 'Oberwolfach Atmosphere' contributed greatly to the success of the conference.

Vancouver, Tokyo, Paderborn, and Leningrad, September 1989.

J.G. Heywood, K. Masuda, R. Rautmann, V.A. Solonnikov.

CONTENTS

Open Problems in the Theory of the Navier–Stokes Equations for Viscous Incompressible Flow

JOHN G. HEYWOOD
University of British Columbia
Vancouver, Canada

The Navier-Stokes equations occupy a central position in the study of nonlinear partial differential equations, dynamical systems, and modern scientific computation, as well as classical fluid dynamics. Because of the complexity and variety of fluid dynamical phenomena, and the simplicity and exactitude of the governing equations, a very special depth and beauty is expected in the mathematical theory. Thus, it is a source of pleasure and fascination that many of the most important questions in the theory remain yet to be answered, and seem certain to stimulate contributions of depth, originality and influence, far into the future. It is the purpose of this article to discuss some of these questions. Many others are described elsewhere in these proceedings. The variety of these questions and of other present researches, too numerous to be fully represented here, suggests a vast and growing range of interesting problems for the future.

Global existence. Without doubt, today's premiere problem in this subject area, brought to prominence by early works of Leray [77], Hopf [56], Kiselev and Ladyzhenskaya [66], and Lions and Prodi [78], is to show that a solution, which is smooth at one instant of time, cannot develop a singularity at a later time. From this would follow the global existence of smooth solutions. It should be emphasized that what is needed now is not necessarily a new existence theorem, in the sense of a new construction of the solution, or new treatment of the functional analysis, but only *an appropriate a priori estimate for smooth solutions*, to serve in a continuation argument via well known local existence theorems. For this, we seem to need a deeper understanding of the physics of fluid motion.

To be precise in describing what is needed, consider the initial boundary value problem for flow in a smoothly bounded three dimensional domain Ω :

$$(1) \qquad \boldsymbol{u}_t + \boldsymbol{u}\cdot\nabla\boldsymbol{u} = -\nabla p + \Delta\boldsymbol{u} + f \quad and \quad \nabla\cdot\boldsymbol{u} = 0\,, \quad for \;\; x\in\Omega,\;\; t>0,$$

$$(2) \qquad \boldsymbol{u}(x,0) = \boldsymbol{u}_0(x) \quad for \;\; x\in\Omega, \qquad and \qquad \boldsymbol{u}(x,t) = b(x,t) \quad for \;\; x\in\partial\Omega,\;\; t>0.$$

For simplicity, assume that the prescribed boundary values b and forces f are zero, and that the initial velocity $\boldsymbol{u}_0$ is smooth and solenoidal, and vanishes on $\partial\Omega$. Then, by a well known local existence theorem, there exists a fully classical solution $\boldsymbol{u}, p \in C^\infty(\overline{\Omega}\times(0,T))$, on a time interval $[0,T)$, with T bounded below in terms of the Dirichlet norm $\|\nabla \boldsymbol{u}_o\|$ of the initial data, i.e., the $L^2(\Omega)$-norm of $\nabla \boldsymbol{u}_o$. A simple proof of this is given in [44], [45]; important prior results on classical solutions were given by Ito [59], Fujita and Kato [34], Solonnikov [96], Ladyzhenskaya [72], and others. It follows that to infer the global existence of smooth solutions, it is enough to obtain an estimate for $sup_{t\geq 0}\|\nabla \boldsymbol{u}(t)\|$, *in terms of the data, for solutions which may be assumed to be smooth.* Alternatively, it would suffice to find a bound for $sup_{\Omega\times[0,\infty)}|\boldsymbol{u}|$, or for any one of certain other quantities.

It is easy to give the flavor of known estimates, for either two or three dimensional domains, by referring to the energy identity

$$\frac{1}{2}\|\boldsymbol{u}(s)\|^2 + \int_0^s \|\nabla \boldsymbol{u}\|^2\,dt = \frac{1}{2}\|\boldsymbol{u}_o\|^2, \tag{3}$$

together with the differential inequality

$$\frac{d}{dt}\|\nabla \boldsymbol{u}\|^2 \leq \begin{cases} c\,\|\nabla \boldsymbol{u}\|^4, & \text{for } n=2, \\ c\,\|\nabla \boldsymbol{u}\|^6, & \text{for } n=3. \end{cases} \tag{4}$$

The latter, due to Prodi [89], is proved using Sobolev's inequality and elliptic estimates for the steady Stokes equations, after taking the $L^2(\Omega)$-inner-product of (1) with $P\Delta\boldsymbol{u}$, where P denotes L^2-projection onto the L^2-completion of the solenoidal test functions. For a hypothetically singular solution $\boldsymbol{u}$, with first singularity at some time t^*, one can plot $\|\nabla \boldsymbol{u}(t)\|^2$ in comparison with solutions of the differential equation that corresponds to the differential inequality (4), i.e., $\varphi'(t) = c\,\varphi^2(t)$, if $n=2$, or $\varphi'(t) = c\,\varphi^3(t)$, if $n=3$:

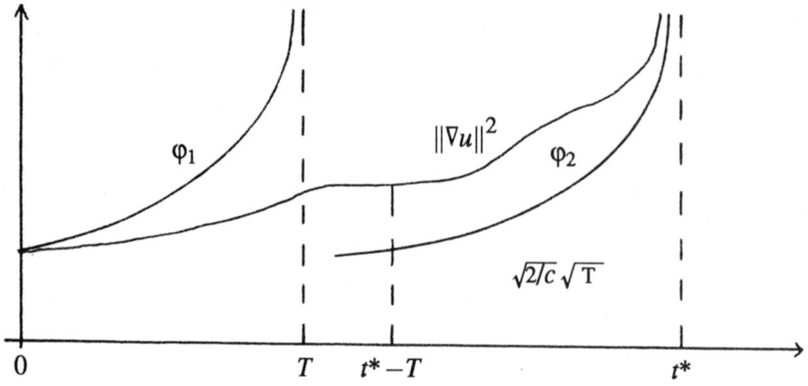

The solution φ_1, chosen to satisfy $\varphi_1(0) = \|\nabla \boldsymbol{u}_o\|^2$, is continuous and bounds $\|\nabla \boldsymbol{u}(t)\|^2$ on an interval $[0,T)$. The graph of the solution φ_2, chosen to have its singularity at t^*, lies below the graph of $\|\nabla \boldsymbol{u}(t)\|^2$, so that the integral of φ_2 over the interval (t^*-T, t^*) is less than or equal to that of $\|\nabla \boldsymbol{u}(t)\|^2$. By a simple calculation, one sees that this integral of φ_2 is infinite if $n = 2$, and equal to $\sqrt{2/c}\sqrt{T}$ if $n = 3$. Thus, in view of (3), a first singularity is impossible in the two dimensional case, and impossible in the three dimensional case if $\frac{1}{2}\|u_0\|^2 < \sqrt{2/c}\sqrt{T}$. In these cases the development of a singularity requires more than the available energy; and a simple modification of the argument provides an explicit bound for $\|\nabla u(t)\|$, in terms of the data, for all $t \geq 0$. For several other variants of the basic estimates see Serrin [94] and Kato [64].

Lest one be inclined to simply assume the existence of smooth solutions, it should be pointed out that any explicit estimate for the continuous dependence of solutions, *on the data*, seems to require the same type of a priori estimate as a global existence theorem. Similarly, taking a computational viewpoint, one may consider that any fluid motion is a combination of motions on different scales: big eddies, small eddies, extremely small eddies, and that the motion on each scale affects that on the others. While numerical computations can be justified in principle, by simply assuming that only a small amount of energy passes into the very small scales of motion (i.e., by assuming regularity), any actual determination of how small the mesh size must be, requires again this same type of a priori estimate.

The global estimate for two dimensional flow is based on dimensional analysis, in the application of Sobolev's inequality in deriving (4). As this does not suffice in three dimensions, it is natural to seek more detailed information about the way energy moves through the spectrum. In this regard, there are interesting recent results on the decay of solutions, as $t \to \infty$. Schonbek [92] has proven, for $\Omega = R^3$ and $u_0 \in L^2 \cap L^1$, that the L^2-norm $\|u(t)\|$ tends to zero as $t \to \infty$, by showing that the energy does not go too low in the spectrum as the solution decays. Foias and Saut [31] considered bounded domains, giving among other things a new proof that smooth solutions cannot come to rest in finite time, by proving that the energy does not go too high in the spectrum as the solution decays.

Another observation which can be made, following a remark in [44, p. 32], is that in a developing singularity, energy cannot go directly from low in the spectrum to high in the spectrum, bypassing the mid ranges. For simplicity, suppose that the initial energy is in relatively low modes; say $P_M u_0 = u_0$, where P_M denotes L^2-projection onto the first M eigenfunctions of the Stokes operator $P\Delta$. Let $E = \|u_0\|^2$, and $A = \lambda_M E$, where λ_M is the Mth eigenvalue of the Stokes operator (thus, $\|\nabla u_0\|^2 \leq A$). *Then there exists* N, *such that* $\|\nabla u(t)\|^2 < 2A$ *for all* $t \geq 0$, *if* $(P_N - P_M)u(t) = 0$ *for all* $t \geq 0$. This is a consequence of a well known estimate for the regularity of solutions, frequently used in [47-51]. Namely, for any $A, T > 0$, there exists $F > 0$, such that for any $t > T$, $sup_{[0,t)}\|\nabla u\|^2 \leq 2A$ implies $sup_{[T,t)}\|P\Delta u\|^2 < F$. Here, we take A as above, and choose T to be the time at which the solution of $\varphi' = c\varphi^3$, satisfying $\varphi(0) = A$, reaches the value $2A$. Choosing N to satisfy $\lambda_{N+1}^{-1} F < A$, let t^* be the first time such that $\|\nabla u(t^*)\|^2 = 2A$. Then, $\|\nabla u(t^*)\|^2 < \|\nabla P_M u(t^*)\|^2 + \|\nabla(I-P_N)u(t^*)\|^2 \leq \lambda_M E + \lambda_{N+1}^{-1}\|P\Delta u(t^*)\|^2 < 2A$.

Qualitative properties of steady solutions in unbounded domains. Frequently, existence theorems provide solutions without giving much information about their properties. Consider the "exterior stationary problem" for flow past an obstacle. One seeks a solution $\boldsymbol{u}(x)$, $p(x)$, in the open complement Ω of a smoothly bounded set Ω^c, tending to a prescribed constant limit $\boldsymbol{b}_\infty$ at infinity:

(5) $$\boldsymbol{u}\cdot\nabla\boldsymbol{u} = -\nabla p + \Delta\boldsymbol{u} \quad \textit{and} \quad \nabla\cdot\boldsymbol{u} = 0 \,, \quad \textit{for } x\in\Omega,$$

(6) $$\boldsymbol{u}(x) = 0 \;\; \textit{for } x\in\partial\Omega \,, \qquad \boldsymbol{u}(x)\to\boldsymbol{b}_\infty \;\text{ as } |x|\to\infty.$$

Leray [76] proved the existence of a solution, in the three dimensional case, by a topological argument based on an a priori estimate for the solution's Dirichlet norm $\|\nabla\boldsymbol{u}\|^2$. The argument was completed, and improved and simplified, by Finn [24], Ladyzhenskaya [69], Fujita [33], and the author [44]. Yet it gives very little information about the solution. Is the solution unique if $|\boldsymbol{b}_\infty|$ is small? Is it stable? Is the wake energy $\frac{1}{2}\|\boldsymbol{u}-\boldsymbol{b}_\infty\|^2$ finite, or infinite? Does the net force $\boldsymbol{F}$ on the obstacle satisfy the expected energy relation, $\boldsymbol{F}\cdot\boldsymbol{b}_\infty = \|D\boldsymbol{u}\|^2$, where the right side is the rate of energy dissipation due to the deformation $D\boldsymbol{u} \equiv \frac{1}{2}(\partial u_i/\partial x_j + \partial u_j/\partial x_i)$? Does the solution possess a reasonable downstream wake structure? These questions were taken up and largely answered by Finn in a profound series of works [24-28]. Finn showed that solutions which decay like $|x|^{-\alpha}$, as $|x|\to\infty$, for some $\alpha > \frac{1}{2}$, allow a potential theoretic representation in terms of the fundamental solution of Oseen's linearization about $\boldsymbol{b}_\infty$. He showed that this implies the qualitative properties he sought, in particular the existence of a parabolic wake structure, and he gave a new existence theorem for such solutions using potential theoretic methods. Subsequently, in another of the great works on the subject, Babenko [6] succeeded in showing that every solution of the exterior stationary problem which possesses a finite Dirichlet norm, in particular Leray's solution, decays like $|x|^{-\alpha}$ in the manner specified above, and thus enjoys the properties for such solutions obtained by Finn.

Another very notable success in the pursuit of qualitative properties is contained in the works of Amick and Fraenkel on steady flow through pipes and two dimensional channels, [1], [3], [32]. However, generally speaking, rigorous qualitative results are lacking for most problems. For example, the existence of a steady jet of prescribed net flux through a hole in a wall was proven in [43], [74]. Yet, very little is known about the solution, in particular about its uniqueness, stability, and decay as $|x|\to\infty$. The existence theorem provides a classical solution with a bounded Dirichlet norm. But, a mere bound for the Dirichlet norm implies only very weak results about the solution's decay as $|x|\to\infty$. While a decay rate like $\boldsymbol{u}\sim|x|^{-2}$ would be consistent with the condition $\nabla\cdot\boldsymbol{u} = 0$, suggesting that probably $\boldsymbol{u}\in L^2(\Omega)$, and almost certainly $\boldsymbol{u}\in L^3(\Omega)$, the relevant three dimensional Sobolev inequality is $\|\boldsymbol{u}\|_{L^6(\Omega)} \le c\,\|\nabla \boldsymbol{u}\|$. A bound for the L^6-norm does not suffice even for proving the uniqueness and stability of the solution which is expected when the prescribed flux is very small. Indeed, the difference $\boldsymbol{w} = \boldsymbol{v} - \boldsymbol{u}$ of two solutions satisfies $\boldsymbol{w}\cdot\nabla\boldsymbol{u} + \boldsymbol{v}\cdot\nabla\boldsymbol{w} = -\nabla q + \Delta\boldsymbol{w}$. If one knew that solutions belonged to $L^3(\Omega)$, one could justify multiplying by $\boldsymbol{w}$ and integrating over Ω , obtaining $\|\nabla\boldsymbol{w}\|^2 = -(\nabla q + \boldsymbol{w}\cdot\nabla\boldsymbol{u} + \boldsymbol{v}\cdot\nabla\boldsymbol{w}\,,\,\boldsymbol{w}) \le c\,\|\boldsymbol{u}\|_{L^3(\Omega)}\|\nabla\boldsymbol{w}\|^2$, and thus uniqueness for small data. But, it is not known whether solutions of the problem belong to $L^3(\Omega)$.

Steady flow in unbounded domains seems to present even greater difficulty in two dimensions than three. This is due mainly to the fact that for functions defined in R^2, a finite Dirichlet norm does not imply the existence of a generalized limit at infinity, as it does in three dimensions. The basic function space used in treating steady problems is the completion $J_0(\Omega)$ of the solenoidal test functions in the Dirichlet norm. In three dimensions, elements of $J_0(\Omega)$ tend to zero at infinity. But in two dimensions they can tend to nonzero limits at infinity, or even grow at infinity; see [42]. Consider the implications of this for Leray's method of solving the exterior stationary problem (5), (6). It consists of first finding solutions $\boldsymbol{u}_R$ in annular regions $\Omega_R = \{x \in \Omega : |x| < R\}$, satisfying $\boldsymbol{u}_R = \boldsymbol{b}_\infty$ on the outer boundary, and then passing to a limit, $\boldsymbol{u}_R(x) \to \boldsymbol{u}(x)$, in bounded subregions, as $R \to \infty$. A limit $\boldsymbol{u}$ exists (for a sequence $R_i \to \infty$) and satisfies the Navier-Stokes equations, in either two or three dimensions. And in either case, $\boldsymbol{u}(x) - \boldsymbol{b}_\infty$ behaves like an element of $J_0(\Omega)$ at infinity. While this can be used to show that $\boldsymbol{u}(x) \to \boldsymbol{b}_\infty$ at infinity in the three dimensional case, the question is still unresolved in two dimensions. The situation can be compared with that of trying to construct a harmonic function which vanishes on the unit circle and tends to the value one at infinity. The functions $u_R(x) = \log x / \log R$ are harmonic, zero on the unit circle, and equal to one on the circle of radius R. As $R \to \infty$, they tend in every bounded region to the harmonic limit $u(x) \equiv 0$, losing the desired limit at infinity. Of course, there is no harmonic function which vanishes on the unit circle and tends to a nonzero limit at infinity. Similarly, there is no solution of the linear Stokes equations $\Delta\boldsymbol{u} - \nabla p = 0$, $\nabla\cdot\boldsymbol{u} = 0$, which vanishes on the unit circle and tends to a nonzero limit at infinity (that's Stokes' paradox; see [19], [42]). But might it be that the full nonlinear Navier-Stokes equations do possess such solutions? In fact it was shown by Finn and Smith [29] that they do, at least for small data, by an argument completely different (and far more involved) than Leray's. The question of whether Leray's construction works remains a very interesting one, with obvious practical implications. Amick [2] has recently obtained a beautiful argument, drawing upon ideas of Gilbarg and Weinberger [37], showing that the solution obtained by Leray's construction does in fact tend to *some nonzero* limit at infinity. Yet the question remains of whether the limit achieved is the same as that prescribed on the outer boundaries of the annular regions, in the construction of the solution.

Problems with infinite energy. Most problems of real practical or scientific interest, like those above, concern flows which are driven by nonhomogeneous boundary values, or pressure drops, in situations which are usually best modeled in unbounded domains. To extend the usual existence theory to the case of nonhomogeneous boundary values, one usually seeks the solution in the form $\boldsymbol{u} = \boldsymbol{v} + \boldsymbol{b}$, where $\boldsymbol{b}$ is a solenoidal continuation of the boundary values into the domain, obtained by some ad hoc construction, and $\boldsymbol{v}$ is a "correction term", found by the same argument as used to prove the existence theorem for homogeneous data, but now applied to the slightly more complicated equation

$$(7) \qquad \boldsymbol{v}_t + \boldsymbol{v}\cdot\nabla\boldsymbol{v} + \boldsymbol{v}\cdot\nabla\boldsymbol{b} + \boldsymbol{b}\cdot\nabla\boldsymbol{v} = -\nabla p + \Delta\boldsymbol{v} + (\boldsymbol{f} + \Delta\boldsymbol{b} - \boldsymbol{b}_t - \boldsymbol{b}\cdot\nabla\boldsymbol{b}) .$$

(The terms $\boldsymbol{v}_t$ and $\boldsymbol{b}_t$ are absent in the steady case.) Of course, the correction term $\boldsymbol{v}$ must be sought in the function space associated with the basic existence theorem. If one has in mind the usual existence theorems via energy estimates, this means that for steady problems $\boldsymbol{v}$ is sought in the completion $J_0(\Omega)$ of the solenoidal test functions in the Dirichlet norm $\|\nabla\cdot\|$, while for nonsteady problems $\boldsymbol{v}(t)$ is sought in the completion $J_1(\Omega)$ of the solenoidal test functions in the $W_2^1(\Omega)$-norm $(\|\cdot\|^2 + \|\nabla\cdot\|^2)^{1/2}$.

How many problems are really accessible by this method? Among problems for steady flow, those considered in the last section should probably be regarded as rare examples of its success. Success is achieved because the solution which is sought has a finite Dirichlet norm. If the expected solution has an infinite Dirichlet norm, then, either the preliminary construction of $\boldsymbol{b}$ must approximate the far field extremely well, to be "correctable" by a term with finite Dirichlet norm, or one must seek new existence theorems for $\boldsymbol{v}$, in wider function spaces.

One very notable success with this type of difficulty concerns steady flow of prescribed net flux through an infinite tube. If the infinite ends of the tube are cylindrical, and the prescribed net flux is small, $\boldsymbol{b}$ can be set equal to Poiseuille flow in the far field, and there is no great difficulty in proving the existence of a solution (and further qualitative properties have been given by Amick [1]). But what if the tube is allowed to bend and vary in some uniform but irregular way, so that one cannot hope to construct a function $\boldsymbol{b}$ that approximates the solution within a term of finite Dirichlet norm? Ladyzhenskaya and Solonnikov [75] solved this problem in a beautiful way, proving existence (even for large data) by using a variant of Saint-Venant's principle to obtain *local bounds* for the Dirichlet integral. Most problems with this type of difficulty, however, seem beyond our present reach. One, for example, of considerable interest in engineering, concerns flow onto the leading edge of a semi-infinite plate. The boundary layer approximations for such flow seem to be too crude to serve as the basis for an existence theorem.

To solve time dependent problems by the standard methods, obtaining $v(t) \in J_1(\Omega)$, requires the construction of a function b that approximates the solution u, at every time t, within a finite amount in the L^2-norm, as well as in the Dirichlet norm. There are some interesting problems which can be approached naturally in this way, among them the acceleration of self propelled bodies studied in [40], and the a posteriori error estimates for numerical methods given in [49]. However, in general, the requirements just mentioned severely restrict the applicability of the method.

Consider, for example, the "starting problem" for flow past an obstacle. That is, consider finding the nonstationary flow about a body which is gradually accelerated (by an external force) from rest to some slow terminal velocity $-b_\infty$. It is to be assumed that the surrounding fluid fills all R^3 exterior to the body, that it is initially at rest, and that it adheres to the boundary. Taking the prescribed velocity of the body to be $-\mu(t)b_\infty$, with $\mu(t)$ growing from zero to a final value of one, the problem can be posed in a fixed domain by introducing the fictitious force $f = \mu_t(t)b_\infty$, and prescribing a corresponding velocity at infinity, $u(x,t) \to \mu(t)b_\infty$, as $|x| \to \infty$. One would then like to establish the global existence of a smooth solution $u(x,t)$, and show that it converges as $t \to \infty$, to the solution $\tilde{u}(x)$ of the corresponding exterior stationary problem. The difficulty is due to the fact (proved by Finn [26]) that the steady solution $\tilde{u}$ has an infinite wake energy, $\|\tilde{u} - b_\infty\| = \infty$, while on the other hand, the desired solution u will surely satisfy, for all time, $\|u(t,\cdot) - \mu(t)b_\infty\| < \infty$. Thus, if one attempts to solve the problem by setting $b = \mu(t)\tilde{u}$, the corresponding term $v(t)$ cannot be found in $L^2(\Omega)$. Alternatively, one might take b to be a solenoidal cut-off function which equals $\mu(t)b_\infty$ in a neighborhood of infinity and vanishes at the boundary. Then $\|v(t)\|$ will be always finite, but will necessarily grow with time beyond any finite bound, so that eventually the energy which is present will exceed that for which we know how to prove the smoothness of the solution. A "weak" solution can be obtained, but it appears impossible to prove its convergence to steady state without proving its regularity. Because the global existence of a smooth solution may depend upon showing that v remains small in some sense, and because it can be expected that $\|\nabla(u(\cdot,t) - \tilde{u})\| \to 0$, as $t \to \infty$, the first alternative of setting $b = \mu(t)\tilde{u}$ would seem to have some advantages (since then $\|\nabla v(t)\|$ should remain finite and tend to zero), if only one had an existence theory for non-square-summable $v(t) \in J_0(\Omega)$. In fact, several variants of such an existence theory were developed in [41], [44], [79]. It is noteworthy that the analysis of [41], [79] involved the introduction of new function spaces whose characterizations in classical terms led to unexpected consequences regarding the proper posing of problems in unbounded domains, given in [43]. However, to date, the methods of [41], [44], [79] have not been successfully applied to the "starting problem". Convergence to steady state was obtained for the linearized problem, but because of the infinite energy present in the term v (even though it is probably smoothly distributed over a wide area), only a local existence theorem could be obtained for the nonlinear problem.

This difficulty has led Bemelmans [12] to develop a nonstationary existence theory based on a combination of potential theoretic and semigroup methods, drawing ideas from the works of Finn [28], Fujita and Kato [34], von Wahl [106], and others. Bemelmans proved global existence for nonstationary perturbations of Finn's steady solution, provided the initial values belong to the domain $D(A^{\delta})$ of a suitable fractional power of Oseen's operator, and are small in a suitable norm. The spaces $D(A^{\delta})$ have not yet been fully characterized in classical terms. But, what is very remarkable is that they apparently contain functions with infinite $L^2(\Omega)$-norms. The existence theorem of Bemelmans is a rare instance of a *global* existence theorem in which the "correction term" $v(t)$, supplied by the functional analysis, has an infinite energy. Existence theorems of Kato [64] and of Giga and Miyakawa [35] now share this property, but concern only the Cauchy problem, $\Omega = R^3$. Regarding the application of Bemelmans' result to the starting problem, there is still a difficulty in determining whether, after the initial period of acceleration, the nonsteady and steady solutions differ by a term belonging to $D(A^{\delta})$. Bemelmans outlined a proof of this in [13], but the manner of acceleration was not specified, and the details of the proof have not appeared.

Recent progress can be reported for a different but related problem, which was solved by a special construction of b, making possible the finding of $v(t) \in J_1(\Omega)$ by standard energy estimates. Consider two dimensional flow through a channel Ω bounded by the two branches of a hyperbola. Suppose the fluid adheres to the walls of the channel, and let it be initially at rest. Seek, then, a solution $u(t)$ carrying a prescribed net flux $F(t)$ through the constriction (F is assumed to be smooth and satisfy $F(0) = 0$). The difficulty is that in two dimensions a solenoidal function like $u(t)$, which carries nonzero flux through the constriction, cannot belong to $L^2(\Omega)$, or hence to $J_1(\Omega)$; it can decay at best like $|x|^{-1} \notin L^2(\Omega)$. Thus, to find $v(t) \in J_1(\Omega)$, $b(t)$ must carry the prescribed flux through the construction, and must approximate the solution extremely well in the far field. In [53], a suitable flux carrier $b(t)$ was found by a special "boundary layer" construction. However, even though the global existence of a smooth solution was established, it remains an open question whether the solution is stable and converges to steady state if $F(t)$ is brought to a small constant value.

Proper posing of problems in unbounded domains. Consider slow steady two dimensional flow around a corner. Say Ω is the region expressed in polar coordinates r, θ by $r > 0$, $\alpha < \theta < 2\pi - \alpha$, for some $\alpha \in [0, \pi)$. One could approach the problem by solving the Navier-Stokes equations numerically in a finite region $\Omega_R = \Omega \cap \{ |x| < R \}$, prescribing some more or less ad hoc boundary values on the artificial boundary. But might the nature of the solution near the origin be highly dependent upon the particular choice of artificial boundary values, and/or might there be no reasonable limit of the solutions near the origin, as $R \to \infty$, so that the "pure corner effect" one seeks does not exist? These questions are analogous to that of whether Leray's construction solves the exterior stationary problem, and it is typically to answer just such questions that one considers problems in unbounded domains.

To discuss one specific case, take $\alpha = 0$, so that the x-axis represents a semi-infinite plate. We seek flow at the edge of a large horizontal plate as it moves slowly downwards (infinitely large plate, infinitely slowly). There is a corresponding potential flow, $\boldsymbol{v} = \nabla\varphi$, where $\varphi = A r^{1/2} \cos\theta/2$, with arbitrary A. One can construct a kind of pseudo boundary layer approximation $\boldsymbol{b}$, by replacing the values of $\boldsymbol{v}$ inside some chosen streamline with those of a solenoidal function which matches $\boldsymbol{v}$ along the streamline and vanishes on $\partial\Omega$. This can be done in such a way that $\|\nabla \boldsymbol{b}\| < \infty$. Thus, while highly speculative (the author would conjecture against it), it seems just plausible that there might exist a nontrivial solution $\boldsymbol{u}$ of the Navier-Stokes equations, with a finite Dirichlet norm, whose far field values become close to those of the potential flow along every ray $\theta \neq 0$, and which can be obtained by an analog of Leray's construction, taking the values of $\boldsymbol{b}$ on $\partial\Omega_R$ as prescribed boundary values for solutions $\boldsymbol{u}_R$ defined in Ω_R.

It would be interesting to prove the existence or nonexistence of such solutions. Existence, if that turned out to be the case, would be extremely interesting from the point of view of uniqueness theory. Recall that it was shown in [43] that a key relationship between the geometry of a domain and the uniqueness of flow in the usual energy classes depends upon the dimension of certain quotient spaces, $\overset{*}{J_1}(\Omega)/J_1(\Omega)$ for nonsteady solutions, and $\overset{*}{J_0}(\Omega)/J_0(\Omega)$ for steady solutions, where $\overset{*}{J_1}(\Omega)$ consists of the solenoidal functions belonging to $\overset{\circ}{W}{}_2^1(\Omega)$, and $\overset{*}{J_0}(\Omega)$ consists of the solenoidal functions belonging to the completion $\overset{\circ}{W}(\Omega)$ of $C_0^\infty(\Omega)$ in the Dirichlet norm. When the quotients are nontrivial, solutions of the Stokes or Navier-Stokes equations (belonging to $\overset{*}{J_1}(\Omega)$ in the nonsteady case, and $\overset{*}{J_0}(\Omega)$ in the steady case) can be driven by pressure drops, and thus pressure drops (or alternative flux conditions) must be included as part of the prescribed data in order to determine a solution uniquely (as in studying a jet of fluid flowing through an aperture in a wall). Briefly, the explanation for this is that the pressure term in the energy identity

$$(8) \qquad \frac{1}{2}\frac{d}{dt}\|\boldsymbol{u}\|^2 + \|\nabla \boldsymbol{u}\|^2 = -(\nabla p, \boldsymbol{u}) - (\boldsymbol{u}\cdot\nabla\boldsymbol{u}, \boldsymbol{u})$$

vanishes automatically if and only if $\boldsymbol{u}$ is the limit of solenoidal test functions. For the domain Ω under present consideration, it is known that $\overset{*}{J_0}(\Omega) = J_0(\Omega)$. This was proven in [73]; for further studies of these spaces, and further references, see also [16], [62], [81], [82], [97], [99]. Because the spaces are equal, there can be no nontrivial solution $\boldsymbol{u} \in \overset{*}{J_0}(\Omega)$

of the steady Stokes equations (for a given domain, equality of the spaces is equivalent to uniqueness for the Stokes equations; [43]). However, this does not automatically exclude the possible existence of a solution of the Navier-Stokes equations, driven through the inertial term, by the infinite kinetic energy present in the flow. In this respect the situation might be analogous to that of the nonunique two dimensional swirling flows given in the preface of Ladyzhenskaya's book [72]. And also analogous to the Stokes paradox for flows past bodies, in that there might be solutions of the nonlinear problem, but not of the linear problem. The existence of such solutions would raise many further questions about the dependence on the angle α, and about the possibility of prescribing various features of the flow at infinity.

Any solutions representing flow around corners in unbounded domains would add very substantially to the collection of known steady solutions of the Navier-Stokes equations which describe "pure effects". The ones known to date are, first, the classical Poiseuille and Couette flows, and the two dimensional swirling flows just mentioned. Then, there are the similarity solutions describing flow toward a stagnation point or above a rotating disc in a half space, or from a submerged jet in R^3, or radially to or from the vertex of a sector. These all have singularities either at the origin or at infinity. Finally, there are the solutions provided by modern existence theory, describing steady flows past obstacles, steady jets through apertures and channels, and steady Taylor vortices between rotating cylinders. One might also list here some solutions which have been obtained recently, describing flows with free boundaries under the influence of surface tension; see [14], [68], [98].

Energy estimates in annular regions. This is another long standing problem which seems to call for good physical insight. Consider the initial boundary value problem (1), (2), with nonhomogeneous boundary values, in a smoothly bounded two or three dimensional domain. Provided the boundary values remain smooth and bounded for all time (in your choice of norm), it seems plausible that the energy should also remain bounded for all time, say $\limsup_{t\to\infty} \|\boldsymbol{u}(t)\| \leq B$, with B depending only on the domain and the norm of the boundary values. Indeed, this was proved by Hopf [55] in 1941, but under an additional restriction, which he called special attention to in [57], and which may not be natural and necessary for the result. This is that the net flux across each component of $\partial\Omega$ must be small (there is no restriction on the size of the boundary values when the boundary is connected). In the simplest case for which the problem is open, let Ω be the annular region between two circles, $1 < r < 2$, and take as boundary values the restriction of $\boldsymbol{b} = F\,\nabla(\log r)$ to the boundary. Then, if $|F|$ is large, and if one takes a smooth nonsymmetric perturbation of $\boldsymbol{b}$ as the initial value $\boldsymbol{u}_0$, it is not known whether $\|\boldsymbol{u}(t)\|$ remains bounded as $t \to \infty$. Multiplying (7) by $\boldsymbol{v}$, and integrating over Ω, one easily sees that $\|\boldsymbol{v}(t)\|$ grows at most exponentially with time, from which one can ultimately infer the global existence of a classical solution (since the domain is two dimensional). If $\|v(t)\|$ does tend to infinity with time, then so does the rate of energy dissipation. But that could conceivably be supported by a growing pressure drop from one component of the boundary to the other. Hopf's result depends upon constructing a special solenoidal continuation $\tilde{b}$ of the boundary values into Ω, such that $|(\varphi\cdot\nabla\tilde{b},\varphi)| \le \frac{1}{2}\|\nabla\varphi\|^2$, for all solenoidal test functions φ. However, it has been shown by Takeshita [101], through a symmetrization argument, that no such $\tilde{b}$ can exist in the present situation.

Convergence to Euler flow as the viscosity tends to zero. Insert a coefficient of kinematic viscosity ν before the viscous term on the right of equation (1). It is a famous question whether, as $\nu \to 0$, the solution of the Navier-Stokes problem (1), (2) tends to the solution of the corresponding problem for the Euler equations, in which the viscous term is omitted from equation (1) and the boundary condition is weakened to $u(x,t)\cdot n = 0$, where n is the outer normal to $\partial\Omega$. For the case $\Omega = R^3$, and for an interval of time depending on the initial values but not the viscosity, the problem was solved in works of Swann [100], and Kato [63]. But the problem with boundaries has remained a famous challenge. There is now a major new development reported by Asano [5], at this conference, who has solved the problem in a half space. The key to his analysis is to construct solutions of the Navier-Stokes equations incorporating an appropriate boundary layer approximation, of Prandtl type. Thus, the result is also important as a step in the rigorous justification of classical boundary layer theory. However, the problem still remains open for more general boundaries, as Asano's construction uses the assumption that the boundary is a plane.

Vortex methods. A solenoidal velocity field u which is defined in all R^3 and decays at infinity can be determined in terms of its vorticity $\omega \equiv \nabla\times u$ by the relation

(9) $$u(x,t) = \int_{R^3} K(x-y)\times\omega(y,t)\,dy\ ,$$

where $K(x) = (-1/4\pi)(x/|x|^3)$. Taking the curl of the Navier-Stokes equations, the vorticity is seen to satisfy

(10) $$\omega_t + u\cdot\nabla\omega = \omega\cdot\nabla u + \Delta\omega\ .$$

It is apparent that the initial value problem (9), (10), with a prescribed solenoidal initial vorticity ω_0, is formally equivalent to the initial value problem (1), (2) in R^3.

There is a great interest in this formulation because it provides the basis for numerical simulations designed to approximate solutions of the Euler equations, and of the Navier-Stokes equations at high Reynolds numbers. The idea is particularly simple for the two dimensional Euler equations. In this case, $\omega = \partial u_2/\partial x_1 - \partial u_1/\partial x_2$, $K = (-x_2, x_1)/2\pi|x|^2$, and (10) becomes $\omega_t + u\cdot\nabla\omega = 0$. This last equation implies the constancy of vorticity along particle paths, determined as solutions of the ordinary differential equations

$$x' = u(x,t) . \tag{11}$$

In numerical simulations, the vorticity is assumed to be concentrated in a finite number of "vortex blobs", which then move along associated particle paths. At each time step, the blobs advance in accordance with (11) and the velocity field is updated in accordance with (9). This method is extended to the Navier-Stokes equations by adding a step of random walk to the particle's position, to simulate viscosity. If boundaries are present, additional vortex blobs are continually introduced at the boundary to satisfy the no slip condition. The principal effort in this problem area is being directed toward the design of increasingly effective numerical methods and to the study of their convergence. For a selection of some of the important papers in this area, and further references, see Chorin [20], Hald [38], Beal and Majda [11], Anderson and Greengard [4], and Benfatto and Pulvirenti [15]. Another very interesting development is the recent proof of existence theorems for the system (9), (10), with initial vorticities concentrated as measures along vortex rings and filaments, by Cottet and Soler [22], and Giga and Miyakawa [35].

Continuity properties of weak solutions, and the strong energy inequality. Consider again the initial boundary value problem (1), (2) in a smoothly bounded three dimensional domain Ω. For simplicity, suppose the boundary values and forces are zero, and that the initial values are smooth. Since, to date, a fully regular solution (defined for all $t > 0$) has been obtained only for small data, there is a great interest in less regular "weak solutions", the existence of which has been established. The simplest method of obtaining a weak solution is Hopf's [56]; see [46]. He constructed a sequence of smooth Galerkin approximations $\boldsymbol{u}^n$, the nth of which satisfies the weak form of the Navier-Stokes equations,

$$\int_0^s [(\boldsymbol{u},\varphi_t) - (\nabla \boldsymbol{u},\nabla\varphi) - (\boldsymbol{u}\cdot\nabla \boldsymbol{u},\varphi)]\, dt = (\boldsymbol{u}(s),\varphi(s)) - (\boldsymbol{u}_0,\varphi(0)) , \tag{12}$$

for any combination $\varphi(x,t)$ of the first n basis functions, and also the energy identity

$$\tfrac{1}{2}\|\boldsymbol{u}(s)\|^2 + \int_{s_1}^s \|\nabla \boldsymbol{u}\|^2\, dt = \tfrac{1}{2}\|\boldsymbol{u}(s_1)\|^2 , \tag{13}$$

for all $0 \le s_1 \le s$. A subsequence (again denoted by $\boldsymbol{u}^n$) was shown to converge to a limit $\boldsymbol{u}$, weakly in $L^2(\Omega)$ for every $t \ge 0$ (strongly for $t = 0$), weakly in $L^2(0,\infty;\, J_1(\Omega))$, and strongly in $L^2(0,\infty;\, L^2(\Omega))$. As a consequence, $\boldsymbol{u}$ satisfies (12) for all smooth solenoidal φ which vanish on $\partial\Omega\times[0,\infty)$, and is *weakly continuous* in $L^2(\Omega)$ as a function of time.

Finally, from the strong convergence in $L^2(0,\infty;\, L^2(\Omega))$, one obtains $\boldsymbol{u}^n \to \boldsymbol{u}$ strongly in $L^2(\Omega)$, for almost all $t > 0$ (taking a further subsequence). Thus, one can pass to a limit from the identities (13) for the approximations u^n, using strong convergence on the right and weak convergence on the left, to obtain for u the energy inequality

$$\tag{14} \frac{1}{2}\|u(s)\|^2 + \int_{s_1}^{s} \|\nabla u\|^2 dt \le \frac{1}{2}\|u(s_1)\|^2,$$

for almost all $s_1 \ge 0$ (including $s_1 = 0$), and for all $s > s_1$. It's curious, that so far as is known, there may be losses in the kinetic energy which are not accounted for by the dissipation of kinetic energy into heat, and the norm $\|u(s)\|$ may even fail to be continuous. The same problem persists in all other known constructions of globally defined weak solutions.

It is interesting to reflect on some of the pathological behavior which might occur, consistent with what is known about weak solutions. Suppose u is expressed in a Fourier series, $u(x,t) = \Sigma c_i(t)\, a_i(x)$, say in terms of the eigenfunctions of the Stokes operator. Then, the weak continuity of u simply means that each coefficient $c_i(t)$ is continuous and that $\Sigma c_i^2(t)$ remains bounded. Let us consider what might happen at an isolated singular instant of time (assuming there is one). Of course, as the singularity is approached, some part of the energy must be passed up the spectrum, so that $\|\nabla u(t)\|^2 \equiv \Sigma \lambda_i c_i^2(t) \to \infty$, where the λ_i's are the eigenvalues of the Stokes operator. Could it be that all the energy is passed up the spectrum, and lost, at such a singularity? In fact, it is an open problem to show that a weak solution cannot come to a complete motionless stop at (and following) a singular instant of time. Even more pathologically, could the solution's total energy be passed up the spectrum as the singularity is approached, and then some of it return, coming back down the spectrum, so that the weakly continuous solution u would be identically zero for just one instant of time (each continuous coefficient $c_i(t)$ vanishing at this one instant)? Such behavior is consistent with what we currently know about weak solutions. However, the energy inequality (14) does rule out the possibility of *essential* "jump ups" in the energy; that is, for every $t > 0$, $ess\ inf_{s \to t-} \|u(s)\| \ge sup_{s \ge t} \|u(s)\|$.

In introducing the concept of a weak solution, Leray [77] included the nonexistence of essential jump ups in the energy as part of his definition. And he used this property in proving his famous theorem about a weak solution's "epochs of regularity" (though this can now be proved independently [46]). Leray, however, dealt only with the case $\Omega = R^3$. When Hopf later considered arbitrary domains, he dropped the condition on energy jump ups from his definition, proving (14) only for $s_1 = 0$. The holding of (14) for almost all $s_1 > 0$ is now frequently referred to as the "strong energy inequality". It was proved for bounded domains by Ladyzhenskaya [72]. However, as pointed out by Masuda [84], her proof does not extend to unbounded domains. This is because Hopf's construction (unlike Leray's) provides strong $L^2(0,\infty;\, L^2(\Omega'))$ convergence of the approximations only for bounded subdomains $\Omega' \subset \Omega$. Conceivably, some part of the energy in the approximations can run away to spatial infinity over whole intervals of time (that is, the energy in successive approximations might move successively further away), and then suddenly return, causing an essential jump up in the energy of the solution. Because of this, it is unknown whether Hopf's solutions satisfy the strong energy inequality in unbounded domains.

There has recently been a major success in the effort to find alternative constructions of weak solutions, free of essential jump ups in the energy. For exterior domains, the existence of a weak solution satisfying the strong energy inequality (11) is now proven in the works of Miyakawa and Sohr [85], and of Sohr, von Wahl and Wiegner [95]. However, the existence of such solutions in domains with noncompact boundaries is still unknown. In view of the way the pressure can interact with distant boundaries, it may be very difficult to properly localize the energy, as needed for the result.

Localization of the energy is a recurring problem with the Navier-Stokes equations. Scheffer [91] and Caffarelli, Kohn and Nirenberg [18] have proven partial regularity results for weak solutions, consisting of bounds for the Hausdorff dimension and measure of the set of possible singularities in space-time. These results seem particularly notable for their dependence upon estimates for the localization of the energy, as exemplified by several estimates given in the introduction of [18]. Even if weak solutions are ultimately proved to be smooth, these estimates will remain of lasting interest.

Navier-Stokes equations combined with other phenomena. Motivated by practical applications, it is important to consider the Navier-Stokes equations as a part of larger models incorporating additional physical phenomena. Heat convection is often modeled by an incompressible flow, by taking the external force to be a vector field aligned with the gravitational field, proportional to the temperature. The temperature must then satisfy an additional convection diffusion equation and appropriate boundary conditions. Similarly, in magneto-hydrodynamics, the external force is taken to be $\boldsymbol{f} = \boldsymbol{J} \times \boldsymbol{B}$, where $\boldsymbol{J}$ is the electric current density and $\boldsymbol{B}$ is the magnetic field. These become additional unknowns, governed by coupling the Navier-Stokes equations with Maxwell's equations. As it becomes better developed, the theory of magneto-hydrodynamics is expected to become rich in its own special qualitative features. Another problem of great current interest is the solution of the Navier-Stokes equations in domains with free boundaries governed by surface tension, as in studying surface waves (see Beal [9]), or flow in a neighborhood of the junction of a free surface with a rigid boundary (see Kroner [68]), or the shape of rising bubbles (see Bemelmans [14] and Solonnikov [98]). Concerning the latter, it would be interesting to study a bubble's loss of rotational symmetry and surface smoothness as its size is increased (as observed in the bubble experiments at science museums; J.-C. H. & F. H.).

Stability. Through the work of many authors, among the earliest of whom were Prodi, Velte, Yudovich, Iooss and Kirchgässner (see [89], [105], [110], [58], [65], [60] for representative papers and further references), the theory of stability and bifurcation has been successfully extended from the theory of ordinary differential equations, to applications in the study of steady and time periodic solutions of the Navier-Stokes equations. The objective is to explain, in the case of steady data, the existence of multiple steady and time periodic solutions, and the exchanges of stability between them as the data is varied. While the abstract principles of this analysis have been generally established, the application to specific problems remains difficult, depending upon a spectral analysis of linearizations about solutions which are not usually explicitly known. See the papers by Nagata [87] and Okamoto [88] (these proceedings), and the papers cited there, for recent results explaining phenomena in the exchange of stabilities in the Rayleigh-Benard and Taylor experiments.

Results on the stability of Finn's steady solutions are given in [40], [83], [44, p. 674], [67], [12] and [86]. The works of Bemelmans [12] and Mizumachi [86] are particularly noteworthy, using, respectively, methods of Finn and Babenko to show that the qualitative properties of the wake are preserved under nonstationary perturbations. For the future, one of the major challenges in this area is to extend spectral stability analysis to problems in unbounded domains, and particularly to the study of bifurcations from Finn's steady solutions, as proposed by Babenko [7], [8]. A landmark problem awaiting analysis is that of proving, completely and rigorously, the existence of two dimensional time periodic flows representing von-Kármán vortex shedding behind a cylinder.

Mindful of the experiments of K. G. Roesner (shown in movies at this conference), in which one sees apparently stable steady and time periodic solutions of extreme complexity, it will be recognized that many stable situations are complex beyond hope of rigorous spectral analysis. Yet, regardless of this, the *apparent* stability of such solutions seems reason enough to expect their faithful replication in numerical experiments. This matter was taken up in joint papers with Rannacher [47], [49]. There, we have formulated simple definitions of stability, seemingly descriptive of many of the principal types of stability which are observed, and shown, among other things, the persistence of these stabilities in different norms, and under progressively wider classes of perturbations, including the numerical discretization of the equations. In particular, error estimates were obtained which are uniform over arbitrarily long periods of time.

It is an interesting question whether a stability theory, analogous to that in [47], can be developed for flows in unbounded domains. For this, it seems necessary to first understand the decay properties of solutions of the Cauchy problem, $\Omega = R^n$. There are major recent results in this area, including the solution of Leray's [77] long standing problem of proving L^2-decay, as $t \to \infty$, for solutions of the three dimensional Cauchy problem. This matter was studied simultaneously, using different methods, by Schonbek [92], [93] and Masuda [84]. Schonbek proved decay, with explicit decay rates, for data $u_0 \in L^2 \cap L^p$, $1 \leq p < 2$. Masuda solved the problem for data $u_0 \in L^2$, as proposed by Leray. In this case an explicit decay rate is impossible, as was later shown in [93]. For further results on the decay of solutions of the n-dimensional Cauchy problem, in the L^2-norm and in other L^p-norms, see Kato [64], Kajikiya and Miyakawa [61], Wiegner [109], and references given there. Algebraic L^2-decay for exterior domains has been just recently proven in a major new work of Borchers and Miyakawa [17], drawing upon results of Giga and Sohr [36].

Partial stability. In trying to describe the types of stability one commonly sees in the motions of real fluids, it will be recognized that there are often minor, small scale, seemingly random irregularities, even when the large scale features of a flow are fairly stable. There were notable examples of this in the experiments of Roesner. Strictly speaking, such flow is not stable if, as often seems the case, there are chaotic ambient flutterings of a definite finite magnitude. The usual stability theory is an all or nothing proposition. Yet, the "partial stability" one sees gives reason to think that the main features of such flows can be replicated in numerical experiments, and indeed they are. In collaboration with Rannacher [47], [49], we introduced a notion of partial stability which we call "contractive stability to a tolerance", and showed its persistence under numerical discretization of the equations, thus obtaining long time error estimates "to a tolerance" for numerical simulations of such flow. However, it remains a very interesting problem to find satisfactory explanations for partially stable motions. Probably, small scale irregularities are sometimes merely the result of local perturbations of an otherwise stable large scale motion in which the viscous term dominates the inertial, in accordance with linear spectral analysis. The perturbations might be due, for example, to discontinuities in the boundary values of a physical experiment, or to additional superimposed forces with small supports. In other instances, large scale stability might be inherited from the stability of a neighboring Euler flow, particularly if irregularities are exiting across a downstream boundary. For works treating the stability of Euler flow, see [10], [54], [80], [108]. Related situations in unbounded domains may call for a partial stability analysis in terms of spatially weighted norms. There are yet other situations, such as Taylor cells with turbulent cores, which could perhaps be explained by a linear spectral analysis of averaged equations for turbulent flow. As yet, there does not seem to be a mathematical demonstration, by theory or example, of any of these possibilities.

Attractors. The question of whether the Navier-Stokes equations are capable of describing turbulence seems to have been convincingly answered by a sharpened awareness that chaos is one of the generic types of behavior common even for solutions of ordinary differential equations. The significance of this point of view was first brought clearly into focus by Ruelle and Takens [90]. It is now believed that turbulence in fluids is closely related to the chaotic motions which one can easily see and understand in low dimensional systems of ordinary differential equations, like the Lorenz equations. Conceptually, the Navier-Stokes equations can be considered as an infinite system of ordinary differential equations in phase space, by expanding solutions in Fourier series. Through examples like the Lorenz equations, it is clear that there is no inherent contradiction between the existence of long time random aspects in the behavior of solutions, and the holding of smoothness, uniqueness, and continuous dependence theorems. The chaos results from the exponential departure, along certain directions, of neighboring trajectories which can later fold back upon each other, in other directions. The salient features of the equations needed for an understanding of this are brought together in the definition of "dissipative dynamical system". That the Navier-Stokes equations define such a system is a consequence of a general energy bound and the smoothing property of the equations (a global smoothing property is needed, which for now has to be assumed in three dimensions; see [48, theorems 2.2 - 2.5]).

Of course, there is order to be found in the chaos. If one considers the surface of an appropriate large ellipsoid in R^3, as it evolves under the dynamics of the Lorenz system, it may be verified that it is mapped into itself, and that the enclosed volume shrinks exponentially in time. Thus, the forward evolution of its closure defines a compact "attractor", of Lebesgue measure zero. It is invariant, both forwards and backwards in time, and all trajectories converge to it, as $t \to \infty$. The structure of this attractor is quite complicated. It has fractal cross sections, and of course there is a chaotic behavior of the trajectories which lie on it. Early proofs of the existence of an analogous attractor for the Navier-Stokes equations go back to Foias and Prodi [30], and Ladyzhenskaya [70],[71]. Recent results about the properties of this attractor can be found in Constantin and Foias [21] and Temam [104]. Observing that for a given bound on the regularity of solutions, all but a finite number of orthogonal directions in phase space must shrink, exponentially, under evolution by the Navier-Stokes equations, it is shown in these works that the volume of any finite dimensional hypersurface of sufficiently high dimension must shrink. Consequently, the attractor is shown to have finite fractal Hausdorff dimensions.

What kind of applications do these concepts have? One thing which can be shown, similarly to the error analysis given in [49] for "partially stable" solutions which are "contractively stable to a tolerance", is that upon discretization of the equations, the discrete attractor must closely neighbor the smooth attractor; see Hale, Lin and Raugel [39]. Still, this is far from an adequate demonstration that discrete solutions faithfully replicate the important properties of smooth solutions. The attractor we've referred to, and for which this result is proved, sometimes called the "maximal attractor", is too large, being the limit of a kind of envelope containing all the system's stationary and periodic orbits, and invariant manifolds, etc., unstable as well as stable. Further, a proper characterization of the important properties of chaotic solutions which should be preserved under discretization, must include, among other things, something about the statistics of the time spent on different parts of the attractor, described in the form of invariant measures. Conjectures about how such measures may vary under perturbations must take into account the limitations of simple examples like o.d.e. flow on a torus. Recent progress in this area is reviewed by Eckmann and Ruelle in [23].

Beyond these matters, the most important problem concerning attractors seems to be to make convincing contact with observable properties of fluid motions, preferably measurable properties. As a step toward this goal, we are studying pressure drops in turbulent pipe flow [52]. One preliminary result is a bound for the pressure gradient required to drive a turbulent flow of prescribed net flux. Another natural objective is to try to connect the study of attractors with explanations for the partial stability of particular flows. Seeking other phenomena which might be explained through the study of attractors, a good candidate might be the visual patterns one seems to so easily recognize in homogeneous turbulence.

References

[1] AMICK, C. J., Properties of steady Navier-Stokes solutions for certain unbounded channels and pipes, *Nonlinear Anal., Theory, Methods & Appl.*, **6** (1978), 689-720.

[2] AMICK, C. J., On Leray's problem of steady flow past a body in the plane, *Acta Math.*, **161** (1988), 71-130.

[3] AMICK, C. J. & FRAENKEL, L. E., Steady solutions of the Navier-Stokes equations representing plane flow in channels of various types, *Acta Math.* **144** (1980), 83-152.

[4] ANDERSON, C. & GREENGARD, C., On vortex methods, *SIAM J. Numer. Anal.*, **22** (1985), 413-440.

[5] ASANO, K., Zero-viscosity limit of the incompressible Navier-Stokes equations, (preprint 1988).

[6] BABENKO, K. I., On stationary solutions of the problem of flow past a body by a viscous incompressible fluid, *Mat. Sb.*, **91** (1973), 3-26.

[7] BABENKO, K. I., On properties of steady viscous incompressible fluid flows, Proc. IUTAM Symp. 1979, *Approximation Methods for Navier-Stokes Problems*, R. Rautmann, ed., Springer-Verlag, *Lecture Notes in Mathematics*, **771** (1980), 12-42.

[8] BABENKO, K. I., Spectrum of the linearized problem of flow of a viscous incompressible liquid round a body, *Sov. Phys. Dokl.*, 27 (1) (1982), 25-27.

[9] BEAL, J. T., Large-time regularity of viscous surface waves, *Arch. Rational Mech. Anal.*, **84** (1984), 307-352.

[10] BEAL, J. T. & MAJDA, A., Vortex methods for flow in two and three dimensions, *Contemp. Math.*, **28** (1984), 221-229.

[11] BEAL, J. T. & MAJDA, A., High order accurate vortex methods with explicit velocity kernels, *J. Comput. Phys.*, **58** (1985), 188-208.

[12] BEMELMANS, J., Eine Außenraumaufgabe für die instationären Navier-Stokes-Gleichungen, *Math. Z.*, **162** (1978), 145-173.

[13] BEMELMANS, J., Navier-Stokes equations in exterior domains, Rendiconti del Seminario Matematico e Fisico di Milano, Vol. L (1980), Tipografia Fusi - Pavia, **4** (1982), 91-100.

[14] BEMELMANS, J., Gleichgewichtsfiguren zäher Flüssigkeiten mit Oberflächenspannung, *Analysis,* 1 (1981), 241-282.

[15] BENFATTO, G. & PULVIRENTI, M., Convergence of Chorin-Marsden product formula in the half-plane, *Comm. Math. Phys.*, **106** (1986), 427-458.

[16] BOGOVSKII, M. E., On the L_p-theory of the Navier-Stokes system for unbounded domains with noncompact boundaries, *Soviet Math. Dokl.*, 22 (1980), 809-814.

[17] BORCHERS, W. & MIYAKAWA, T., Algebraic L^2 decay for Navier-Stokes flows in exterior domains, (preprint 1989).

[18] CAFFARELLI, L., KOHN, R. & NIRENBERG, L., Partial regularity of suitable weak solutions of the Navier-Stokes equations, *Comm. Pure Appl. Math.*, **35** (1982), 771-831.

[19] CHANG, I-D. & FINN, R., On the solutions of a class of equations occurring in continuum mechanics, with application to the Stokes paradox, *Arch. Rational Mech. Anal.*, **7** (1961), 388-401.

[20] CHORIN, A. J., Numerical study of slightly viscous flow, *J. Fluid Mech.*, **57** (1973), 785-796.

[21] CONSTANTIN, P. & FOIAS, C., *Navier-Stokes Equations*, Univ. Chicago Press, Chicago and London, 1988.

[22] COTTET, G.-H. & SOLER, J., Three-dimensional Navier-Stokes equations for singular filament data, (preprint).

[23] ECKMANN, J.-P. & RUELLE, D., Ergodic theory of chaos and strange attractors, *Rev. Modern Phys.*, **57** (1985), 617-656.

[24] FINN, R., On steady state solutions of the Navier-Stokes partial differential equations, *Arch. Rational Mech. Anal.*, 3 (1959), 381-396.

[25] FINN, R., Estimates at infinity for stationary solutions of the Navier-Stokes equations, *Bull. Math. Soc. Sci. Math. Phys. R. P. Roumaine*, **3** (51) (1959), 387-418.

[26] FINN, R., An energy theorem for viscous fluid motions, *Arch. Rational Mech. Anal.*, **6** (1960), 371-381.

[27] FINN, R., On the steady state solutions of the Navier-Stokes equations. III, *Acta. Math.,* **105** (1961), 197-244.

[28] FINN, R., On the exterior stationary problem for the Navier-Stokes equations, and associated perturbation problems, *Arch. Rational Mech. Anal.* **19** (1965), 363-406.

[29] FINN, R. & SMITH, D. R., On the stationary solution of the Navier-Stokes equations in two dimensions, *Arch. Rational Mech. Anal.*, **25** (1967), 26-39.

[30] FOIAS, C. & PRODI, G., Sur le comportement global des solutions nonstationaires des equations de Navier-Stokes en dimension 2, *Rend. Semin. Matem. Univ. Padova*, **39** (1967), 1-34.

[31] FOIAS, C. & SAUT J. C., Asymptotic behavior, as $t \to \infty$ of solutions of Navier-Stokes equations and nonlinear spectral manifolds, *Indiana Univ. Math. J.*, **33** (1984), 459-477.

[32] FRAENKEL, L. E. & EAGLES, P. M., On a theory of laminar flow in channels of certain class. II, *Math. Proc. Camb. Phil. Soc.*, **77** (1975), 199-224.

[33] FUJITA, H., On the existence and regularity of the steady-state solutions of the Navier-Stokes equation, *J. Fac. Sci. Univ. Tokyo, Sect. I A Math.* **9** (1961), 59-102.

[34] FUJITA, H. & KATO, T., On the Navier-Stokes initial value problem, I., *Arch. Rational Mech. Anal.*, 16 (1964), 269-315.

[35] GIGA, Y. & MIYAKAWA, T., Navier-Stokes flow in R^3 with measures as initial vorticity and Morrey Spaces, Comm. PDE, (to appear).

[36] GIGA, Y. & SOHR, H., On the Stokes operator in exterior domains, *J. Fac. Sci. Univ. Tokyo, Sect. I A Math.*, **36** (1989), 103-130.

[37] GILBARG, D. & WEINBERGER, H. F., Asymptotic properties of Leray's solution of the stationary two-dimensional Navier-Stokes equations, *Russian Math. Surveys*, **29** (1974), 109-123.

[38] HALD, O., On the convergence of vortex methods. II, *SIAM J. Numer. Anal.*, **16** (1979), 726-755.

[39] HALE, J. K., LIN, X.-B. & RAUGEL, G., Upper semicontinuity of attractors for approximations of semigroups and partial differential equations, *Math. Comp.*, **50** (1988), 89-123.

[40] HEYWOOD, J. G., The exterior nonstationary problem for the Navier-Stokes equations, *Acta Math.*, **129** (1972), 11-34.

[41] HEYWOOD, J. G., On nonstationary Stokes flow past an obstacle, *Indiana Univ. Math. J.*, **24** (1974), 271-284.

[42] HEYWOOD, J. G., On some paradoxes concerning two-dimensional Stokes flow past an obstacle, *Indiana Univ. Math. J.*, **24** (1974), 443-450.

[43] HEYWOOD, J. G., On uniqueness questions in the theory of viscous flow, *Acta Math.*, **136** (1976), 61-102.

[44] HEYWOOD, J. G., The Navier-Stokes equations: On the existence, regularity and decay of solutions, *Indiana Univ. Math. J.*, **29** (1980), 639-681.

[45] HEYWOOD, J. G., Classical solutions of the Navier-Stokes equations, Proc. IUTAM Symp. 1979, *Approximation Methods for Navier-Stokes Problems*, R. Rautmann, ed., Springer-Verlag, *Lecture Notes in Mathematics*, **771** (1980), 235-248.

[46] HEYWOOD, J. G., Epochs of regularity for weak solutions of the Navier-Stokes equations in unbounded domains, *Tôhoku Math. J.*, **40** (1988), 293-313.

[47] HEYWOOD, J. G. & RANNACHER, R., An analysis of stability concepts for the Navier-Stokes equations, *J. Reine Angew. Math.*, **372** (1986), 1-33.

[48] HEYWOOD, J. G. & RANNACHER, R., Finite element approximation of the nonstationary Navier-Stokes problem, I: Regularity of solutions and second-order error estimates for spatial discretization, *SIAM J. Numer. Anal.*, **19** (1982), 275-311.

[49] HEYWOOD, J. G. & RANNACHER, R., Finite element approximation of the nonstationary Navier-Stokes problem, II: Stability of solutions and error estimates uniform in time, *SIAM J. Numer. Anal.*, **23** (1986), 750-777.

[50] HEYWOOD, J. G. & RANNACHER, R., Finite element approximation of the nonstationary Navier-Stokes problem, III: Smoothing property and higher order error estimates for spatial discretization, *SIAM J. Numer. Anal.*, **25** (1988), 489-512.

[51] HEYWOOD, J. G. & RANNACHER, R., Finite element approximation of the nonstationary Navier-Stokes problem, IV: Error analysis for second order time discretization, *SIAM J. Numer. Anal.* (to appear).

[52] HEYWOOD, J. G. & RANNACHER, R., Pressure drops in turbulent pipe flow, (to appear).

[53] HEYWOOD, J. G. & SRITHARAN, S. S., Justification of a boundary layer approximation for nonstationary two dimensional jets, (to appear).

[54] HOLM, D. D., MARSDEN, J. E., RATIU, T., & WEINSTEIN, A., Nonlinear stability of fluid and plasma equilibria, *Phys. Rep.*, **123** (1985), 1-116.

[55] HOPF, E., Ein allgemeiner Endlichkeitssatz der Hydrodynamik, *Math. Ann.*, **117** (1940/41), 764-775.

[56] HOPF, E., Über die Anfangswertaufgabe für die Hydrodynamischen Grundgleichungen, *Math. Nachr.*, **4** (1950/51), 213-231.

[57] HOPF, E., On nonlinear partial differential equation, *Lecture series of symposium on partial differential equations*, Univ. of Kansas, (1951), 1-32.

[58] IOOSS, G., Existence et stabilité de la solution périodique secondaire intervenant dans les problemes d'évolution du type Navier-Stokes, *Arch. Rational Mech. Anal.*, **47** (1972), 301-329.

[59] ITO, S., The existence and the uniqueness of regular solution of nonstationary Navier-Stokes equation, *J. Fac. Univ. Tokyo, Sect. I A Math.* **9** (1961), 103-140.

[60] JOSEPH, D. D., *Stability of Fluid Motions*, Springer-Verlag, Berlin, Heidelberg, New York, 1976.

[61] KAJIKIYA, R. & MIYAKAWA, T., On L^2-decay of weak solutions of the Navier-Stokes equations in R^n, *Math. Z.*, **192** (1986), 135-148.

[62] KAPITANSKII, L. V. & PILETSKAS, K. I., On spaces of solenoidal vector fields and boundary value problems for the Navier-Stokes equations in domains with noncompact boundaries, Proc. Steklov Inst. Math., Issue 2 (1984), 3-34.

[63] KATO, T., Nonstationary flows of viscous and ideal fluids in R^3, *J. Func. Anal.*, **9** (1972), 296-305.

[64] KATO, T., Strong L^p-solutions of the Navier-Stokes equations in R^n, with applications to weak solutions, *Math. Z.*, **187** (1984), 471-480.

[65] KIRCHGASSNER, K., Bifurcation in nonlinear hydrodynamic stability, *SIAM Rev.*, **17** (1975), 652-683.

[66] KISELEV, A. A. & LADYZHENSKAYA, O. A., On the existence and uniqueness of the solution of the nonstationary problem for a viscous incompressible fluid, *Izv. Akad. Nauk SSSR Ser. Mat.* **21** (1957),655-680.

[67] KNIGHTLY, G., Some decay properties of solutions of the Navier-Stokes equations, Proc. IUTAM Symp. 1979, *Approximation Methods for Navier-Stokes Problems*, R. Rautmann, ed., Springer-Verlag, *Lecture Notes in Mathematics*, **771** (1980), 287-298.

[68] KRONER, D., Asymptotic expansions for a flow with a dynamic contact angle, (these proceedings).

[69] LADYZHENSKAYA, O. A., Investigation of the Navier-Stokes equations in the case of stationary motion of an incompressible fluid, *Uspekhi Mat. Nauk* (in Russian), **14** (1959), 75-97.

[70] LADYZHENSKAYA, O. A., A dynamical system generated by the Navier-Stokes equations, *Zap. Nauch. Semin. Lening. Otd. Mat. Inst. Steklov*, **27** (1972), 91-115.

[71] LADYZHENSKAYA, O. A., Mathematical analysis of Navier-Stokes equations for incompressible liquids, *Ann. Rev. of Fluid Mech.*, **7** (1975), 249-272.

[72] LADYZHENSKAYA, O. A., *The Mathematical Theory of Viscous Incompressible Flow*, Second Edition, Gordon and Breach, New York, 1969.

[73] LADYZHENSKAYA, O. A. & SOLONNIKOV, V. A., Some problems of vector analysis and generalized formulations of boundary-value problems for the Navier-Stokes equations, *J. Sov. Math.*, **10** (1978), 257-285.

[74] LADYZHENSKAYA, O. A. & SOLONNIKOV, V. A., On the solvability of boundary and initial-boundary value problems in regions with noncompact boundaries, *Vestn. Leningr. Univ.* (in Russian), **13** (1977), 39-47.

[75] LADYZHENSKAYA, O. A. & SOLONNIKOV, V. A., Determination of the solutions of boundary value problems for stationary Stokes and Navier-Stokes equations having an unbounded Dirichlet integral, *J. Sov. Math.*, **21** (1983), 728-761.

[76] LERAY, J., Étude de diverses équations intégrales non linéaires et de quelques problèmes que pose l'hydrodynamique, *J. Math. Pures Appl.*, **12** (1933), 1-82.

[77] LERAY, J., Sur le mouvement d'un liquide visqueux emplissant l'espace, *Acta Math.*, **63** (1934), 193-248.

[78] LIONS, J. L., & PRODI, G., Un théorème d'existence et unicité dans les équations de Navier-Stokes en dimension 2, *C. R. Acad. Sci. Paris*, **248** (1959), 3519-3521.

[79] MA, C.-M., A uniqueness theorem for Navier-Stokes equations, *Pacific J. Math.*, **93** (1981), 387-405.

[80] MARSDEN, J. E. & RATIU, T., Nonlinear stability in fluids and plasmas, *Sem. on New Results in Nonlinear P.D.E.*, A. Tromba ed., Vieweg (1987), 101-134.

[81] MASLENNIKOVA, V. N. & BOGOVSKII, M. E., Sobolev spaces of solenoidal vector spaces, *Sibirian Math. J.*, **22** (1982), 399-420.

[82] MASLENNIKOVA, V. N. & BOGOVSKII, M. E., Approximation of potential and solenoidal vector fields, *Sibirian. Math. J.*, 24 (1983), 768-787.

[83] MASUDA, K., On the stability of incompressible viscous fluid motions past objects, *J. Math. Soc. Japan*, 27 (1975), 294-327.

[84] MASUDA, K., Weak solutions of the Navier-Stokes equations, *Tôhoku Math. J.*, **36** (1984), 623-646.

[85] MIYAKAWA, T. & SOHR, H., On energy inequality, smoothness and large time behavior in L^2 for weak solutions of the Navier-Stokes equations in exterior domains, *Math. Z.*, **199** (1988), 455-478.

[86] MIZUMACHI, R., On the asymptotic behavior of incompressible viscous fluid motions past bodies, *J. Math. Soc. Japan*, **36** (1984), 497-522.

[87] NAGATA, W., Symmetry-breaking effects of distant sidewalls in Rayleigh-Benard convection, (these proceedings).

[88] OKAMOTO, H., Applications of degenerate bifurcation equations to the Taylor problem and the water wave problem, (these proceedings).

[89] PRODI, G., Theoremi di tipo locale per il sistema di Navier-Stokes e stabilita della soluzioni stazionarie, *Rend. Sem. Mat. Univ. Padova,* **32** (1962), 374-397.

[90] RUELLE, D. & TAKENS, F., On the nature of turbulence, *Comm. Math. Phys.*, **20** (1971), 167-192.

[91] SCHEFFER, V., Partial regularity of solutions to the Navier-Stokes equations, *Pacific J. Math.*, **66** (1976), 535-552.

[92] SCHONBEK, M. E., L^2 decay for weak solutions to the Navier-Stokes equations, *Arch. Rational Mech. Anal.*, **88** (1985), 209-222.

[93] SCHONBEK, M. E., Large time behavior of solutions to the Navier-Stokes equations, *Comm. PDE*, 11 (1986), 733-763.

[94] SERRIN, J., The initial value problem for the Navier-Stokes equations, *Nonlinear Problems*, R. Langer ed., University of Wisconsin Press, Madison, (1963), 69-98.

[95] SOHR, H., WAHL, W. VON & WIEGNER, M., Zur Asymptotik der Gleichungen von Navier-Stokes, *Nachr. Akad. Wiss. Göttingen,* **3** (1986), 45-59.

[96] SOLONNIKOV, V. A., On differential properties of the solutions of the first boundary-value problem for nonstationary systems of Navier-Stokes equations, *Trudy Mat. Inst. Steklov,* **73** (1964), 221-291.

[97] SOLONNIKOV, V. A., On the solvability of boundary and initial-boundary value problems for the Navier-Stokes system in domains with noncompact boundaries, *Pacific J. Math.*, **93** (1981), 443-458.

[98] SOLONNIKOV, V. A., Solvability of the problem of evolution of an isolated volume of viscous, incompressible capillary fluid, *J. Sov. Math.*, **32** (1986), 223-228.

[99] SOLONNIKOV, V. A. & PILETSKAS, K. I., Certain spaces of solenoidal vectors and solvability of the boundary value problem for the Navier-Stokes system of equations in domains with noncompact boundaries, *J. Sov. Math.*, **34** (1986), 2101-2111.

[100] SWANN, H., The convergence with vanishing viscosity of nonstationary Navier-Stokes flow to ideal flow in R^3, *Trans. Amer. Math. Soc.*, **157** (1971), 373-397.

[101] TAKESHITA, A., A remark on Hopf's Inequality, *Preprint Series, Nagoia Univ. Coll. Gen. Educ., Dep. Math.* (1982), to appear in *Pacific J. Math.*.

[102] TEMAM, R., *Navier-Stokes Equations*, Revised Edition, North-Holland, Amsterdam, New York, Oxford, 1984.

[103] TEMAM, R., *Navier-Stokes Equations and Nonlinear Functional Analysis*, SIAM, Philadelphia 1983.

[104] TEMAM, R., *Infinite-Dimensional Dynamical Systems in Mechanics and Physics,* Springer-Verlag, Berlin, Heidelberg, New York, 1988.

[105] VELTE, W., Stabilität und Verzweigung stationärer Lösungen der Navier-Stokesschen Gleichungen beim Taylorproblem, *Arch. Rational Mech. Anal.*, **22** (1966), 1-14.

[106] WAHL, W. VON, Gebrochene Potenzen eines elliptischen Operators und parabolische Differentialgleichungen in Räumen Hölderstetiger Funktionen, *Nachr. Akad. Wiss. Göttingen, Math.-Phys. Kl. II*, 11 (1972), 231-258.

[107] WAHL, W. VON, *The Equations of Navier-Stokes and Abstract Parabolic Equations*, Vieweg, Braunschweig/Wiesbaden, 1985.

[108] WAN, Y. H. & PULVIRENTI, M., Nonlinear stability of circular vortex patches, *Comm. Math. Phys.*, **99** (1985), 435-450.

[109] WIEGNER, M., Decay results for weak solutions of the Navier-Stokes equations on R^n, *J. London Math. Soc.*, (2) **35** (1987), 303-313.

[110] YUDOVICH, V. I., On the stability of self-oscillations of a fluid, *Sov. Math. Dokl.*, **11** (1970), 1543-1546.

NAVIER-STOKES EQUATIONS FROM THE POINT OF VIEW OF THE THEORY OF ILL-POSED BOUNDARY VALUE PROBLEMS

A.V.Fursikov
Department of Mechanics and Mathematics, Moscow University,
Lenin Hills, 119899 Moscow, USSR

The problem of existence "in the large" and of uniqueness of solutions of initial-boundary value problem for three-dimensional nonstationary Navier-Stokes equations is one of fundamental mathematical problems of fluid mechanics. There were many attempts to solve this problem in the classical formulation, i.e. to prove its unique solvability for arbitrary data from an appropriate functional space.Up to now these attempts were unsuccessful, therefore it seams reasonable to change the formulation of the problem and to consider it from the point of view of the theory of ill-posed boundary value problems. This point of view is quite appropriate, because it is known from physical experiments, that for large Reynolds numbers the fluid motion can change greatly under small perturbations of the data. Therefore we cannot be absolutly shure, that initial-boundary value problem for the Navier-Stokes equations with large Reynolds numbers is well-posed.

An abstract ill-posed problem has the from

$$\Phi u = f, \tag{1}$$

were $f \in F$ is a datum and $u \in U$ is an unknown element, U, F are Banach spaces, $\Phi: U \to F$ is a mapping. The specific character of this problem consists in the following. The solution of the equation (1) is unique in the space U, but it does not exist for all $f \in F$. Let F_1 be the set of the data $f \in F$ for which exists a solution $u \in U$ of the problem (1) (F is called the set of solvability of the problem (1)). The study of the question concerning the unique solvability of an ill-posed problem (1) is reduced to the investigation of the properties of the set of solvability F_1. We are interested to know how "large" is the set F_1. More preciesly, in the case of the Navier-Stokes equations and of their statistical analogs (the chain of moment equations, the Hopf equation) we shall be interested in the question of the density of the set F_1 in the functional space F. In §1 we consider the case, when (1) is the three-dimensional Navier-Stokes equations with homogeneous boundary conditions, and with fixed initial value and right hand side (r.h.s.) f. The theorem from [1] [2] asserts that in this case the set F_1 is dense in F relative to a certain norm, which is weaker than the norm in the space F.

In §2 we study the case, when (1) is the Hopf equation or the chain of moment equations, corresponding to the three dimensional Navier-Stokes equations with the zero r.h.s. In this case f is the initial value. The density F_1 in F is established (see [3]). We observe that in the case of the Hopf equation the problem is considered in the spaces of analytic functionals.

When the investigation of ill-posed equations is connected with the numerical solution, the conditionally well-posed formulation of a boundary value problem is used often (see [4], [5]). In this formulation it is supposed that we know only the approximation f_ε of the datum f (the assumption about the exact knowledge of f is natural only in the case of well-posed problems). In this formulation the assumption about existence of the exact solution u, satisfying a certain estimate (condition), is of principal importance. It is needed to find the approximation of solution of the problem (1) and to estimate the error between the exact solution and its appoximation. In §3 the conditionally well-posed formulation of the boundary value problem for three-dimensional Navier-Stokes system is discussed in details. The estimate of the error between the exact solution and its appoximation and the method of construction of approximation are presented.

§1. The unique solvability of three-dimensional Navier-Stokes system for the dense set of right hand sides

In the bounded domain $\Omega\in R^3$ with the boundary $\partial\Omega$ of class C^∞ we consider the Navier-Stokes system:

$$\frac{\partial u\ (t,x)}{\partial t} + \sum_{j=1}^{3} u_j \frac{\partial u}{\partial x_j} - \Delta u + \nabla p(t,x)=f(t,x) \qquad (1.1)$$

$$\mathrm{div} u(t,x)=0 \qquad (1.2)$$

where $t\in[0,T]$, $T>0$ is a fixed number or $T=\infty$, $x=(x_1,x_2,x_3)\in\Omega$, $u=(u_1,u_2,u_3)$, $f=(f_1,f_2,f_3)$, div f=0. We suppose that

$$u|_{t=0} = u_0(x) \tag{1.3}$$

and

$$u|_{\partial\Omega} = 0 \tag{1.4}$$

In this section the theorem of the unique solvability of the problem (1.1)-(1.4) is formulated, when r.h.s.f belongs to some dense set in an appropriate functional space. First of all we introduce functional spaces and write the problem (1.1)-(1.4) in an abstract form. This form was used in many papers. It is convenient also for our purposes too. Set

$$V = \{v(x) \in C_0^{\infty}(\Omega): \operatorname{div} v(x) = 0\}$$

$$H^0 = \text{the closure of } V \text{ in } (L_2(\Omega))^3 \tag{1.5}$$

$$H^1 = \text{the closure of } V \text{ in } (W_2^1(\Omega))^3 \tag{1.6}$$

$$H^2 = H^1 \cap (W_2^2(\Omega))^3 \tag{1.7}$$

where $W_2^k(\Omega)$ is the Sobolev space of functions, defined in Ω and square summable together with their derivatives up to order k. Let

$$\pi: (L_2(\Omega))^3 \to H^0$$

be the orthogonal projection. Set $A = -\pi\Delta$. Clearly $A: H^0 \to H^0$ is a self-adjoint positive operator with the domain $D(A) = H^2$, which has a discrete spectrum $0 < \lambda_1 \le \lambda_2 \le \ldots \lambda_k \to \infty$, $k \to \infty$. The set of its eigenvectors $\{e_k\}$ forms an orthonormal basis in H^0. Now, we introduce the spaces

$$H^{\alpha} = \{v = \sum_{k=1}^{\infty} v_k e_k: \|v\|_{\alpha}^2 = \sum_{k=1}^{\infty} \lambda_k^{\alpha} |v_k|^2 < \infty\} \tag{1.8}$$

with $\alpha \in R$. For $\alpha = 0,1,2$ the space (1.8) coincides, respectively, with the spaces (1.5), (1.6) or (1.7) . Finally, let

$$L_p^{\alpha} = L_p(0,T;H^{\alpha}), \quad \alpha \in R, \quad 1 \le p \le \infty \tag{1.9}$$

We shall look for the solution u of the problem (1.1)-(1.4) in the space

$$H^{1,2} = \{u \in L_2^2 : \dot{u} \in L_2^0\}, \qquad (1.10)$$

where

$$\dot{u} = \partial u(t,x)/\partial t$$

It is well known (see for instance [2], p.33) that in the space (1.10) the problem (1.1)-(1.4) is equivalent to the following Cauchy problem for the ordinary differential equation with operator coefficients

$$Q(u) = \dot{u} + Au + B(u) = f \qquad (1.11)$$

$$\gamma_0 u = u_0 \qquad (1.12)$$

Here $Au = -\pi\Delta u$, $\hat{B}(u,v) = \pi(\sum_{i=1}^{3} u_i \frac{\partial v}{\partial x_i})$, $B(u) = \hat{B}(u,u)$, and γ_0 is the operator of restriction of the function $u(t)=u(t,\cdot)$ at $t=0$: $\gamma_0 u=u(0)$.

We consider the question of the unique solvability of the problem (1.11), (1.12)for a dense set of right hand sides. It is known that the operator

$$(Q,\gamma_0):H^{1,2} \to L_2^0 \times H^1$$

is continuons, (Q is the operator from (1.11)). We take arbitrary initial value $u_0 \in H^1$ in (1.12) and we set

$$U_{u_0} = \{u \in H^{1,2} : \gamma_0 u = u_0\}$$

We denote by F_{u_0} the image of U_{u_0} under Q:

$$F_{u_0} = QU_{u_0} \qquad (1.13)$$

It is evident that $F_{u_0} \subset L_2^0$. According to the definition, the set F_{u_0} consists of those and only those r.h.s.f of the problem (1.11), (1.12), for which there exists the solution $u \in H^{1,2}$ of this problem. It is unique in the space $H^{1,2}$ (see, for example [2], p.31), therefore F_{u_0} is such a set of r.h.s., that the problem (1.11), (1.12) has the unique solution in $H^{1,2}$. The following theorem concerning properties of the set F_{u_0} holds.

Theorem 1.1. The set F_{u_0} is open in the topology of L_2^0 and dense everywhere in L_2^0 with respect to the norm of L_p^{-1}, where $1\le p<4/(5-2l)$ for $1/2<l<3/2$ and $p=2$ for $l>3/2$.

The proof of the theorem is given in [1], [2]. We note, that the definition of F_{u_0} in [1] differs from definition in this paper, and it was proved in [1] that F_{u_0} is open in the topolody of $L_2^{-1/2}$.

Remark 1.1 H.Sohr and W.van Wahl in [6] extended the result of theorem 1.1 in the following way. Let $R(U_0) = F_{u_0}\cap L_p(0,T;(L_p(\Omega))^3)$, where $2\le p<\infty$, $u_0\in H^4$, $0<T<\infty$. Then $R(U_0)$ is dense everywhere in $L_p(0,T;(L_p(\Omega))^3)$ with respect to the norm of the space $L_s(0,T;(L_q(\Omega))^3)$, where s, $q\in(1,\infty)$ and $4<2/s+3/q$.

Let us change the formulation of the problem. Assume now the r.h.s.f in (1.1) is fixed (for instance, $f=0$) and let

$$V_f= \{u_0\in H^1:\ \exists \text{ solution } u\in H^{1,2} \text{ of the problem (1.11), (1.12)}\} \qquad (1.14)$$

It is rather important to investigate the properties of the set V_f. For instance, is the set V_f dense everywhere with respect to some norms? This problem is still open. But for statistical analogous of the three-dimension Navier-Stokes system there exists the answer to this question. This result will be presented in the following section.

§2. Unique solvability of the chain of moment equations and of the Hopf equation for a dense set of initial values

Consider the Navier-Stokes equations with periodic boundary conditions, i.e in the problem (1.1)-(1.4) we replace (1.4) by the condition

$$u(t, \ldots x_i \ldots) = u(t, \ldots x_i + 2\pi \ldots),\ i=1,2,3 \qquad (2.1)$$

In this case the spaces H^α, $\alpha\in R$ are defined by the formula

$$H^\alpha =\{v=(v^1,v^2,v^3):\ v \text{ is periodic, } \operatorname{div} v=0,\ \int_\Omega v\,dx=0,\ \|v\|_\alpha<\infty\} \qquad (2.2)$$

where $\Omega = \{x = (x_1,x_2,x_3):0<x_i<2\pi,\ i=1,2,3\}$ is the cube of periods,

$$\|\bar{\bar{v}}\|_\alpha^2 = \sum_{\xi\in Z^2\setminus\{0\}} |\xi|^{2\alpha}|\hat{u}(\xi)|^2,$$

and $\hat{u}(\xi)$ is a ξ-th Fourier coefficient of the function $\overset{=}{V}(x)$; div $\overset{=}{V}$ and $\int v dx$ in (2.2) are understood in the sence of distributions. As in §1, the problem (1.1)-(1.3), (2.1) is written in the form (1.11). Here we shall consider the case f=0:

$$\dot{u} + Au + B(u) = 0 \qquad (2.3)$$

The chain of moment equations arises in the statistical hydromechanics. In the statistical approach to the of equation (2.3), we suppose, that instead of concrete initial value u_0 there is a statistical distribution of initial values, i.e. the probability measure $\mu(du_0)$, defined on the space of initial values. We present a general idea of deduction of the chain of moment equations. Suppose first that $\dim\Omega=2$. Then the problem (2.3), (1.12) has a unique solution. Therefore the operator S_t, assigning a solution $S_t u_0$ to the initial value u_0 is well defined. By means of S_t we can define a statistical solution $\mu(t,du)$ of the Navier-Stokes equations

$$\mu(t,\omega) = \mu_0(S_t^{-1}\omega) \qquad (2.4)$$

where ω is a Borel subset of the space H^0, and S_t^{-1} is pre-image of the set ω : $S_t^{-1}\omega = \{u_0 \in H^0 : S_t u_0 \in \omega\}$.

For any t it is possible to determine moments $M_k(t,x_1 \dots x_k)$ of the measure $\mu(t,du)$ by means of the formula

$$M_k(t,x_1,\dots,x_k) = \int_{H_0} u(x_1)\otimes \dots \otimes u(x_k)\, \mu(du) =$$

$$= \int_{H_0} S_t u_0(x_1) \otimes \dots \otimes S_t u_0(x_k)\, \mu_0(du_0), \qquad (2.5)$$

the identities in (2.5) being understood in the sense of distributions theory (see the details in [2] p.83, and see also the formula (2.19) below). We note, that the moments $M_k(t,x_1,\dots,x_k)$ are determined, if

$$\int \|u_0\|^k \mu_0(du_0) < \infty \qquad \forall k > 0$$

For the deduction of the chain of moment equations it is sufficient now to take the derivative of both sides of the identity (2.5) with respect to t and to express $\frac{d}{dt}S_t u_0$ from the equation (2.3). After some transformations we obtain an infinite chain of equations for the moments $M_k(t,x_1 \dots x_k)$ of a statistical solution $\mu(t,du)$:

$$\dot{M}_k + A_k M_k + B_k M_{k+1} = 0, \quad k = 1,2,... \tag{2.6}$$

Here $\dot{M}_k = \partial M_k/\partial t$, the operator A_k is determined by the formula

$$A_k = \sum_{j=1}^{k} I_{j-1} \otimes A \otimes I_{k-j}, \tag{2.7}$$

where A is operator from (2.3), $I_l = I\otimes ...\otimes I$ (l - times), I is the identity operator. For the definition of the operator B_k from (2.6) we need to introduce the linear operator $\bar{B}$: $H^2 \otimes H^2 \to H^0$ such that

$$B(u) = \bar{B}(u\otimes u) \quad \forall u \in H^2,$$

where B(u) is the nonlinear operator from equation (2.3). The existence of the operator $\bar{B}$ in the case of the Navier-Stokes equations was proved in [2] p.p.115-116 and in [3]. Now the operator B_k from (2.6) is defined by means of the formula

$$B_k = \sum_{j=1}^{k} I_{j-1} \otimes \bar{B} \otimes I_{k-j} \tag{2.8}$$

The system of equations (2.6) is called the chain of moment equations corresponding to the Navier-Stokes system. The details of deduction of the chain of equations see in [2] p.93. In the case dim Ω=3, corresponding to a three-dimensional Navier-Stokes system, we suppose by definition, that if the moments M_k of some measure satisfy the chain of moment equations (2.6), then this measure is a statistical solution. We shall suppose below, that dim Ω=3.

The chain of equations (2.6) is supplied with the initial conditions

$$M_k|_{t=0} = m_k, \quad k=1,2,3,... \tag{2.9}$$

We want to investigate the question of the unique solvability of the problem (2.6), (2.9). First of all we untroduce functional spaces, in which this problem will be studied. Set

$$H^{\alpha}(k) = H^{\alpha}\otimes ...\otimes H^{\alpha} \qquad \text{(k times)} \tag{2.10}$$

where $\alpha \in R$. We define the notion of a symmetric element of the space $H^{\alpha}(k)$. Let $\{e_j\}$ be an orthogonal basis in H^{α}. Then the set of vectors

$$e(\bar{n}^k) = e_{n_1} \otimes \ldots \otimes e_{n_k}, \quad \bar{n}^k \in N^k \tag{2.11}$$

forms an orthogonal basis in $H^{\alpha}(k)$.
Here $\bar{n}^k = (n_1 \ldots n_k)$, $N^k = N \times \ldots \times N$ (N is the set of natural numbers). Therefore for $m \in H^{\alpha}(k)$ there holds the expansion

$$m = \sum_{\bar{n}^k \in N^k} m(\bar{n}^k) e(\bar{n}^k) \tag{2.12}$$

We determine the operator σ of symmetrization of the vector m by the formula

$$\sigma_m = \sum_{\bar{n}^k} \sigma m(\bar{n}^k) e(\bar{n}^k),$$

where

$$\sigma m(\bar{n}^k) = \sigma m(n_1, \ldots, n_k) = \frac{1}{k!} \sum_{(j_1, \ldots, j_k)} m(n_{j_1}, \ldots, n_{j_k})$$

and the summation is taken over all permutations of $(1, \ldots k)$.

Vector $m \in H^{\alpha}(k)$ is called symmetric, if $\sigma m = m$. The subspace of all symmetric vectors of the space $H^{\alpha}(k)$ is denoted by $SH^{\alpha}(k)$. From $SH^{\alpha}(k)$ we shall choose initial values m_k for the problem (2.6), (2.9). We denote

$$H_q^{\alpha}(k) = \{w: A_k^{q/2} w \in H^{\alpha}(k)\},$$

where $\alpha \in R$, $q = \pm 1$, A_k is the operator (2.7). The subspace of symmetric elements of the space $H_q^{\alpha}(k)$ is denoted by $SH_q^{\alpha}(k)$. Let

$$Y^{\alpha}(k) = \{u(t) \in L_2(0,T;SH_1^{\alpha}(k)) : \dot{u}(t) \in L_2(0,T;SH_{-1}^{\alpha}(k))\}$$

In this space we shall look for the component M_k of the solution of the problem (2.6), (2.9). For any $R>0$ we set

$$N_R^\alpha = \prod_{k=1}^{\infty} SH^\alpha(k), \quad \|m\|^2_{N_R^\alpha} = \sum_{k=1}^{\infty} R^{-2k} \|m_k\|^2_{H^\alpha(k)} \tag{2.13}$$

$$Y_R^\alpha = \prod_{k=1}^{\infty} Y^\alpha(k), \quad \|M\|^2_{Y^\alpha(k)} = \sum_{k=1}^{\infty} R^{-2k}\|M_k\|_{Y^\alpha(k)} \tag{2.14}$$

The solution $M = (M_1, M_2, \dots)$ of the problem (2.6), (2.9) will be looked for in the space Y_R^α, and the initial value $m = (m_1, m_2, \dots)$ will be taken from the spase N_R^α. It is possible to show, that the parameter R in N_R^α, Y_R^α is closely connected with the Reynolds number: the set of moments of the fluid flow with a large Reynolds number can belong to Y_R^α only if R is quite large.

Let us formulate the results about the unique solvability of the problem (2.6), (2.9). We begin with the uniqueness theorem.

Theorem 2.1. Let $\alpha>2$, $R>0$. Then the solution $M(t) = \{M_k(t, x_1, \dots x_k), k\in N\}$ of the problem (2.6), (2.9) is unique in the space Y_R^α.

The proof of this theorem is given in [3].

Denote by V_R^α the set of such initial values $m\in N_R^\alpha$, that the problem (2.6), (2.9) has the solution $M\in Y_R^\alpha$.

Theorem 2.2. Let $\alpha>1/2$, $R>0$. Then the set V_R^α is dense in N_R^α(with respect to the norm of N_R^α). Any element $m = (m_1, m_2, \dots)$ such, that only a finite number of components m_k are not equal to zero, belongs to V_R^α.

This theorem was proved in [3] too.

Let us deduce from theorems 2.1 and 2.2 the unique solvability of the Hopf equation for a dense set of initial values. We shall take initial values from the class of analytic functionals, and we shall look for the solution in the class of analytic functionals, depending on the parameter t.

We begin with definitions. We suppose, that all the functional spaces in this section are complex. Let X be a Hilbert space. Function $f: X\to\mathbb{C}$ is analytic at the point $x_0\in X$, if it is bounded in a neighbourhood of a point x_0 and can be developed into a converging series

$$f(x) = f(x_0) + \sum_{k=1}^{\infty} f_k(x-x_0)^k, \tag{2.15}$$

where $f_k z^k = \hat{f}_k(z, \dots, z)$, and $\hat{f}_k: X^k\to\mathbb{C}$ is a k-linear symmetric function.

Function f is analytic on X, if it is analytic at any point $x_0 \in X$ and bounded on any closed bounded set $K \subset X$. By means of such K we define the seminorm

$$|f|_k = \sup_{x \in K} |f(x)| \tag{2.16}$$

We denote by Z(X) the space of analytic functions on X supplied with the topology, that is determined by seminorms (2.16). The linear continuous functional on Z(X) is called an analytic functional. In virtue of definition, there exist such a closed bounded set $K \subset X$ and a constant $C>0$, that

$$|\langle F,f\rangle| \leq C|f|_K \quad \forall f \in Z(X) \tag{2.17}$$

where $\langle F,f\rangle$ is a value of the functional F on $f \in Z(X)$. This K is called a determinant set of an analytic functional F. We introduce the convergence in the space $Z^*(X)$ of analytic functionals in the following way. The sequence $F_n \in Z^*(X)$ converges to $F \in Z^*(X)$, if there exists a closed bounded set $K \subset X$, that is determinant both for F, and for any F_n, and

$$\langle F_n,f\rangle \to \langle F,f\rangle \text{ as } n \to \infty \quad \forall f \in Z(X) \tag{2.18}$$

Now we consider analytic functionals on $Z(H^\alpha)$ where H^α is the space (2.2). For $g_k \in H^{-\alpha}(k)$ (see (2.10)) we introduce the functional

$$g_k \to \langle F, (g_k, \overset{k}{\otimes} u)_{(k)}\rangle$$

where $F \in Z^*(H^\alpha), (.,.)_{(k)}$ is the duality between $H^{-\alpha}(k)$ and $H^\alpha(k)$. Because of (2.17) this functional is continuous and therefore there exists such $m_k \in H^\alpha(k)$, that

$$\langle F, (g_k, \overset{k}{\otimes} u)_{(k)}\rangle = (g_k, m_k)_{(k)} \quad \forall g_k \in H^{-\alpha}(k) \tag{2.19}$$

Vector m_k is called the k-th moment of an analytic functional F. Let $F \in Z^*(H^\alpha)$ and $m = (m_1, m_2 \ldots)$ be its moments. It is always possible to choose a ball

$$B_\xi^\alpha = \{u \in H^\alpha : \|u\|_\alpha \leq \xi\} \tag{2.20}$$

with some $\xi>0$, that is a determinant set K from (2.17). From (2.19), (2.17), (2.20) it follows, that for any $k\in N$

$$|(g_k,m_k)_{(k)}| \le C|(g_k,\overset{k}{\otimes}u)_{(k)}|_{B^{\alpha}_{\xi}} \le C\|g\|_{H^{-\alpha}(k)}\,\xi^k,$$

i.e. $\|m_k\|_{H^{-\alpha}(k)} \le C\xi^k$ and therefore $m = (m_1,m_2,\dots)\in N^{\alpha}_R$ with $R>\xi$.

Suppose now that there are $m = (m_1,m_2,\dots)\in N^{\theta}_R$ and $m_0\in\mathbb{C}$ are given with $R>0$, $\theta>0$. We want to construct such an analytic functional $F\in Z^*(H^{\alpha})$, $\alpha<\theta$, that the equation (2.19) is satisfied for any $g_k\in H^{-\alpha}(k)$ and $k\in N$. In other words, we want to solve the moment problem (2.19) in the class of analytic functionals.

For $f\in Z(H^{\alpha})$ we denote

$$f_k(\bar n^k) = \hat f_k(e_{n_1},\dots,e_{n_k}),$$

where $\hat f_k$ is a k-linear functional, defined by $f_k(x-x_0)^k$ from (2.15) with $x_0=0$, $\bar n^k = (n_1,\dots,n_k)$ and $\{e_j\}$ is an orthogonal basis in H^{α}. We set

$$\langle F,f\rangle = f_0w_0 + \sum_{k=1}^{\infty}\sum_{\bar n^k\in N^k} m_k(\bar n^k)f_k(\bar n^k) \tag{2.21}$$

where $m_k(\bar n^k)$ is determined by m_k from the equality (2.12). As it is proved in the theorem formulated below, for some α the formula (2.21) determines an analytic functional on $Z(H^{\alpha})$.

Theorem 2.3. Let $\theta>\alpha>3/4$. Then (2.21) determines an analytic functional on $Z(H^{\alpha})$, and there exists such $\xi_1>R$, depending on R that

$$|\langle F,f\rangle| \le C\|m\|_{N^{\theta}_R}\,|f|_{B^{\alpha}_{\xi_1}}, \tag{2.22}$$

where B^{α}_{ξ} is the ball (2.20). The solution F of the moment problem is unique in $Z^*(H^{\alpha})$.

The proof of theorem 2.3 is analogous to the proof of theorems 2.1, 2.2 from [8].

Let $\theta>0$, $R>0$ and let $M(t) = (M_1,M_2,\dots)\in Y^{\theta}_R$ be a solution of the problem (2.6), (2.9), where $m = (m_1,m_2,\dots)\in N^{\theta}_R$, and $M_0(t) = m_0 = 1$. By theorem 2.3, for $\theta-\alpha>3/4$ there exists a unique solution $F\in Z^*(H^\alpha)$ of the moment problem (2.19). Moreover, for any $t>0$ the solution $G(t)\in Z^*(H^\alpha)$ of the moment problem is determined uniquely by the set of moments $M(t) = = (M_1(t), M_2(t),\dots)$. Because of (2.22) and of the inclusion $Y_R^\theta\subset C(0,T; N_R^\theta)$ there holds the inclusion $G(t)\in C(0,T; Z^*(H^\alpha))$, where α can be supposed greaten than 1.

Proposition 2.1. Let $M(t)\in Y_R^\theta$ be a solution of the problem (2.6), (2.9), with $m\in N_R^\theta$. Let $G(t)\in C(0,T;Z^*(H^\alpha))$, $F\in Z^*(H^\alpha)$, where $\theta-\alpha>3/4,\alpha>1$, be analytic functionals, whose moments equal correspondingly to $M(t)$, m, and $M_0(t) = m_0 = 1$. Then $G(t),F$ satisty for any $v\in H^1$ the equation

$$\langle G(t),e^{i(\cdot,v)_0}\rangle-\langle F,e^{i(\cdot,v)_0}\rangle + \int_0^t \langle G(\tau),L(\cdot,v)\rangle\, d\tau=0 \qquad (2.23)$$

where

$$L(u,v) = ie^{i(u,v)_0}(Au + B(u),v)_0$$

The equality (2.23) is called the Hopf equation. From theorems 2.1, 2.2, 2.3 and from proposition 2.1 the following result about the unique solvability of the Hopf equation follows. Denote by X_R^θ the family of analytic functionals $G(t)$, depending on the parameter t, whose moments $M(t) = (M_1(t),M_2(t),\dots)$ belong to the space Y_R^θ. Let W^θ is such subset of $Z^*(H^\theta)$, that for any $F\in W^\theta$ there exists the solution $G(t)\in X_R^\theta$ of equation (2.23) with some $R>0$. The set W^θ is called the set of existence of equation (2.23).

Theorem 2.4. Let $\theta>2$. Then

(i). The set of existence W^θ is dense everywhere in the space of initial values $Z^*(H^\theta)$ of equation (2.23)relatively to convergence in the space $Z^*(H^\alpha)$, wich $\theta-\alpha>3/4$.

(ii). The solution $G(t)$ of equation (2.23) is unique in the space X_R^θ.

Let $\mu_0(du_0)$ be a probability measure on H^θ with support in the ball B_R^θ. Then the analytic functional $F\in Z^*(H^\theta)$ is determined by the formula

$$\langle F,f\rangle = \int_{H^\theta} f(u_0)\mu_0(du_0) \qquad (2.24)$$

From results of the paper [3] there follows.

Theorem 2.5. Let $\theta>2$, $R>0$ and let $G(t)\in X_R^\theta$ be a solution of the Hopf equation (2.23) with an initial value $F\in Z^*(H^\theta)$, admitting the representation (2.24). Then for any $t>0$ there exists the probability measure $\mu(t,du)$ with the support in the ball B_R^θ such that

$$\langle G(t),f\rangle = \int f(u)\mu(t,du)$$

The measure $\mu(t,du)$ satisfies (2.4) and is called a spatial statistical solution of the equation (2.3).

§3. Conditionally well-posed statement of the boundary-value problem for three-dimensional Navier-Stokes equations

The conditionally well-posed statement of boundary-value problems has a great significance ([4],[5]) in the theory of ill-posed problems. We know, that solutions of evolutional Navier-Stokes equations are very unstable relative to fluctuations of data, when the Reynolds number is quite large. As it was pointed out, for large Reynolds numbers we have no absolute confidence that the boundary value problem for three-dimensional Navier-Stokes equations is well-posed. Therefore it seems expedient to consider a conditionally well-posed statement of the boundary value problem for the Navier-Stokes equations.

Let us consider the problem (1.1), (1.2), (1.4) in a bounded domain $\Omega\subset R^3$ with a smooth boundary $\partial\Omega$. More precisely, we consider equation (1.11) with r.h.s.$f=0$:

$$\dot{u} + Au + B(u) = 0 \tag{3.1}$$

Instead of condition (1.12) we assume that

$$\|\gamma_0 u - u_0\|_0<\varepsilon , \tag{3.2}$$

where $u_0\in H^0$ is a given function, ε is a given positive number (by the sense of the problem it must be small). Next we suppose, that

$$\|u\|_{L_2^2} \le M \tag{3.3}$$

where $M>0$ is given number, $L_2^2 = L_2(0,\infty;H^2)$ is defined in (1.9). (We shall consider the solution $u(t,\cdot) = u(t)$ for $t\in(0,\infty)$.)

We shall look for functions, satisfying (3.1)-(3.3), in the space $H^{1,2}$. It is evident that the function, satisfying (3.1)-(3.3), exists not for any u_0,ε,M. Therefore we must assume, that the following condition for u_0,ε,M is satisfied.

Condition 3.1. There exists a function $u\in H^{1,2}$, satisfying (3.1)-(3.3).

The equation (3.1) and inequalities (3.2), (3.3) are called conditional well-posed statement of the boundary-value problem for the Navier-Stokes equations. The ideas leading to this statement are as follows. Suppose, that we are studying the evolution of a fluid flow, and we are able to measure the vector field of velocities at the initial moment. As a result of this measurement we have $u_0(x)$. Any measurement is approximate, therefore it is natural to write the initial condition in the form (3.2), where ε is the accuracy of a measuring instrument. It is assumed that some global characteristic of the flow is known, which is described by means of inequality (3.3). The condition of the form (3.3) is quite important for the conditional well-posed statement of a problem. We are not obliged to require the boundedness of the solution just in the norm L_2^2, but it is necessary, that the solution of the Cauchy problem for equation (3.1) would be unique in the class of functions, satisfying (3.3). The problem under consideration describes a real physycal process, therefore the condition 3.1 concerning the existence of the solution is natural.

All these considerations are also valid for the case, when equation (3.1) is well-posed, and its solution changes slightly under small changes of the initial value (for example, if $\|u_0\|_1$ is small). But in this case it is better to abstract oneself from errors of measurements and to replace the condition (3.2) by a traditional condition (1.12). Then the condition (3.3) becames unnecessary, because the solution $\hat{u}$ of the problem (3.1), (1.12) is determined uniquely by the initial value u_0, and $\hat{u}$ differs slightly from the vector field u, satisfying (3.1)-(3.3). Therefore $\hat{u}$ satisfies to (3.3) automatically (at least approximatly). Replacement of the condition (3.2) by (1.12) leads to wrong results in the case, when equation (3.1) is ill-posed, or its solution changes considerably with small fluctuations of initial values. Indeed, in this case it is possible, that the problem (3.1), (1.12) has no smooth solution, or this solution does not satisfy the inequality (3.3) and its L_2^2-norm is greater more than M.

Let us now discuss the question: what does it mean to solve the problem (3.1)-(3.3). We recall, that the existence of a solution, i.e.

of a function satisfying condition (3.1)-(3.3) is postulated. On the other hand, we cannot hope, that a function, satisfying (3.1)- (3.3) is unique. The aim of the investigation of the problem (3.1)-(3.3) consists in the construction of an approximate solution.

To this end it is necessary to show first of all that the difference $w = u_1 - u_2$ of any two functions u_1, u_2, satisfying (3.1)-(3.3), is small, i.e. that

$$\|w(t)\| \leq C(t,\varepsilon), \text{ where } \forall t \in R \; C(t,\varepsilon) \to 0, \text{ when } \varepsilon \to 0$$

Secondly it is necessary to present the method of construction of a function, satisfying (3.1)-(3.3).

Let us solve the problem (3.1)-(3.3).

Lemma 1.1. Let $w = u_1 - u_2$, where $u_i \in H^{1,2}$, $i=1,2$ are functions, satisfying conditions (3.1)-(3.3). Then

$$\|w(t)\|_0 \leq C(t,\varepsilon) \tag{3.5}$$

with

$$C(t,\varepsilon) = \min \, (2de^{-\lambda_1 t}, \; 2\varepsilon \exp \, [cd^{1/2} M^{3/2} \lambda_1^{-1/4} (1-e^{-2\lambda_1 t})^{1/4}] \tag{3.6}$$

where λ_1 is the number from (1.8), $d = \|u_0\|_0 + \varepsilon$, and C is a constant, depending only on the domain Ω.

The proof of this lemma is based on two well-known inequalities. Firstly, for any solution u(t) of the equation (3.1) we have

$$\|u(t)\|_0^2 \leq e^{-2\lambda_1 t} \|\gamma_0 u\|_0^2 \tag{3.7}$$

Secondly, for the difference $w = u_1 - u_2$ of the two solutions (3.1) the following estimate holds:

$$\|w(t)\|_0^2 \leq \|w(0)\|_0^2 \exp \, (C \int_0^t \|u_1\|_{3/2}^2 \, d\tau) \tag{3.8}$$

From (3.7) and (3.8) the inequality (3.6) easily follows.

Now we give the method of construction of the function $u(t,\cdot)$, satisfying conditions (3.1)-(3.3). We consider the extremal problem

$$\|\gamma_0 u - u_0\|_0 \to \inf \tag{3.9}$$

$$\dot{u} + Au + B(u) = 0 \tag{3.10}$$

$$\|u\|_{L_2^2} \leq M \tag{3.11}$$

A function $u \in H^{1,2}$, satisfying (3.10), (3.11), is called an admissible element of the problem (3.9)-(3.11). The set of admissible elements is denoted by Z.

The function $\hat{u} \in Z$ such that

$$\|\gamma_0 \hat{u} - u_0\|_0 = \inf_{u \in Z} \|\gamma_0 u - u_0\|$$

is called the solution of the problem (3.9)-(3.11).

Theorem 3.1. The solution $\hat{u}$ of the problem (3.9)-(3.11) exists.

The proof is analogous to the proof of theorem 3.1 from [7]. First, we prove the solvability of the problem in the space $Y = \{y \in L_2^2 : \dot{y} \in L_2^{-1}\}$. Now we conclude from the inclusion $\hat{u} \in Y$ where $\hat{u}$ is a solution of the problem (3.9)-(3.11), that the $\hat{u} \in H^{1,2}$.

Proposition 3.1. The solution $\hat{u}$ of the problem (3.9)-(3.11) satisfies conditions (3.1)-(3.3).

Proof. We have to prove, that $\hat{u}$ satisfies (3.2). Let $v \in Y$ be a function, satisfying (3.1)-(3.3). It exists in virtue of condition 3.1. Evidently, $v \in Z$ and therefore

$$\|\gamma_0 \hat{u} - u_0\|_0 \leq \|\gamma_0 v - u_0\|_0 \tag{3.12}$$

g.e.d.

The solution $\hat{u}$ of the problem (3.9)-(3.11) is called quasisolution of the problem (3.1)-(3.3).

Remark 3.1. As it was pointed out the condition 3.1 does not hold for all u_0, ε, M. But in virtue of theorem 3.1 the solution $\hat{u}$ of the problem (3.9)-(3.11) exists for any u_0, ε, M.

It is possible to use $\hat{u}$ for check-up of condition 3.1 for arbitrary u_0, ε, M. Indeed, $\hat{u}$ satisfies (3.12) for any solution v of (3.1)-(3.3). Therefore the condition 3.1 is equivalent to inequality $\|\gamma_0 \hat{u} - u_0\| \leq \varepsilon$.

Note, that it is in final form and that no similar paper has been or is being submitted elsewhere.

References

[1] Fursikov A.V. Control problems and theorems, concerning the unique solvability of a initial boundary value problem for the three-dimensional Navier-Stokes and Euler equations. -Math USSR Sbornik, 43, (2), (1982), 251-273

[2] Vishik M.I., Fursikov A.V. Mathematical problems of statisfical hydromechanics - Kluwer academic publishers, Dordrecht, Boston, London, 1988

[3] Fursikov A.V. On the uniqueness of solutions of the chain of moment equations, corresponding to the three dimensional Navier-Stokes system. Mat.Sb. 134, (4), (1987), 472-495 (in Russian)

[4] Tikhonov A.N., Arsenin V.Ia. Methods of solution of ill-posed problems, Nauka, Moscow, 1979 (in Russian)

[5] Lavrentiev M.M., Romanov V.G., Shishatskiy S.P. Ill-posed problems of mathematical physic and analysis. Nauka, Moscow, 1980 (in Russian)

[6] Sohr H., von Wahl W. Generic solvability of the equations of Navier-Stokes. Hirosima Math. J. 17, (3), 613-625

[7] Fursikov A.V. Some problems of the theory of optimal control of nonlinear systems with distributed parameters. Proc. of I.G.Petrovskiy's sem.9, (1983), 167-189 (in Russian)

[8] Fursikov A.V. Analytic functionals and unique solvability of quasilinear dissipative systems for almost all initial conditions. Trans. Moscow Math.Soc. 1987, 1-55

(received February 27, 1989).

ON THE STATISTICAL APPROACH TO THE NAVIER-STOKES EQUATIONS

A.V. Fursikov
Department of Mechanics and Mathematics, Moscow University, Lenin Hills, 119899 Moscow, USSR

This article is a short review of mathematical results in the statistical hydromechanics. The details may be found in [1]. That book contains a number of the proofs. For the latest results see [6,7,8].

1. The viscous incompressible fluid flow is described by the Navier-Stokes equations

$$\frac{\partial u(t,x)}{\partial t} + \sum_{j=1}^{n} u^j \frac{\partial u}{\partial x^j} = \nu \Delta u - \nabla p(t,x) + g(t,x), \ \operatorname{div} u = 0 \tag{1}$$

where $u(t,x) = (u^1,\dots,u^n)$ is velocity, $p(t,x)$ is pressure, $g(t,x) = (g^1,\dots,g^n)$, $\nu>0$, $t\in[0,T]$, $x = (x^1,\dots,x^n)\in\Omega\subset R^n$. Here $T>0$ is fixed and Ω is a bounded domain with a boundary $\partial\Omega\in C^\infty$. The dimension n of Ω equals 2 or 3. We suppose that

$$u(t,x) = 0, \quad x\in\partial\Omega \tag{2}$$

$$\gamma_0 u(x) = u_0(x) \tag{3}$$

where γ_τ is the restriction operator for the function $u(t,x)$ to $t=\tau$: $\gamma_\tau u(x) = u(\tau,x)$. Set $V = \{u(x)\in(C_0^\infty(\Omega))^n : \operatorname{div} v = 0\}$,

H^0 = the closure of V in $(L_2(\Omega))^3$, $\|u\|_{H^0} = \|u\|_{(L_2(\Omega))^3} \equiv \|u\|$

H = the closure of V in $(W^1(\Omega))^3$, $\|u\|_{H^1} = \|u\|_{(L_2(\Omega))^3} \equiv \|u\|_1$

where $W_2^k(\Omega)$ is a Sobolev space of functions, square summable in Ω together with their derivatives up to order k. Let $\pi : (L_2(\Omega))^3 \to H^0$ be an orthogonal projection. Applying π to the both parts of the first equation in (1), we obtain the equation without p:

$$\dot{u} + Au + B(u) = f \tag{4}$$

where $\dot{u} = \partial u/\partial t$, $A = -\nu\pi\Delta$, $B(u) = \pi\sum_1^n (u^j \frac{\partial u}{\partial x^j})$, $f = \pi g$. We set

$$Z = C(0,T;\ (W_2^{-s}(\Omega))^3) \cap L_\infty(0,T;H^0) \cap L_2(0,T;H^1)$$

where $s > 1 + \frac{n}{2}$ is fixed, $W_2^{-s}(\Omega)$ is the Sobolev space with the negative index of the smoothness (See the definition in [2]). The existence of the solution $u(t,x) \in Z$ of the problem (4),(3) was proved for any $u_0 \in H^0$, $f \in L_2(0,T;H^0)$, when $n = 2$ or 3. For $n = 2$ the uniqueness of the solution $u \in Z$ of the problem (4),(3) is proved, but for $n = 3$ the uniqueness in the space Z is not obtained up to now. The solution of the equation (4) will be called the individual solution.

2. The statistical approach to the solution of hydromechanical problems was originated by Reynolds, Taylor, Kolmogorov and other authors. E.Hopf was the first who obtained the equation for the characteristic functional of statistical solution giving the probability description of fluid flows [3]. There is much information about statistical hydromechanics with detailed review of literature in the books by A.S.Monin and A.M.Jaglom [4]. Ĉ.Foiaş investigated in [5] the questions of existence and uniqueness of spatial statistical solutions. Space-time and homogeneous statistical solutions were constructed in [1].

Denote $\mathcal{B}(x)$ the σ-algebra of Borel subsets of a linear topological space X. The statistical approach to problem (4),(3) implies that instead of the fixed initial value u_0 in (3) we set the random vector field u_0. It means that we set the probability measure $\mu_0(\omega_0)$, $\omega_0 \in \mathcal{B}(H^0)$, determining the distribution of probabilities of the random field u_0. In this case the solution of the problem (4),(3) is the random vector field, and its distribution of probabilities $P(\omega)$, $\omega \in \mathcal{B}(Z)$ is called the space-time statistical solution. Thus, if the probability measure $P(\omega)$, $\omega \in \mathcal{B}(Z)$ is concentrated on the individual solutions of (4) and connected with the initial measure μ_0 by the equality

$$P(\{u(t,x) : \gamma_0 u \in \omega_0\}) = \mu(\omega_0) \qquad \forall\ \omega_0 \in \mathcal{B}(H^0)$$

it is called the space-time statistical solution of equation (4).

In [1] the following theorem was proved.

Theorem 1. Let

$$\int \|u_0\|^2 \mu_0(du_0) < \infty \tag{5}$$

Then there exists a space-time statistical solution $P(\omega)$, $\omega \in \mathcal{B}(Z)$, satisfying the energy estimate

$$\int \Big(\sup_{t\in[0,T]} \|u(t,\cdot)\|^2 + \int_0^T \|u(t,\cdot)\|_1^2 dt\Big) P(du) \le c\Big(1 + \int \|u_0\|^2 \mu(du_0)\Big) \qquad (6)$$

If n = 2, then the space-time statistical solution P is uniquelly determined by the initial measure μ_0, satisfying (5). The existence of a space-time statistical solution P may be proved when the initial measure μ_0 does not satisfy condition (5) of the finiteness of mean energy, i.e. we have (see [6]) an

Theorem 2. For arbitrary probability measure μ_0 on $\mathcal{B}(H^0)$ there exists the space-time statistical solution P defined on $\mathcal{B}(Z)$.

3. Let $\mu(t, \omega_0)$ be a family of probability measures, depending on the parameter t, which are the restriction of a space-time statistical solution P to the moment t

$$\mu(t,\omega_0) = \gamma_t^* P(\omega_0) = P(\gamma_t^{-1}\omega_0) \qquad \forall\, \omega_0 \in \mathcal{B}(H^0) \qquad (7)$$

The energy inequality for $\mu(t, \omega_0)$ is easily deduced from (6):

$$\int \|v\|^2 \mu(t,dv) + \nu \int_0^t\int \|v\|_1^2 \mu(\tau,dv)\, d\tau \le \int \|v\|^2 \mu_0(dv) + \frac{1}{\nu}\int_0^t \|f(\tau,\cdot)\|^2 d\tau \qquad (8)$$

Let (u,v) be the scalar product in $(L_2(\Omega))^n$, and

$\chi(t,w) = \int \exp\{i(u,v)\}\mu(t,dv)$

be a characteristic functional of $\mu(t,dv)$. The family $\mu(t,dv)$ satisfies the Hopf equation

$$\chi(t,w) - \chi_0(w) = \int_0^t\int \exp\{i(v,w)\}\,(L(\tau,v),w)\,\mu(\tau,dv)\,d\tau \qquad (9)$$

for arbitrary $t \in [0,T]$, $w \in H^0 \cap (W_2^s(\Omega))^n$, $s > 1 + \frac{n}{2}$, where

$$\chi_0(w) = \int \exp\{i(v,w)\}\mu_0(dv) \qquad (10)$$

is the characteristic functional of an initial measure, and

$$(L(t,v),w) = i\Big[\sum_{j=1}^{n} (\partial w/\partial x^j, v^j v) + \nu(\Delta w, v) + (w, f(t,\cdot))\Big]$$

A family of measures $\mu(t, \omega_0)$, $t \in [0,T]$, $\omega_0 \in \mathcal{B}(H^0)$ is called a spatial statistical solution with an initial measure $\mu_0(\omega_0)$, if $\mu(t,dv)$ satisfies the energy estimate (8) and the Hopf equation (9).

The following result of Ĉ.Foiaş [5] is one of the first rigorous results in statistical hydromechanics.

Theorem 3. Let the initial measure $\mu_0(dv)$ satisfy condition (5). Then the spatial statistical solution exists.

The following uniqueness theorem holds.

Theorem 4. Let $n = 2$ and $\mu(t,dv), t \in [0,T]$ be a spatial statistical solution satisfying the inequality

$$\int_0^T \int \|u\|^2 e^{C\|u\|^4} \mu(t,du)\,dt < \infty \tag{11}$$

where $C>0$ is a constant, depending on T and other data of problem (4). Then $\mu(t,dv)$ is uniquelly determined by the initial measure $\mu_0(du_0)$.

This theorem was proved by Ĉ.Foiaş [5] (see also [1]). Note that the existence of spatial statistical solutions satisfying (11) can be proved if the following condition on the initial measure holds

$$\int \|u\|^2 e^{N\|u_0\|^4} \mu_0(du_0) < \infty$$

Here $N>0$ is a sufficiently large number defined by the constant C from (11). Recently V.I.Gishlarkaev proved a more general theorem of uniqueness of statistical solutions.

Theorem 5. Let $n = 2$ and $\mu(t,dv), t \in [0,T]$ be a spatial statistical solution satisfying the inequality

$$\exists\, C, C_0 \;\; \forall R>0 \quad \int_{\Lambda_R} \|v\|^2 \mu(t,dv) + \int_0^t \int_{\Lambda_R} \|v\|_1^2 \mu(\tau,dv)\,d\tau \leq$$

$$\leq C \int_{\Lambda_{R-C_0}} (\|v\|^2 + 1)\mu_0(dv) \tag{12}$$

where

$$\Lambda_R = \{ v \in H^0 : \|v\| \geq R \}$$

Then $\mu(t,dv)$ is uniquely determined by the initial measure μ_0. The existence of the spatial statistical solution satisfying (12), may be proved if condition (5) is satisfied.

4. Let $n = 2,3$ and $f = 0$ in (4). If the measure μ_0 from (10) satisfies the condition

$$\int \|v\|^2 \mu_0(dv) < \infty \qquad k = 1,2,\ldots \tag{13}$$

then the functional $\chi_0(w)$, $w \in H^0$, has Frechet derivatives of arbitrary order. It is possible to construct the spatial statistical solution $\mu(t,dv)$ with the initial measure μ_0 satisfying (13), such that its characteristic functional $\chi(t,w)$ also has Frechet derivatives an arbitrary order. Variation derivatives of the functional $\chi(t,w)$ at $w = 0$ coincide with moments of the measure $\mu(t,dv)$:

$$\delta^k \chi(t,w)/\delta w^{j_1}(x_1) \dots \delta w^{j_k}(x_k)\Big|_{w=0} = M_k^{j_1,\dots,j_k}(x_1,\dots,x_k) =$$

$$= \int u^{j_1}(x_1) \dots u^{j_k}(x_k)\, \mu(t,dv) \qquad (14)$$

where $j_l = 1,\dots,n;\ l = 1,\dots,k;\ u^j(x)$ are components of the vector $u(x)$, and the integral in (14) is meant in the sense of distributions (see [1], p.83). Values of moments can be measured experimentally, therefore the moments are important objects of statistical hydromechanics.

It follows from the Hopf equation, that the tensors $M_k(t,\bar{x}^k) = M_k^{j_1,\dots,j_k}(x_1,\dots,x_k)$, $j_l = 1,\dots,n;\ l = 1,\dots,k$, where $\bar{x}^k = (x_1,\dots,x_k)$, satisfies an infinite chain of the moment equations

$$M_k(t,\bar{x}^k) + A_k M_k(t,\bar{x}^k) + (B_k M_{k+1})(t,\bar{x}^k) = 0, \quad k = 1,2,\dots \qquad (15)$$

where $A_k = \sum_{j=0}^{k-1} I(j)\otimes A \otimes I(k-j-1)$, $I(j) = I\otimes\dots\otimes I$ (j times)

$I: H^0 \longrightarrow H^0$ is identity, A is the operator from (4), B_k are linear operators defined by B from (4) (see [1]). The presence of M_{k+1} in the k-th equation from (15) is explained by the nonlinearity of the Navier-Stokes equations. Equations (15) are supplied by initial data

$$M_k(t,\bar{x}^k)\Big|_{t=0} = M_k^0(x^k) \quad k = 1,2,\dots \qquad (16)$$

where $M_k^0(\bar{x}^k) = M_k^{0;j_1,\dots,j_k}(x_1,\dots,x_k)$ are the moments of the initial measure μ_0.

The existence theorem for problem (15),(16) and the uniqueness theorem when $n = 2$ may be easily obtained from theorems 3,5. We do not formulate these results. We consider the result of unique solvability of problem (15),(16) for $n = 3$ in case of small Reynolds numbers and the result of analytic dependence of the solutions M_k of problem (15),(16) on the initial moments M_j^0.

Theorem 6. Let the support of an initial measure μ_0 be contained in the ball $Q_\rho = \{u: \|u\|_1 < \rho\}$ of the space H^1 where $\rho = \rho(\nu)$ is sufficiently small. Then problem (15),(16) possesses the unique solution in the class of tensors $M_k(t,\bar{x}^k)$. For any

$t\in[0,T]$ the $M_k(t,\cdot)$ are the moments of a spatial statistical solution $\mu(t,\omega)$, and $\mu(t,\omega)$ is concentrated in the ball Q_{ρ_1} of the space H^1, where $\rho_1 = \rho_1(\rho) \xrightarrow{\rho\to 0} 0$. For an arbitrary k the moment $M_k(t,\bar{x}^k)$ is expanded into a converging series

$$M\ (t,\bar{x}^k) = \sum_{r=k}^{\infty} \Phi_{r,k}(t,M_r^0)(\bar{x}^k)$$

where $\Phi_{r,k}(t,\cdot)$ are linear operators (see [1]).

5. Let consider problem (15),(16) in case $n = 3$ for large Reynolds numbers. We suppose that instead of (2) the periodic conditions

$$u(t,\dots,x^j,\dots) = u(t,\dots,x^j + 2\pi,\dots) \qquad j = 1,2,3,$$

are posed, i.e. $\Omega = R^3/2\pi Z^3$ is a torus. Set

$$H^\alpha = \{v = (v^1,v^2,v^3):\ \mathrm{div}\ v = 0,\ \int_\Omega v dx = 0,\ \|v\|_\alpha < \infty\} \quad \alpha\in R$$

where

$$\|v\|_\alpha^2 = \sum_{\xi\in Z^3\setminus\{0\}} |\xi|^{2\alpha} |\hat{v}(\xi)|^2$$

and $\hat{v}(\xi)$ is the ξ-th Fourier coefficient of the function $v(x)$. In this case, as in case of boundary condition (2), applying orthogonal projection $\pi : (L_2(\Omega))^3 \longrightarrow H^0$ to (1), we obtain (5) and after that, repeating the above considerations, we come to problem (15),(16). Set

$$H^\alpha(k) = H^\alpha \otimes \dots \otimes H^\alpha(k \text{ times}),\quad H_q^\alpha(k) = \{w : A_k^{q/2} w \in H^\alpha(k)\}$$

where $\alpha\in R$, $q = \pm 1$, A_k is the operator from (15). The subspace of symmetric elements of the space $H^\alpha(k)$ (the space $H_q^\alpha(k)$) is denoted by $SH^\alpha(k)$ (respectively by $SH_q^\alpha(k)$). The definition of the symmetric element of the tensor product of Hilbert spaces see in [1] p.185. Let

$$\mathcal{Y}^\alpha(k) = \{u(t)\in L_2(0,T;SH_1^\alpha(k)\ :\ \dot{u}(t)\in L_2(0,T;SH_{-1}^\alpha(k))\}$$

For an arbitrary $R>0$ set

$$N_R^\alpha = \prod_{k=1}^{\infty} SH^\alpha(k), \qquad \|m\|_{N_R^\alpha}^2 = \sum_{k=1}^{\infty} R^{-2k} \|m_k\|_{H^\alpha(k)}^2$$

$$\mathcal{Y}_R^\alpha = \prod_{k=1}^{\infty} \mathcal{Y}^\alpha(k), \qquad \|M\|^2_{\mathcal{Y}_R^\alpha} = \sum_{k=1}^{\infty} R^{-2k} \|M_k\|^2_{\mathcal{Y}^\alpha(k)}$$

For a large R the spaces N_R^α, Y_R^α contain the moments, corresponding to the case of large Reynolds numbers.

Theorem 7. Let $\alpha > 2$, $R > 0$. Then the solution $M(t,\cdot) = \{M_k(t,\bar{x}^k),\ k = 1,2,\dots\}$ of problem (15),(16) is unique in the space Y_R.

The set of initial data $M_k^0 = \{M_k^0(\bar{x}^k), k = 1,2,\dots\} \in N_R^\alpha$, for which the problem (15),(17) possesses a solution $M \in Y_R^\alpha$, is denoted by V_R^α.

Theorem 8. Let $\alpha > 1/2$, $R > 0$. Then V_R^α is dense in N_R^α.

Theorems 7,8 are proved in [8].

Note that theorem 7 was proved without the assumption of the positive definiteness of the moment system $\{M_k\}$. Theorem 8 was also proved in this class. It is possible to deduce the existence and uniqueness of solutions of the Hopf equation (9) for a dense set of initial data from the theorems 7,8. However these solutions would pertain to the class of analytic functionals but not to the class of measures.

6. If statistical characteristics of turbulent flows do not depend on shifts along spatial variables, then such flows are described by the homogeneous statistical solutions. Let

$$\mathcal{H} = \{u(x) = (u^1,\dots,u^n),\ x \in R^n : \operatorname{div} u = 0,\quad \int_\Omega |u(x)|^2 dx < \infty\}$$

where $\Omega \subset R^n$ is an arbitrary bounded domain. The measure $\mu_0(\omega_0)$, $\omega_0 \in \mathcal{B}(\mathcal{H})$ is called homogeneous, if

$$\mu_0(\omega_0) = \mu_0(\hat{h}\omega_0) \quad \forall \omega_0 \in \mathcal{B}(\mathcal{H}),\ \forall h \in R^n$$

where $\hat{h}u(x) = u(x+h)$, $\hat{h}\omega_0 = \{\hat{h}u,\ u \in \omega_0\}$. The energy density $\bar{e}_0 = \int\int_\Omega |u_0(x)|^2 dx\, \mu(du_0)/|\Omega|$ of a homogeneous measure μ (where $|\Omega|$ is the volume of the domain Ω) does not depend on Ω with $|\Omega| < \infty$ and is denoted

$$\bar{e}_0 = \int |u_0(x)|^2 \mu(du_0) \tag{17}$$

The space-time statistical solution P is called homogeneous, if

$$P(\omega) = P(\hat{h}\omega) \quad \forall \omega \in \mathcal{B}(L_2(0,T;\mathcal{H})) \quad \forall h \in R^n$$

where $\hat{h}u(t,x) = u(t,x+h)$.

Theorem 9. Let μ_0 be a homogeneous initial measure with $\bar{e}_0 <$

$<\infty$. Then there exists the homogeneous space-time statistical solution P of the Navier-Stokes equations (1) with $g = 0$, which satisfies the following analog of energy inequality:

$$\int \left(|u(t,x)|^2 + 2\nu \int_0^t |\nabla u(\tau,x)|^2 d\tau \right) P(du) \leq \bar{e}_0, \quad t \in [0,T]$$

where the integral is meant in sense of (17).

Theorem 9 is proved in [1].

From the existence of a homogeneous space-time statistical solution P it is possible to deduce the solvability of the Navier-Stokes system (1),(3) with $g = 0$. Namely, the solvability is proved if initial datum $u_0 = \sum c_k e^{i\xi_k \cdot x}$ is an almost periodic polinomial with $\{c_k\}$ belonging to a certain set of the complete Lebegue measure ([1]). Such initial data in most cases have the infinity energy in R^n : $\|u_0\|_{L_2(R^n)} = \infty$, therefore it is impossible to use classical methods for such initial data.

The tensor $Q(t,y) = \{Q^{ij}(t,y), i,j = 1,\ldots,n\}$, where

$$Q^{ij}(t,y) = \int u^i(t,x+y)u^j(t,x)P(du) \tag{18}$$

is called a correlation tensor of a homogeneous statistical solution P. (The integral in (18) is meant in sense of (17)). The $Q(t,y)$ possesses the following asymptotics as $|y| \longrightarrow 0$.

Theorem 10. For almost all $t \in [0,T]$, the following equality holds

$$Q^{ij}(t,y) = Q^{ij}(x,0) + \sum_{1\leq l\leq n} y^l \int u^i(t,x) \frac{\partial u^j(t,x)}{\partial x^l} P(du) -$$

$$- \frac{1}{2} \sum_{1\leq r,l\leq n} y^r y^l \int \frac{\partial u^j(t,x)}{\partial x^l} \frac{\partial u^j(t,x)}{\partial x^r} P(du) + R^{ij}(t,y),$$

where $R^{ij}(t,y)/|y|^2 \longrightarrow 0$ as $|y| \longrightarrow 0$. In the expansion of Q^{jj}, the terms of the first order in y are absent.

The proof of theorem 10 see in [1].

7. It is well-known that the problem of unique solvability of the three-dimensional Navier-Stokes system for arbitrary data from an appropriate functional space is not solved up to now. Let the right hand side f in (4) be fixed and don't depend on t. We say that a measure $\mu(\omega_0)$, $\omega_0 \in \mathcal{B}(H^1)$, belongs to the class E, if problem (4),(3) has the solution $u(t,x) \in C(0,T;H^1)$ for μ-almost all ini-

tial data u_o. So, problem (4),(3) is uniquely solvable for μ-almost all initial data if $\mu \in E$.

Let $\mu_o(\omega_o)$, $\omega_o \in \mathcal{B}(H^1)$, be an arbitrary measure, satisfying the inequality $\int \exp\|v\| \mu_o(dv) < \infty$, and m_k^o be the moments of the measure μ_o. We shall construct the measure $\mu \in E$, using μ_o. Let $\mathcal{P}_+$ be the class of measures $P(\omega)$, $\omega \in \mathcal{B}(C(0,T;H^1))$, concentrated on individual solutions of problem (4),(3). We consider the extremal problem

$$J(P) = \int \exp\|u\|^2_{L_2(0,T;H^2)} P(du) + \lambda \sum_{k=1}^{\infty} \frac{1}{k!}\|m_k - m_k^o\|_{(k)} \longrightarrow \inf, \ P \in \mathcal{P}_+ \quad (19)$$

where $H^2 = H^o \cap (W_2^2(\Omega))^3$, m_k are the moments of the measure $\gamma_o^* P$, defined in (7), $\|\cdot\|_{(k)}$ is the norm of $(L_2(\Omega^k))^{3^k}$, $\Omega^k = \Omega \times \dots \times \Omega$ (k times). For any $\lambda > 0$ problem (19) has the unique solution $\hat{P}$. Then the measure $\mu = \gamma_o^* \hat{P} \in E$ (see [1], Appendix I).

References

1. Vishik M.I., Fursikov A.V. Mathematical problems of statistical hydromechanics. - Kluwer academic publishers, Dordrecht, Boston, London, 1988.
2. Lions J.-L., Magenes E. Problemes aux limites non homogènes et applications. V.1. Dunod. Paris. 1968.
3. Hopf E. Statistical hydromechanics and functional calculus. - J. Rational Mech. and Anal., 1 (1952), 87-123.
4. Monin A.S., Jaglom A.M. Statistical hydromechanics. V.1. Nauka, Moscow, 1965; V.2. Nauka, Moscow, 1967 (in Russian). English Translation (revised). MIT Press, Cambridge, Mass, 1977.
5. Foiaş C. Statistical study of Navier-Stokes equations I. - Rend. Semin. Mat. Univ. Padova, 1973, v.49, p.9-123.
6. Gishlarkaev V.I. Statistical solutions of the Navier-Stokes equations in the case of infinite mean energy. Moscow Univ. Math. Bull. (1)(1983), 9-13 (in Russian).
7. Gishlarkaev V.I. The theorems of uniqueness of statistical solutions of the Navier-Stokes equations. - Moscow Univ. Math. Bull. (to appear).
8. Fursikov A.V. On the uniqueness of solutions of the chain of moment equations, corresponding to the three-dimensional Navier-Stokes system. Mat. Sb. 134, (4), (1987), 472-495 (in Russian).

(received February 27, 1989.)

Asymptotic expansions for a flow with a dynamic contact angle

DIETMAR KRÖNER
UNIVERSITÄT HEIDELBERG

In a container, which is partially filled with a fluid, there is a free surface, separating the air and the fluid, and a contact line on the wall of this container, where the free surface touches the rigid boundary. In this paper we shall consider only two dimensional problems and in this case the contact line will reduce to two single contact points. The angle between the free surface and the rigid boundary is called contact angle. If the fluid is at rest, the contact angle can be considered as a constant, depending only on the materials of the fluid and the container. If the fluid and the contact points are in motion, the angle will depend on the velocity of the contact points. This situation arises for instance , if the container is beeing filled, or if we push the fluid with a piston through a tube or if we draw off a plate out of a fluid. In this paper we shall study the mathematical model for this case, which consists of a free boundary value problem for the equations of Navier and Stokes.

The static situation was studied by Finn [FI]), Finn and Shinbrot [$FS1$], [$FS2$]. The case where the fluid is in motion and the contact points are fixed was treated by Solonnikov ([$S1$], [$S2$]). The corresponding situation in three dimensions was considered by Mazja, Plamenevskii and Stupyalis [MPS], [MP]. Numerical results has been proven by Cuvelier [$CU1$], [$CU2$] . For problems with free boundaries, which do not touch the rigid boundary we refer to Bemelmans [BE] and Solonnikov [$S3$] .

In this paper we assume, that the fluid is pushed through the container such that the contact points move upwards with a constant velocity S_1 . We shall use a coordinate system which moves with the fluid, such that the origin is equal to the left contact point (see Figure 1).

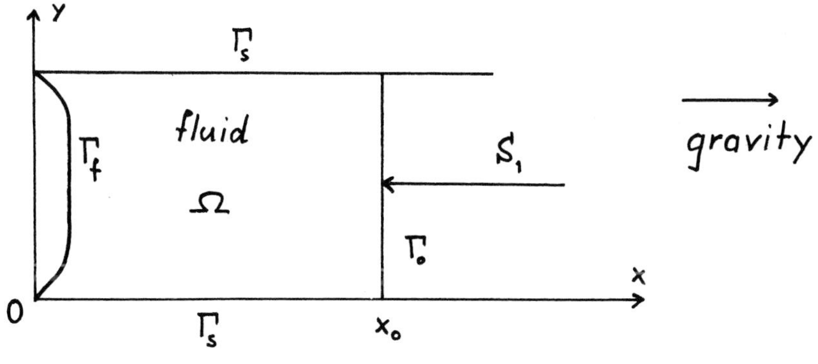

FIGURE 1

In order to describe the mathematical model, let us fix some notations. We assume that the free boundary is described as the graph of a function g with respect to $y \in I :=]0,1[$. Let

$$(1) \qquad \begin{array}{rcl} \Omega & := & \{(x,y) \in I\!R^2 \mid 0 < y < 1, g(y) < x < x_0\} \\ \Gamma_f & := & \{(x,y) \in I\!R^2 \mid 0 < y < 1, x = g(y)\} \\ \Gamma_0 & := & \{(x,y) \in I\!R^2 \mid 0 < y < 1, x = x_0\} \\ \Gamma_s & := & \{(x,y) \in I\!R^2 \mid 0 < x < x_0, y \in \{0,1\}\} \end{array} \quad .$$

The mathematical model for our problem consists of the following free boundary value problem for the incompressible equations of Navier and Stokes.

$$(2) \qquad \begin{array}{rcll} -\nu\Delta v_i + \varrho v \nabla v_i + \partial_i p & = & 0 & \text{in } \Omega \\ \operatorname{div} v & = & 0 & \text{in } \Omega \\ & & & \\ v \cdot n & = & 0 & \text{on } \partial\Omega \\ \nu\partial_n v_\tau + \gamma v_\tau & = & -\gamma S_1 & \text{on } \Gamma_s \\ & & & \\ \tau\sigma_v n & = & 0 & \text{on } \Gamma_f \\ n\sigma_v n & = & -H\kappa + \beta g & \text{on } \Gamma_f \\ & & & \\ \nu\partial_n v_\tau + \gamma_0 v_\tau & = & 0 & \text{on } \Gamma_0 \\ g(1) = 0, \quad g(0) & = & 0 & \end{array} \quad .$$

The meaning of the parameters is as follows: ν denotes the viscosity, ϱ the density, σ_v the stress tensor of the fluid, i.e. $(\sigma_v)_{ij} := -p\delta_{ij} + \nu(\partial_i v_j + \partial_j v_i), i,j = 1,2$, S_1 is the velocity of the contact point with respect to the solid boundary, γ and γ_0 are friction coefficients with respect to the boundaries Γ_s and Γ_0, H is the mean curvature of the free boundary: $H := (g'(1+g'^2)^{-\frac{1}{2}})'$, κ the surface tension, $\beta \in I\!R^+$ the product of the density ϱ and the acceleration of gravity, τ and n are the tangent and normal unit vectors.

S_1 and the parameters of the fluid are given and the velocity v, the pressure p and the parametrization g of the free boundary are to be determined.

The most interesting feature of this model is the slip boundary condition on the fixed boundary Γ_s. In the literature you can find a comprehensive discussion about the correct model for the slip on the solid boundary near the contact point. ([DD], [HS]). Using formal asymptotic expansions it was known, that the classical no-slip condition on Γ_s (i.e. $v_\tau = -S_1$) implies infinite energy and dissipation and therefore doesn't make sense ([DD], [HS]). On the other hand the slip-condition as in (2) turns out to be more successful. *Puchnachev* und *Solonnikov* [PS] have shown, that if one assumes the classical no-slip condition and if the solution v of (2) is in $H^{1,2}(\Omega)$, then the contact angle has to be 0 or π .

In this paper we would like to study the free boundary value problem (2), in particular the asymptotic expansion of the velocity v near the contact point. For the result concerning the asymptotic expansion we assume that there exists already a solution such that the energy and the dissipation is finite and that g is so smooth, that we can define a contact angle. Mathematically this means, that there exists a solution $\{v, p, g\}$ of (2) such that

(3) $$v \in H^{1,2}(\Omega) \quad \text{und} \quad v \quad \text{sufficiently smooth in} \quad \bar{\Omega}\backslash M \qquad ,$$

where $M = \{(0,0),(0,1),(x_0,0),(x_0,1)\}$ is the set of corners of Ω ;

(4) $$g \in C^1(\bar{I}) \quad \text{und} \quad g \quad \text{sufficiently smooth in} \quad I \qquad .$$

Additionally we have to assume that

(5) $$g' \in C_0^{3+\nu}([0,1]), \quad 0 < \nu < 1 \qquad ,$$

(6) $$\| g - g'(0)y \|_{C_1^{4+\nu}([0,a])} \longrightarrow 0 \quad \text{for} \quad a \to 0 \qquad .$$

The weighted Hölder spaces, which we have used here, are defined as follows: Let $\varrho(z) :=$ dist (z, M). Then for $k \in I\!N, s \in I\!R, \nu \in \,]0,1[\,$:

$$C_s^k(\Omega, M) := \left\{ u \mid \| u \|_{C_s^k(\Omega,M)} := \sum_{|\beta| \le k} \sup_z \mid \varrho(z) \mid^{-s+\beta} \mid \partial^\beta u(z) \mid < \infty \right\}$$

$$C_s^{k+\nu}(\Omega, M) := \Big\{ u \in C_s^k(G) \mid \| u \|_{C_s^{k+\nu}(\Omega,M)} := \| u \|_{C_s^k} +$$

$$+ \sum_{|\beta|=k} \sup_z \varrho(z)^{k+\nu-s} \sup_{|z-z'| \le \frac{\varrho(z)}{2}} \frac{\mid \partial^\beta u(z) - \partial^\beta u(z') \mid}{\mid z - z' \mid^\nu} < \infty \Big\} \qquad .$$

If we omit M, we only consider the local situation near the origin and in this case, $\rho(z) = \mid z \mid$.

The result concerning the asymptotic expansion will be given for the streamfunction instead of the velocity, in order to get rid of the pressure. We introduce the streamfunction

ψ such that $\partial_2\psi = v_1$ and $\partial_1\psi = -v_2$ and obtain for ψ and g

(7)
$$\begin{array}{rcll}
\nu\Delta^2\psi &=& \varrho(\partial_2\psi\partial_1\Delta\psi - \partial_1\psi\partial_2\Delta\psi) & \text{in } \Omega \\
\psi &=& 0 & \text{on } \partial\Omega \\
\nu\psi_{nn} + \gamma\psi_n &=& -\gamma S_1 & \text{on } \Gamma_s \\
\psi_{nn} - H\psi_n &=& 0 & \text{on } \Gamma_f \\
e(-\kappa H' + \beta g') &=& \varrho(\partial_2\psi\partial_n\partial_1\psi - \partial_1\psi\partial_n\partial_2\psi) + \nu\partial_n\Delta\psi + 2\nu\partial_\tau\psi_{\tau n} & \text{on } \Gamma_f \\
\nu\psi_{nn} + \gamma_0\psi_n &=& 0 & \text{on } \Gamma_0 \\
g(1) &=& g(0) = 0 & ,
\end{array}$$

where $e = \sqrt{1 + g'(y)^2}$. Notice that $\psi_n, \psi_\tau, \psi_{nn}, \psi_{\tau\tau}, \psi_{\tau n}$ denote partial derivatives in the local coordinate system (n, τ) and $\partial_\tau = e(g'\partial_x + \partial_y), \partial_n = e(g'\partial_y - \partial_x)$. Furthermore we need the weighted Sobolev spaces, which are defined as follows:

$$W_\mu^{k,p}(\Omega, M) := \left\{ u \mid \| u \|^p_{W_\mu^{k,p}(\Omega,M)} := \sum_{|\beta|\le k} \int_\Omega \varrho^{p(\mu-k+\beta)} \mid \partial^\beta u \mid^p < \infty \right\} \quad ,$$

where $k \in I\!N, p \in I\!R^+, \mu \in I\!R$. For the local investigations let $W_\mu^{k,p}(\Omega) := W_\mu^{k,p}(\Omega, \{0\})$. If $p = 2$, we neglect p .

Let (r, φ) denote the polar coordinates in $I\!R^2$. For the streamfunction we get the following asymptotic expansion near the contact point. It's the main result of this paper.

THEOREM 1 (ASYMPTOTIC EXPANSION). *Let $\{v, g\}$ be a solution of (2) in a neighbourhood of the contact point, satisfying (3), (4) (5) and (6). Let φ_0 be the contact angle $\varphi_0 := \frac{\pi}{2} - \arctan g'(0)$ and $k(y)$ as in (11). Then we have $\psi = \psi_{as} + \psi_0$, where*

$$\psi_0 \in W^4_\sigma(\Omega) \quad \text{for some} \quad \sigma > \begin{cases} \frac{3}{2} - \frac{\pi}{2\varphi_0} & \text{if } \; 0 < \varphi_0 \le \frac{\pi}{2} \\ \frac{7}{2} - \frac{3\pi}{2\varphi_0} & \text{if } \; \frac{\pi}{2} < \varphi_0 < \pi \end{cases}$$

and for $0 < r < 1, 0 \le \varphi \le \varphi_0$ and some $s_0, q \in I\!N$:

(8)
$$\begin{array}{rcll}
\psi_{as}(r,\varphi) &=& u_3(r,\varphi) & \text{if } \; 0 < \varphi_0 \le \frac{2\pi}{5} \\
\psi_{as}(r,\varphi) &=& u_3(r,\varphi) + r^{\frac{\pi}{\varphi_0}} \sum_{s=0}^{s_0} \ln^s r \hat{P}_s(r \ln^q r, \varphi) & \text{if } \; \frac{2\pi}{5} < \varphi_0 \le \frac{3\pi}{5} \\
\psi_{as}(r,\varphi) &=& r^{\frac{\pi}{\varphi_0}} \sum_{s=0}^{s_0} \ln^s r P_s(r \ln^q r, \varphi) & \text{if } \; \frac{3\pi}{5} < \varphi_0 < \pi \quad .
\end{array}$$

$u_3 \neq 0$ and u_3 can be computed explicitly. We have

$$u_3(r,\varphi) = \begin{cases} 0(r^2) & \text{if } \ 0 < \varphi_0 < \frac{\pi}{2} \\ 0(r^2 \ln r) & \text{if } \ \varphi_0 = \frac{\pi}{2} \end{cases} .$$

$P_s, \hat{P}_s$ are polynomials in $r \ln^q r$,where the coefficients are smooth functions in φ .

It can be seen very easily, that $u_3 \neq 0$ if $0 < \varphi_0 < \frac{\pi}{2}$. Otherwise the asymptotic expansion of ψ would begin with $r^{\frac{\pi}{\varphi_0}} \ln^s r$. But this means that $| \psi_{yy}(x) | \leq c \mid x |^{\epsilon}$ and $| \psi_y(x) | \leq c \mid x |^{1+\epsilon}$ for $x \in \Gamma_s$ and one $\epsilon > 0$. This is a contradiction to the slip condition $\nu\psi_{yy} + \gamma\psi_y = -\gamma S_1 \neq 0$.

It turns out, that there exists a solution of (2), which satisfies (3), (4), (5) and (6), if S_1 is suffiently small. (see $[KR]$). In the remaining part of this paper we would like to discuss the main ideas of the proof of Theorem 1. For more details we refer to $[KR]$.

Sketch of the proof of Theorem 1: For studying the asymptotic behaviour of the velocity v near the contact points we can restrict ourselves to a neighbourhood of the origin. Then for the streamfunction we have the local situation as shown in Figure 2a.

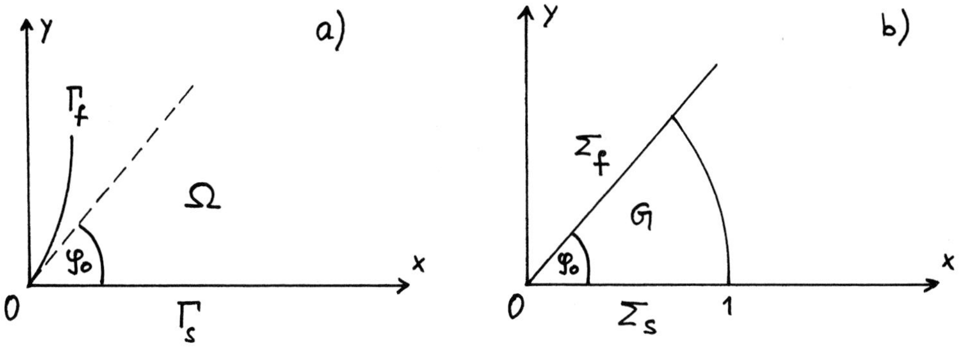

FIGURE 2

It is much more convenient to transform the problem to a domain G with straight boundaries (see Figure 2). Then we shall get a problem on a nice domain but an operator with bad coefficients. Define

(9) $$G := \{(x,y) \mid y > 0, \quad x > g'(0)y\} \cap B_1(0)$$

and

(10) $$u(x,y) := \psi(x - k(y), y) \quad \text{for} \quad (x,y) \in G \quad ,$$

where

$$k(y) := g'(0)y - g(y) \quad . \tag{11}$$

By this transformation Γ_f and Γ_s are transformed to $\Sigma_s := \Gamma_s, \Sigma_f = \{(x,y) \in I\!R^2 \mid x = g'(0)y\}$.

Then for u and g we locally obtain the following free boundary value problem. We can only write down the most important terms. For the complete operator see [KR] .

$$\begin{array}{rcrll} \Delta^2 u + k'^4 \partial_x^4 u + 4k' \partial_y^3 \partial_x u + ... & = & 0 & in & G \\ u & = & 0 & on & \partial G \\ \partial_y^2 u - \gamma \partial_y u + 2k' \partial_x \partial_y u & = & -\gamma S_1 & on & \Sigma_s \\ \partial_n^2 u - 2k' g' \partial_x^2 u + ... & = & 0 & on & \Sigma_f \end{array} \quad . \tag{12}$$

$$-\kappa e H' + \beta e g' - 3g' k' e^3 \partial_x^3 u + ... = 0 \quad \text{on} \quad \Sigma_f \quad . \tag{13}$$

First let us consider the linear main part of the operator in (12):

$$\begin{array}{rcrll} \Delta^2 u & = & 0 & in & G \\ u & = & 0 & on & \partial G \\ \partial_y^2 u & = & -\gamma S_1 & on & \Sigma_s \\ \partial_n^2 u & = & 0 & on & \Sigma_f \end{array} \quad . \tag{14}$$

This linear problem we shall write in the short form

$$Au = F_0 \quad , \tag{15}$$

where A denotes the differential operator including the boundary operators and F_0 the right-hand side in (14). Let us collect the nonlinear parts of the boundary value problem (12) in Nu and the remaining terms like $k'^4 \partial_x^4 u + 4k' \partial_y^3 \partial_x u + ...$ in Pu. Then (12) can be written in the form:

$$Au + Pu + Nu = F_0 \quad . \tag{16}$$

In order to get an asymptotic expansion for the linear problem (14) in the domain G, we apply the following transformation to (14):

a) polar coordinates (r, ϕ) with respect to the origin;

b) $(r, \phi) \to (t, \phi)$, where $r = -lnt$;

c) Fouriertransformation with respect to t .

This idea is due to Kondratev [KO]). Then after carrying out these transformations we end up with the following boundary value problem for an ordinary differential equation for the transformed function u^*:

$$\sum_{i\le 4} a_i(\lambda)\partial_\phi^{(i)} u^* = F^* \tag{17}$$

and additional boundary conditions, which we omit here for simplicity. For the solution operator of (17) we have the following result.

LEMMA 2. *Let $\varphi_0 \notin \{0,\pi\}$ and let $R(\lambda)$ be the solution operator of (17) including the boundary conditions. Then $R(\lambda)$ is a meromorphic function of λ and the poles are*

$$\lambda = -\frac{ik\pi}{\varphi_0}, \quad \lambda = i(2-\frac{k\pi}{\varphi_0}), \quad k \in \mathbb{Z} - \{0\} \qquad \text{if} \quad \varphi_0 \neq \frac{\pi}{2}$$

and

$$\lambda = 2ik, \quad k \in \mathbb{Z} \qquad \text{if} \quad \varphi_0 = \frac{\pi}{2} \quad .$$

Here we have used $\varphi_0 = \frac{\pi}{2} - \arctan g'(0)$.

Now we need some notations:

$$G_\infty := \{(x,y) \in I\!R^2 \mid \quad y \in I\!R^+ \quad \text{and} \quad x > g'(0)y\}$$

$$E(\varphi_0) := \begin{cases} \left\{\lambda \mid \lambda = \frac{k\pi}{\varphi_0} \quad \text{or} \quad \lambda = 2 + \frac{k\pi}{\varphi_0}, \quad k \in \mathbb{Z} - \{0\}\right\} & \text{if} \quad \varphi_0 \neq \frac{\pi}{2} \\ \{\lambda \mid \lambda = 2k, \quad k \in \mathbb{Z}\} & \text{if} \quad \varphi_0 = \frac{\pi}{2} \end{cases} .$$

$$H_\mu^{k,p}(G,M) := W_\mu^{k,p}(G,M) \times W_\mu^{k+4-\frac{1}{p},p}(\Sigma_s \cup \Sigma_f, M) \times W_\mu^{k+2-\frac{1}{p},p}(\Sigma_s, M) \times W_\mu^{k+2-\frac{1}{p},p}(\Sigma_f, \ldots$$

For the local investigations let $H_\mu^{k,p}(\Omega) := H_\mu^{k,p}(\Omega,\{0\})$.

The asymptotic expansion of the solution u of (14) depends on the poles of $R(\lambda)$. The precise statement is given in

THEOREM 3. *Let $F \in H_\alpha^0(G_\infty)$ such that $\alpha \notin E(\varphi_0)$ and $w \in W_\mu^4(G_\infty), \alpha < \mu$ solution of $Aw = F$. Then $w = w_1 + w_2$ such that $w_2 \in W_\alpha^4(G_\infty),$, $Aw_2 = F$ and*

$$w_1(r,\varphi) = \sum_{j=0}^{j_0}\sum_{s=0}^{s_j} r^{\lambda_j}\, ln^s r \psi_{sj}(\varphi) \quad ,$$

where $\{\lambda_j \mid 0 \le j \le j_0\} = E(\varphi_0,\alpha,\mu) := E(\varphi_0) \cap \,]3-\mu, 3-\alpha[$. s_j is the multiplicity of $R(\lambda)$ in λ_j. Furthermore we have the estimate

$$\| w_2 \|_{W_\alpha^4(G_\infty)} \le c_{11}(\| F \|_{H_\alpha^0(G_\infty)} + \| w \|_{W_\mu^4(G_\infty)}) \quad .$$

This result has been proved by Kondratev [*KO*]. Now the question arises if one can generalize this result to the nonlinear problem (16). For studying this, let us consider the regularity of the right-hand side F_0 of (15). We obtain that F_0 is in the weighted space $B^{\nu}_{-2-\epsilon}(G)$ for all $\epsilon > 0$, where

$$B^{k+\nu}_{s-4}(G,M) := C^{k+\nu}_{s-4}(G,M) \times C^{4+k+\nu}_{s}(\Sigma_s \cup \Sigma_f, M) \times C^{k+2+\nu}_{s-2}(\Sigma_s, M) \times C^{k+2+\nu}_{s-2}(\Sigma_f, M)$$

and for the local investigations $B^{k+\nu}_{s}(\Omega) := B^{k+\nu}_{s}(\Omega, \{0\})$. Now let us determine the regularity of u. The assumption (3) implies that

$$\psi \in H^{2,2}(\Omega)$$

and therefore

$$u \in H^{2,2}(G) \quad .$$

Since $u = 0$ on ∂G (see (12)), the Sobolev embedding theorem implies that $u \in C^{0,1}(G)$ and in particular

$$| u(z) | \leq c \, | z |^{1-\epsilon}$$

for all $\epsilon > 0$. In order to estimate higher order derivatives of u, we apply the C^{α}-estimates of Agmon, Douglas, and Nirenberg $[ADN]$) in annular subsets of G. We obtain

$$u \in C^{4+\nu}_{1-\epsilon}(G)$$

and this implies

$$N(u) \in B^{\nu}_{-2-\epsilon}(G)$$

for all $\epsilon > 0$. This is the same regularity, as we have already for F_0. Therefore we can shift $N(u)$ to the right-hand side and we have to consider

$$(18) \qquad Au + Pu = F_1 \quad \text{, where} \quad F_1 := F_0 - N(u) \in B^{\nu}_{-2-\epsilon} \quad .$$

Essentially the leading terms of P consist of differential operators with the same order as Δ^2 and coefficients which behave like k'. Using suitable cut-off functions, which are equal 1 near the contact point and equal 0 outside a small ball around it, we can restrict ourselves to a small neighborhood of the origin. If this neighborhood is small enough, $\| k' \|_{C^{3+\nu}_0}$ will become small (see (6)) and P can be considered as a perturbation of A. In order to use this information for further investigations, we have to apply the inverse operator A^{-1} of A to (18). The existence of A^{-1} follows from

THEOREM 4. *Let $F \in H^0_\alpha(G_\infty)$ such that $3 - \alpha \notin E(\varphi_0)$. Then there exists a unique $w \in W^4_\alpha(G_\infty)$ $Aw = F$. Furthermore the following estimate is valid*

$$\| w \|_{W^4_\alpha(G_\infty)} \leq c_{12}(\| F \|_{H^0_\alpha(G_\infty)}) \quad .$$

Proof: *Kondratev* $[KO]$, Theorem 1.1.

In order to apply A^{-1} to (18), we have to express the regularity of u and F_1 in terms of the weighted Sobolev spaces instead of the weighted Hölder spaces. Applying A^{-1} to (18) we find

$$(id + A^{-1}P)u = A^{-1}F_1 \quad .$$

It can be proved that the smallness of $\| P \|$ (with a suitable norm $\| \,.\, \|$) yields the smallness of $\| A^{-1}P \|$ (with respect to a different norm) and the existence of

$$(id + A^{-1}P)^{-1}.$$

Then formally we infer that

$$u = (id + A^{-1}P)^{-1}w \qquad \text{where} \qquad w := A^{-1}F_1 \tag{19}$$

and therefore

$$Aw = F_1 \quad .$$

For this linear problem we can apply Theorem 3. We have

$$F_1 \in B^{\nu}_{-2-\epsilon} \qquad \text{and} \qquad w \in C^{4+\nu}_{1-\epsilon} \quad .$$

Again we express these regularity properties of w and F_1 in terms of weighted Sobolev spaces and then the asymptotic expansion in Theorem 3 and some additional C^{α}-estimates imply

$$w \in C^{4+\nu}_{1+\epsilon} \quad . \tag{20}$$

and

$$w(r,\varphi) = \sum_{j=0}^{j_0} \sum_{s=0}^{s_j} r^{\lambda_j} \, ln^s r \psi_{sj}(\varphi) + \quad \dots \quad , \tag{21}$$

where we have for the leading term $\lambda_0 > 1$. The asymptotic expansion (21) cannot be used to prove a similar one for u , since there is still the operator $(id + A^{-1}P)^{-1}$ in (19), but in any case (20) implies that

$$u \in C^{4+\nu}_{1+\epsilon} \quad . \tag{22}$$

It turns out that this is the most important step to improve the regularity of u from

$$u \in C^{4+\nu}_{1-\epsilon} \quad \text{to} \quad u \in C^{4+\nu}_{1+\epsilon} \quad .$$

To see this, let us consider the third boundary condition (13) on the free boundary Σ_f. This is a boundary value problem for an ordinary differential equation for g' or k' respectively. The coefficients depend on u and higher derivatives of u. Using (13) and (22) we can state the following result:

THEOREM 5 (REGULARITY OF THE FREE BOUNDARY). *Suppose that the assumptions of Theorem 1 are satisfied. Then $k \in C^4_{3-\mu_0}(I)$ for almost all μ_0 such that*

$$\mu_0 > \begin{cases} 3 - \frac{\pi}{\varphi_0} & \text{if } \ \frac{\pi}{2} < \varphi_0 < \pi \\ 1 & \text{if } \ 0 < \varphi_0 \le \frac{\pi}{2} \end{cases} \quad .$$

Proof: see $[KR]$ Theorem 6.5.

Theorem 5 implies, that $k \in C^4_{2-\delta}(I)$ for all $\delta > 0$. This means in particular $| k^{(j)}(y) | \leq c | y |^{2-\delta-j}$ and $k \in C^{2-\delta}(\bar{I})$ for all $\delta > 0$.

Therefore we can control the regularity of k' in the operator P much better then before and we infer that $k'\partial_x^4 u$ and P respectively lie in $B^{\nu}_{-2-\epsilon}$. This is the same regularity as we have for F_1. This means, that we can put P to the right-hand side in (18) to obtain

$$Au = F_2 \quad ,\text{where} \quad F_2 = F_1 - Pu \in B^{\nu}_{-2-\epsilon} \quad . \tag{23}$$

Then Theorem 3 can be applied directly to (23) to derive the asymptotic expansion as stated in Theorem 1.

References

ADN. Agmon, S., Douglis, A., Nirenberg, L., *Estimates near the boundary for solutions of elliptic partial differential equations satisfying general boundary conditions. I.*, Comm. Pure Appl. Math. **XII** (1959), 623 - 727.

BE. Bemelmans, J., *Gleichgewichtsfiguren zäher Flüssigkeiten mit Oberflächenspannung*, Analysis **1** (1981), 241 - 282.

CU1. Cuvelier, C., *A capillary free boundary problem governed by the Navier-Stokes equations*, Comp. Math. Appl. Mech. Eng. **48** (1985), 45 - 80.

CU2. Cuvelier, C., *On the solution of capillary free boundary problems governed by the Navier-Stokes equations*, preprint.

DD. Dussan V., E.B., Davis, S.H., *On the motion of a fluid-fluid interface along a solid surface*, J. Fluid Mech. **65** (1974), 71 - 95.

FI. Finn, R., *Equilibrium capillary surfaces*, Grundlehren d. mathem. Wissenschaften, Berlin, Heidelberg, New York, Tokyo (1986).

FS1. Finn, R., Shinbrot, M., *The capillary contact angle, I: The horizontal plane and stick-slip motion*, Preprint.

FS2. Finn,R., Shinbrot, M., *The capillary contact angle, II: The inclined plane*, Preprint.

HS. Huh, C., Scriven, L.E., *Hydrodynamic model of steady movement of a solid / liquid / fluid contact line*, J. Colloid and Interface Science (1971), 35 - 101.

KO. Kondratev, V.A., *Boundary problems for elliptic equations in domains with conical or angular points*, Trans. Moscow Math. Soc. **16** (1967), 227 - 313.

KR. Kröner, D., *Asymptotische Entwicklungen für Strömungen von Flüssigkeiten mit freiem Rand und dynamischem Kontaktwinkel*, Preprint 809, SFB 72 , Bonn. 1986.

MP. Maz'ja, V.G., Plamenevskii, V.A., *A problem on the motion of a fluid with a free surface in a container with piecewise smooth walls*, Sov. Math. Dokl. **21** (1980), 317 - 219.

MPS. Maz'ja, V.G., Plamenevskii, B.A., Stupyalis, L.I., *The three-dimensional problem of steady-state motion of a fluid with a free surface*, Amer. Math. Soc. Transl. **123** (1984), 171 - 268.

PS. Pukhnachev, V.V., Sołonnikov, V.A., *On the problem of dynamic contact angle*, PMM USSR **46** (1983), 771 - 779.

S1. Solonnikov, V.A., *Solvability of a problem on the plane motion of a heavy viscous incompressible copillary liquid partially filling a container*, Math. USSR Izvestija **14** (1980), 193 - 221.

S2. Solonnikov, V.A., *On the Stokes equation in domains with non smooth boundaries and on viscous incompressible flow with a free surface. In: Nonlinear partial differential equations and their applications*, College de France, Seminar III, ed. by H. Brezis and J.L. Lions, Boston (1982), 340 - 423.

S3. Solonnikov, V.A., *Solvability of the problem of evolution of an isolated volume of viscous, incompressible capillary fluid*, J. Sov. Math. (1986).

(received February 21, 1989.)

This paper is in its final form and no similar paper has been or is being submitted elsewhere

NONCOMPACT FREE BOUNDARY PROBLEMS FOR THE NAVIER-STOKES EQUATIONS

Konstantin Pileckas
Institute of Mathematics and Cybernetics of the Academy of Sciences of the Lithuanian SSR,
Akademijos str. 4, Vilnius, Lithuanian SSR, USSR

Free boundary problems for the Navier-Stokes system have been the subject of many investigations. Such problems were considered by V.V. Pukhnachov [15, 16], V.G. Osmolovskii [10], O.A. Ladyzhenskaya and V.G. Osmolovskii [8], V.A. Solonnikov [20-24], V.Ia. Rivkind [17, 18], V.G. Maz'ja, B.A. Plamenevskii and L.I. Stupelis [9], I.B.Erunova [7], D. Socolescu [19], I. Bemelmans [4-6], I. Beale [2, 3], G. Allain [1], K. Pileckas [11-14] and others.

This paper deals with two related free boundary problems for the Navier-Stokes equations. Namely, at first the solvability of a noncompact free boundary problem is studied and then the asymptotics of a thin film flow is constructed.

1. Noncompact free boundary problem. Consider the problem of a plane steady flow of a heavy viscous incompressible capillary fluid which flows out of a slit and spreads over an infinite plate directed under an angle α to the horizon. It is assumed that the motion of the fluid is a result of Poiseille flow with a fixed flux F and of the motion of the bottom with a constant speed R. The vector of velocity v, the presure p and the free boundary Γ, tending to a line at the infinity, are sought for.

Denote by Ω the domain occupied by the fluid and choose the coordinate system (x_1, x_2) with x_1-axis coinciding with the bottom, i.e. the bottom S_1 is given by the equation $S_1 = \{(x_1, x_2) = x : -\infty < x_1 < \infty,\ x_2=0\}$. Let S_2 be a nonmoving upper wall of the slit. There are two possibilities:

i) S_2 consists of two semilines $L^{(1)} = \{x : x_1 < -1,\ x_2=h_o\}$ and $L^{(2)} = \{x : x_1=0,\ x_2 > h_o+1\}$, joined by a smooth curve;

ii) $S_2 = \{x : x_1 < 0,\ x_2=h_o\}$.

In the case i) the contact angle θ between the free boundary and the wall S_2 is prescribed and the contact point is unknown.

Contrary, in the case ii) we prescribe the contact point $(0, h_0)$ (the edge of the wall S_2) and the contact angle is unknown.

The vector of velocity $v=(v_1, v_2)$ and the pressure p satisfy in Ω the Navier-Stokes system

$$-\nu\nabla^2 v+(v\cdot\nabla)v+\nabla p=-\nabla G, \qquad \nabla\cdot v=0 \tag{1.1}$$

and the boundary conditions

$$v|_{S_1}=(R, 0), \; v|_{S_2}=(0, 0), \; v\cdot n|_\Gamma=0, \quad \tau\cdot T(v)n|_\Gamma=0 \tag{1.2}$$

Here $\nabla=(\partial/\partial x_1, \partial/\partial x_2)$, $G=g(-x_1\sin\alpha+x_2\cos\alpha)$, g is the acceleration of the gravity, ν is a coefficient of the viscosity, n and τ are unit vectors directed along the normal and the tangent to Γ, T(v) is the matrix with the elements $T_{ij}(v)=\partial v_i/\partial x_j+\partial v_j/\partial x_i$, $i, j=1, 2$.

The free boundary Γ is a priori unknown and satisfies the equation

$$K(x)=\sigma^{-1}(-p(x)+n\cdot T(v)n)|_\Gamma, \tag{1.3}$$

where K(x) is the curvature of Γ and σ is a coefficient of the surface tension.

In the case i) the free boundary Γ can be projected one to one onto the bottom S_1 and hence we shall look for Γ in the form

$$\Gamma=\{x : x_2=\Psi(x_1), \; x_1\in(0,\infty)\} \tag{1.4}$$

Then the equation (1.3) is equivalent to the following:

$$\frac{d}{dx_1}\left(\frac{\Psi'(x_1)}{(1+\Psi'(x_1)^2)^{1/2}}\right)=\sigma^{-1}(-p(x)+n\cdot T(v)n)\Big|_{x_2=\Psi(x_1)} \tag{1.5$_1$}$$

Prescribing the contact angle θ one gets the boundary condition for the function Ψ in the point $x_1=0$:

$$\Psi'(x_1)(1+\Psi'(x_1)^2)^{-1/2}\Big|_{x_1=0}=\cos\theta \tag{1.6$_1$}$$

Moreover, since the free boundary Γ tends to a line at the infinity, one also gets the condition

$$\lim_{x_1 \to \infty} \Psi(x_1) = h_* \qquad (1.7_1)$$

Note that the height h_* of the free boundary at the infinity cannot be prescribed a priori. It is a solution of the problem.

Consider the equation (1.3) in the case ii). Since in this case the contact angle is unknown, it can happen that the free boundary Γ has a turning point and it is impossible to express Γ in the form (1.4). Hence one has to seek for Γ in the parametric form

$$\Gamma = \{(x_1, x_2) : x_1 = x(s),\ x_2 = y(s),\ s \in (0, \infty)\} \qquad (1.8)$$

Then, instead of (1.5_1), one gets a system of differential equations

$$\begin{aligned} &\varphi'(s)\,(x'(s)^2 + y'(s)^2)^{-1/2} = \sigma^{-1}\,(-p(x) + n \cdot T(v)n)\big|_{\Gamma}, \\ &y'(s)\,(x'(s)^2 + y'(s)^2)^{-1/2} - \sin\varphi(s) = 0, \qquad (1.5_2) \\ &x'(s)\,(x'(s)^2 + y(s)^2)^{-1/2} - \cos\varphi(s) = 0, \end{aligned}$$

where $\varphi(s)$ is an angle between the vectors (1,0) and τ.

The boundary conditions (1.6_1), (1.7_1) in the case ii) take the form

$$x(0) = 0, \quad y(0) = h_0\,, \qquad (1.6_2)$$

$$\lim_{s \to \infty} x(s) = \infty\,, \quad \lim_{s \to \infty} y(s) = h_* \qquad (1.7_2)$$

To finish the mathematical formulation of the problem, one has to prescribe the full flux $Q = F + Rh_0/2$ of the fluid:

$$\int_0^{h_0} v_1(x_1, x_2)\, dx_2 = Q \qquad (1.9)$$

The problem (1.1), (1.2), (1.5_1)-(1.7_1), (1.9) for $\alpha = 0$, $F > 0$, $R > 0$ has been studied by the author in [11]. The problem (1.1), (1.2), (1.5_2)-(1.7_2), (1.9) for the case when the fluid moves only under gravitation (i.e. when $R = 0$, $F = g \sin\alpha\, h_0^3/12$) was considered in the joint papers by the author and M. Specovius-Neugebauer [12]. The general case was the subject of the papers [13, 14].

Experimental results for this problem were obtained by A. Stücheli [25].

Considering the problem (1.1), (1.2), (1.5)-(1.7), (1.9) we use the method belonging to V.V. Pukhnachov [15, 16]. This method consists in the following. Firstly, we consider the auxilliary problem (1.1), (1.2), (1.9), assuming that the free boundary Γ is known. Then a change of a solution of this problem under a small variation of the free boundary is found out. Finally, for small data the free boundary problem (1.5)-(1.7) is reduced to the contraction operator equation. Hence, the solution of the full problem is found by the method of successive approximations.

In order to formulate the main theorem we introduce weighted Hölder spaces. Denote by $C^s_r(\Omega;\mu)$, $r\in(0,s)$, $\mu>0$, a space of functions defined in Ω and provided with the norm

$$\|u\|_{C^s_r(\Omega;\mu)} = \|u\|_{C^s_r(\Omega_0)} + \|u\exp(\mu|x_1|)\|_{C^s(\Omega^+\cup\Omega^-)},$$

where $\Omega_0 = \{x\in\Omega : |x_1| < 2\}$, $\Omega^\pm = \{x\in\Omega : \pm x_1 > 1\}$,

$\|\cdot\|_{C^m(\cdot)}$ is the norm in Hölder space C^m,

$$\|u\|_{C^s_r(\Omega_0)} = \|u\|_{C^r(\Omega_0)} + \sum_{r<|\gamma|<s} \sup_{x\in\Omega_0} (\varrho(x)^{|\gamma|-r}|D^\gamma u(x)|) +$$

$$+\sum_{|\gamma|=[s]} \sup_{x,y\in\Omega_0} (\min(\varrho(x)^{s-r}, \varrho(y)^{s-r}) \frac{|D^\gamma u(x) - D^\gamma u(y)|}{|x-y|^{s-[s]}},$$

$\varrho(x) = \operatorname{dist}(x, M)$, $M = S_2\cap\Gamma$ is a contact point.

For $r<0$ the norm of the space $C^s_r(\Omega;\mu)$ is defined by the same formula but without the first term in the right-hand side. Note that for $|\gamma|>[r]$ the derivatives $D^\gamma u$ of functions $u\in C^s_r(\Omega;\mu)$ can have singularities at the contact point M. Moreover, elements u of the space $C^s_r(\Omega;\mu)$ vanish exponentially as $|x_1|\to\infty$.

<u>Theorem.</u> Let the cubic equation

$$g\sin\alpha\, h^3 + 3R\nu h - 3\nu Q = 0 \tag{1.10}$$

have a positive root h_* satisfying the condition $|h_0 - h_*| < 2/g\,\sigma^{-1}\cos\alpha$. Then for sufficiently small $|Q|$, $|R|$ and α the problem (1.1), (1.2), (1.5)-(1.7), (1.9) has a solution (v, p, Γ).

The number of solutions (v, p, Γ) is equal to the number of positive roots of the equation (1.10). The solution (v, p) has the asymtotic expansion

$$v(x) = V(x) + u(x), \quad p(x) = Q(x) + q(x), \tag{1.11}$$

where $V(x) = \zeta(-x_1)v^{(-)}(x) + \zeta(x_1)v^{(+)}(x)$, $Q(x) = -g\cos\alpha(x_2-h_*)+$ $+\zeta(-x_1)(g\sin\alpha - 12\nu F/h_o^3)x_1$, ζ is a smooth cut off function such that $\zeta(t) = 1$ for $t \geqslant 2$ and $\zeta(t) = 0$ for $t \leqslant 1$, $u \in C_\delta^{2+\delta}(\Omega;\mu)$, $\nabla q \in C_{\delta-2}^{\delta}(\Omega;\mu)$, $\delta \in (0, 1)$, $\mu \in (0, \mu_*)$,

$$\begin{aligned} v_1^{(-)}(x) &= R(h_o-x_2)/h_o + 6Fx_2(h_o-x_2)/h_o^3, \quad v_2^{(-)}(x) \equiv 0, \\ v_1^{(+)}(x) &= R + g\sin\alpha\, x_2(2h_* - x_2)/2\nu, \quad v_2^{(+)}(x) \equiv 0 \end{aligned} \tag{1.12}$$

Moreover, the free boundary Γ exponentially tends to the line $\{x : x_2 = h_*\}$ as $x_1 \to \infty$.

Note that considering the auxiliary problem (1.1), (1.2), (1.9) at first we assume the curve Γ to be given and the height h_* of Γ at the infinity to be an arbitrary positive number. In this case the asymptotics of the pressure $p(x)$ as $x_1 \to \infty$ is the following

$$p(x) \sim \left(g\sin\alpha - 3\nu(F + Rh_o/2 - Rh_*)/h_*^3\right)x_1 - g\cos\alpha(x_2 - h_*) \tag{1.13}$$

On the second step of Pukhnachov's scheme one has to substitute (1.13) into the right-hand side of the equation (1.5_1) or (1.5_2). These equations are solvable only if their right sides are bounded as $x_1 \to \infty$. In order to satisfy this condition it is necessary to eliminate the first singular term in (1.13). Physically it means that the pressure $p(x)$ remains bounded as $x_1 \to \infty$. Hence, h_* has to be a positive root of the cubic equation (1.10).

In cases:

1) $\alpha = 0$, sign $R = -$ sign Q;
2) $\alpha > 0$, $Q \leqslant 0$, $R \geqslant 0$;
3) $\alpha = 0$ and one of the numbers Q or R is zero

the equation (1.10) does not have positive roots and therefore the problem (1.1), (1.2), (1.5)-(1.7), (1.9) does not have solutions with a free boundary tending to a line.

If 4) $\alpha > 0$, $Q < 0$, $R < 0$, then the equation (1.10) has two positive roots and accordingly the problem (1.1), (1.2), (1.5)-(1.7), (1.9) has two solutions for arbitrarily small data.

In all other cases the positive root of (1.10) and the solution of problem (1.1), (1.2), (1.5)-(1.7), (1.9) is unique.

2. Asymptotics of a thin film flow. The results of this section were obtained jointly with S.A. Nazarov. Consider the flow of a viscous incompressible fluid in a thin domain $\widetilde{\Omega}_\varepsilon$, which is bounded by the bottom $\widetilde{S}_1 = \{x : x_2 = 0\}$, nonmoving upper walls $\widetilde{S}^+_{2\varepsilon}$, $\widetilde{S}^-_{2\varepsilon}$, $\widetilde{S}^\pm_{2\varepsilon} = \{x : \pm x_1 > d, \ x_2 = \varepsilon \widetilde{h}^\pm_0\} \cup \{x : x_1 = \pm d, \ x_2 > \varepsilon \widetilde{h}^\pm_0\}$, and the free boundary $\widetilde{\Gamma}_\varepsilon = \{x : x_2 = \varepsilon \widetilde{H}(x_1), \ x_1 \in (-d, d)\}$. It is assumed that the bottom $\widetilde{S}_1$ is inclined at an angle α to the horizon and moves with a constant speed $\widetilde{A}_0$. Moreover, the full flux Q of the fluid is prescribed.

Construct the asymptotics with respect to a small parameter ε of the solution $(\widetilde{v}, \widetilde{p}, \widetilde{H})$. It is more convenient to introduce the nondimensional functions $v = dQ^{-1}\widetilde{v}$, $p = (\rho Q \nu)^{-1} d^2 \widetilde{p}$, $H = d^{-1}\widetilde{H}$ and nondimensional coordinates $y = x_1 d^{-1}$, $z = x_2 d^{-1}$ *). Then (v, p, H) satisfy in the domain $\Omega_\varepsilon = d^{-1}\widetilde{\Omega}_\varepsilon$ the following nondimensional form of the free boundary problem for the Navier-Stokes system

$$-\nabla^2 v + \mathrm{Re}(v \cdot \nabla) v + \nabla p = -\mathrm{Re}\mathrm{Fr}^{-1} e^{(\alpha)}, \quad \nabla \cdot v = 0, \tag{2.1}$$

$$v\big|_{S_1} = (A_0, 0), \quad v\big|_{S^+_{2\varepsilon} \cup S^-_{2\varepsilon}} = (0, 0), \tag{2.2}$$

$$v \cdot n\big|_{\Gamma_\varepsilon} = 0, \quad \tau \cdot T(v) n\big|_{\Gamma_\varepsilon} = 0, \tag{2.3}$$

$$\varepsilon \mathrm{Re}\mathrm{We}^{-1} [H'(\varepsilon, y)(1 + \varepsilon^2 H'(\varepsilon, y)^2)^{-1/2}]' = (-p + n \cdot T(v) n)\big|_{\Gamma_\varepsilon}, \tag{2.4}$$

$$\pm \varepsilon H'(\varepsilon, \pm 1)(1 + \varepsilon^2 H'(\varepsilon, \pm 1)^2)^{-1/2} = \cos\theta, \tag{2.5}$$

$$\int_0^{\varepsilon h^\pm_0} v_1(\varepsilon, y, z)\, dz = 1. \tag{2.6}$$

Here ρ is the density of the fluid, $e^{(\alpha)} = (-\sin\alpha, \cos\alpha)$, $A_0 = \widetilde{A}_0 Q^{-1} d$, θ is a given contact angle, S_1, $S^\pm_{2\varepsilon}$, Γ_ε are the curves $\widetilde{S}_1$, $\widetilde{S}^\pm_{2\varepsilon}$, $\widetilde{\Gamma}_\varepsilon$

---

*) The nondimensional parameter ε is assumed to be such that $\max(\widetilde{h}^+_0, \widetilde{h}^-_0) = d$.

in the nondimensional coordinates, $Re = \nu^{-1}Q$, $Fr = Q^2 g^{-1} d^3$, $We = \rho Q^2 \sigma^{-1} d^{-1}$ are Reynolds, Frude and Weber numbers.

Suppose that the following conditions

$$A_0 = \varepsilon^{-1} a_0, \quad Re = \varepsilon^{m} R_0, \quad m > 1, \quad ReWe^{-1} = \varepsilon^{-2} \beta_0,$$
$$ReFr^{-1} = \varepsilon^{-4} \varkappa_0, \quad \sin\alpha = \varepsilon \alpha_0 \tag{2.7}$$

are valid. Consider at first the region $|y| < 1$. We shall seek for the asymptotics of the solution (v, p, H) in the form

$$v(\varepsilon, y, z) \sim \sum_{j=1}^{\infty} \varepsilon^{j-2} v^{(j)}(y, \zeta), \quad p(\varepsilon, y, z) \sim - ReFr^{-1}(y,z)\cdot e_{+}^{(\alpha)}$$
$$+ \varepsilon^{-3} q^{(0)}(y) + \sum_{j=1}^{\infty} \varepsilon^{j-3} q^{(j)}(y, \zeta), \quad H(\varepsilon, y) \sim \sum_{j=0}^{\infty} \varepsilon^{j} H_j(y), \tag{2.8}$$

where $\zeta = \varepsilon^{-1} z$. Substituting the series (2.8) into the Navier-Stokes equations (2.1) and into the boundary conditions (2.2), (2.3) and collecting the coefficients for the same degrees of ε, one gets the system

$$\frac{\partial^2 v_1^{(1)}}{\partial \zeta^2}(y, \zeta) = \frac{\partial q^{(0)}}{\partial y}(y), \quad \frac{\partial^2 v_2^{(1)}}{\partial \zeta^2}(y, \zeta) = \frac{\partial q^{(1)}}{\partial \zeta}(y, \zeta),$$
$$\frac{\partial v_2^{(1)}}{\partial \zeta}(y, \zeta) = 0, \tag{2.9}$$

$$v_1^{(1)}(y, 0) = a_0, \quad v_2^{(1)}(y, 0) = 0, \tag{2.10}$$

$$v_2^{(1)}(y, H_0(y)) = 0, \quad \frac{\partial v_1^{(1)}}{\partial \zeta}(y, H_0(y)) = 0 \tag{2.11}$$

Deriving the boundary conditions (2.11) we expand the function v into Taylor series by the variable ζ at the point $\zeta = H_0(y)$. The solution of the problem (2.9)-(2.11) is as follows

$$v_1^{(1)}(y, \zeta) = \frac{1}{2} \frac{dq^{(0)}}{dy}(y)\, \zeta (\zeta - 2 H_0(y)) + a_0, \quad v_2^{(1)}(y, \zeta) \equiv 0,$$
$$q^{(1)}(y, \zeta) = q_0^{(1)}(y), \tag{2.12}$$

where the function $q^{(0)}$ is still unknown and $q_0^{(1)}$ is an arbitrary function of y.

Similarly, for the functions $v^{(2)}$, $q^{(2)}$ one gets the problem

$$\frac{\partial^2 v_1^{(2)}}{\partial \zeta^2}(y,\zeta) = \frac{dq_o^{(1)}}{dy}(y), \frac{\partial^2 v_2^{(2)}}{\partial \zeta^2}(y,\zeta) = \frac{\partial q^{(2)}}{\partial \zeta}(y,\zeta),$$
$$\frac{\partial v_2^{(2)}}{\partial \zeta}(y,\zeta) = -\frac{\partial v_1^{(1)}}{\partial y}(y,\zeta) \tag{2.13}$$

$$v_1^{(2)}(y, 0) = v_2^{(2)}(y, 0) = 0 \tag{2.14}$$

$$v_2^{(2)}(y, H_o(y)) = v_1^{(1)}(y, H_o(y))H_o'(y), \frac{\partial v_1^{(2)}}{\partial \zeta}(y,H_o(y)) =$$
$$= -q^{(0)\prime}(y)H_1(y) \tag{2.15}$$

One can easily see that the problem (2.13)-(2.15) has a solution if and only if the following condition

$$\frac{\partial}{\partial y}\int_0^{H_o(y)} v_1^{(1)}(y,\zeta)\, d\zeta = 0 \tag{2.16}$$

is valid. Substituting the formulas (2.12) into (2.16), we derive the equation for the function $q^{(0)}$:

$$\frac{d}{dy}\left(H_o(y)^3 \frac{dq^{(0)}}{dy}(y)\right) - 3a_o \frac{dH_o}{dy}(y) = 0, \quad |y| < 1 \tag{2.17}$$

Now consider the equation (2.4). Repeating the above used reasons one obtains

$$-\varkappa_o H_o(y) + q^{(0)}(y) = -\alpha_o \varkappa_o y \tag{2.18}$$

Therefore, to define the functions H_o and $q^{(0)}$, we have to solve the system of equations (2.17), (2.18), which is an analogue of the Reynolds equation in the case of a flow with a free boundary. The boundary conditions for this system will be found below.

In the regions $y > 1$ and $y < -1$ the problem can be solved exactly, then one gets the following formulas

$$p(\varepsilon, y, z) = -\text{ReFr}^{-1}(y, z)\cdot e^{(\alpha)} + \varepsilon^{-3} q^{(0)}(y), \tag{2.19}$$
$$v(\varepsilon, y, z) = \varepsilon^{-1} v^{(1)}(y,\zeta),$$

where

$$q^{(0)}(y) = -6(h_o^{\pm})^3 (2 - a_o h_o^{\pm})\, y + q_o^{(0,\pm)}, \quad q_o^{(0,\pm)} = \text{const},$$

$$v_1^{(1)}(y,\zeta) = \frac{q^{(0)\prime}(y)}{2}\, \zeta(\zeta - h_o^{\pm}) + a_o (h_o^{\pm})^{-1}(h_o^{\pm} - \zeta), \quad \pm y > 1, \tag{2.20}$$

$$v_2^{(1)}(y,\zeta) = 0.$$

As usual, the conjugate conditions for the asymptotic expansions of the solution in different regions are obtained considering the boundary layers. Moreover, the function H given by the series (2.8) cannot satisfy the boundary condition (2.5). For this purpose the boundary layers are also neseccary. Consider the neighbourhood of the point $y = -1$. Let us introduce the "strech" variables $(\xi_1, \xi_2) = \varepsilon^{-1}(y + 1, z)$. We shall seek for the solution of a boundary layer type in the form

$$v(\varepsilon, y, z) \sim \sum_{j=1}^{\infty} \varepsilon^{j-2} u^{(j)}(\xi), \quad p(\varepsilon, y, z) \sim \varepsilon^{-3} r^{(0)}(\xi) +$$

$$+ \sum_{j=1}^{\infty} \varepsilon^{j-3} r^{(j)}(\xi), \quad H(\varepsilon, y) \sim \sum_{j=0}^{\infty} \varepsilon^{j} G_j(\xi_1). \tag{2.21}$$

Rewrite the problem (2.1)-(2.5) in the variables ξ and substitute the series (2.21) into the equations obtained. Collecting the coefficients for the same degrees of ε one gets

$$r^{(0)}(\xi) = -\varkappa_o \xi_2 + r_o^{(0)}, \quad r_o^{(0)} = \text{const}, \tag{2.22}$$

$$\left(G_o'(\xi_1)\left(1 + G_o'(\xi_1)^2\right)^{-1/2}\right)' - \beta_o^{-1}\varkappa_o \left(G_o(\xi_1) - h_*^-\right) = 0, \quad \xi_1 \in (0,\infty)$$

$$G_o'(0)\left(1 + G_o'(0)^2\right)^{-1/2} = \cos\theta, \tag{2.23}$$

where

$$h_*^- = r_o^{(0)} \varkappa_o^{-1} \tag{2.24}$$

$$-\nabla^2 u^{(1)} + \nabla r^{(1)} = (\varkappa_o \alpha_o, 0), \quad \nabla \cdot u^{(1)} = 0$$

$$u^{(1)}(\xi_1, 0) = (a_o, 0), \quad u^{(1)}(\xi)\Big|_{S^-} = (0,0), \quad u^{(1)} \cdot n\Big|_{\xi_2 = G_o(\xi_1)} = 0,$$

$$\tau \cdot T(u^{(1)})\, n\Big|_{\xi_2 = G_o(\xi_1)} = 0, \quad \int_0^{h_*^-} u_1^{(1)}(\xi_1, \xi_2)\, d\xi_2 = 1. \tag{2.25}$$

Problem (2.25) has to be solved in the domain Ω^- bounded by a line $\{\xi : \xi_2 = 0\}$, contour $S^- = \{\xi : \xi_1 < 0,\ \xi_2 = h_o^-\} \cup \{\xi : \xi_1 = 0,\ \xi_2 > h_o^-\}$ and a curve $\Gamma^- = \{\xi : \xi_2 = G_o(\xi_1),\ \xi_1 \in (0, \infty)\}$.

It is well known that the problem (2.23) has a unique solution $G_o(\xi_1)$ such that

$$G_o(\xi_1) \sim h_*^- + O(\exp(-\sqrt{\varkappa_o \beta_o^{-1}}\,\xi_1)) \text{ as } \xi_1 \to \infty.$$

Theorem. If h_*^- is a positive root of the cubic equation

$$\alpha_o \varkappa_o h^3 + 3a_o h - 3 = 0, \tag{2.26}$$

then the problem (2.25) has a unique solution $(u^{(1)}, r^{(1)})$ such that $r^{(1)}$ is bounded as $\xi_1 \to \infty$. For this solution there holds the representation

$$u^{(1)}(\xi) = \zeta(-\xi_1)\, u^{(1,-)}(\xi) + \zeta(\xi_1)\, u^{(1,+)}(\xi) + W^{(1)}(\xi),$$
$$r^{(1)}(\xi) = \zeta(-\xi_1)\,(\alpha_o \varkappa_o - 2c^{(1,-)})\,\xi_1 + s^{(1)}(\xi), \tag{2.27}$$

where

$$u_1^{(1,+)}(\xi) = c^{(1,+)}\,\xi_2(\xi_2 - 2h_*^-) + a_o, \quad u_2^{(1,+)}(\xi) = 0,$$

$$u_1^{(1,-)}(\xi) = c^{(1,-)}\,\xi_2(\xi_2 - h_o^-) + a_o (h_o^-)^{-1}(h_o^- - \xi_2), \quad u_2^{(1,-)}(\xi) = 0, \tag{2.28}$$

$$c^{(1,+)} = \frac{3}{2(h_*^-)^3}(a_o h_*^- - 1), \quad c^{(1,-)} = \frac{3}{(h_o^-)^3}(a_o h_o^- - 2),$$

ζ is a cut off function, $W^{(1)}$ and $s^{(1)}$ vanish exponentially as $|\xi_1| \to \infty$ and may have singularities at the angular points $M_1 = (0, h_o^-)$ and $M_2 = S^- \cap \Gamma^-$.

This theorem follows from the results of section 1. The construction of the boundary layer near the point y = -1 will be finished if we take h_*^- as a positive root of (2.26) and then calculate the constant $r_o^{(0)}$ from (2.24).

Similarly, the boundary layer near the point y = 1 can be constructed. Note only that the height h_*^+ of the free boundary Γ^+ at the infinity is also defined as a root of the same cubic equation (2.26) and, hence $h_*^- = h_*^+ = h_*$.

Let us find the conjugate conditions near the points y = -1 and y = 1.

The height h_* of the free boundary at the infinity (in the boundary layer solutions) has to coincide with $H_o(\pm 1)$. Hence we get the following conditions

$$H_o(1) = H_o(-1) = h_* \tag{2.29}$$

Rewrite the expansions for the function p in the regions $|y| < 1$ and $|y| > 1$ in the coordinates (ξ_1, ξ_2) and compare them with the corresponding boundary layer expansions. As a results, we obtain the conditions

$$\pm \alpha_o + \varkappa_o^{-1} q^{(0)}(\pm 1) = h_*, \quad q^{(0)\prime}(\pm 1) = -\varkappa_o \alpha_o \tag{2.30}$$

Integrating the equation (2.17) we get

$$H_o^3(y) \frac{dq^{(0)}}{dy}(y) - 3a_o H_o(y) + c = 0 \tag{2.31}$$

By virtue of (2.29), (2.30) and the cubic equation (2.26) we find $c = 3$. Using now (2.18) we derive the final equation for H_o:

$$\begin{gathered} \varkappa_o H_o^3(y) H_o'(y) = \varkappa_o \alpha_o H_o^3(y) + 3a_o H_o(y) - 3 \\ H_o(-1) = H_o(1) = h_* . \end{gathered} \tag{2.32}$$

It can be easily proved that the unique solution of (2.32) is $H_o(y) \equiv h_*$. The function $q^{(0)}$ can be uniquely determined from the equation (2.31) and the first of the boundary conditions (2.30).

The asymptotic expansions of the solution in different zones can be joined with the help of cut off functions. Analogously the following membres of the asymptotics can be found.

Let us assume that we have constructed N members of the asymptotics. Let $V^{(N)}$, $P^{(N)}$ and $H^{(N)}$ be partial sums of the asymptotical series. Let

$$W_N = v - V^{(N)}, \quad S_N = p - P^{(N)}, \quad K_N = H - H^{(N)},$$

where (v, p, H) is the exact solution of the problem. It is proved that the norms of W_N, εS_N, $\varepsilon^{-2} K_N$ in the corresponding spaces are $O(\varepsilon^{N-1})$ as $\varepsilon \to 0$.

References

1. Allain G., Small time existence for the Navier-Stokes equations with a free surface. Ecole Polytechnique. Rapport interue, 1985, v. 135, p. 1-24.

2. Beale J.T., The initial value problem for the Navier-Stokes equations with a free boundary. Comm. Pure Appl. Math., 1980, v. 31, p. 359-392.

3. Beale J.T., Large-time regularity of viscous surface waves. Arch. Rat. Mech. Anal., 1984, v. 84, p. 307-352.

4. Bemelmans J., Liquid drops under the influence of gravity and surface tension. Manuscripta Math., 1981, v. 36. No 1, p. 105-123.

5. Bemelmans J., Free boundary problems for the stationary Navier-Stokes equations. Asterisqué, 1984, No 18, p. 115-123.

6. Bemelmans J., Equilibrium figures of rotating fluids. Z. angew. Math. und Mech., 1982, v. 62, No 5, p. 273.

7. Erunova I.B., Solvability of a free boundary problem for two fluids in a container. Vestnik Leningrad. Univ. Math., Mech., Astr., 1986, No 2, p. 9-16 (in Russian).

8. Ladyzhenskaya O.A., Osmolovskii V.G., The free surface of a layer of fluid over a solid sphere. Vestnik Leningrad. Univ., Math., Mech., Astr., 1976, No 13, p. 25-30 (in Russian).

9. Maz'ya V.G., Plamenevskii B.A., Stupelis L.I., Three-dimensional problem with a free boundary. Diff equations and their applications. Inst. of Math. and Cybern. Acad. Sci. Lith. SSR, Vilnius, 1979, v. 23, p. 9-153 (in Russian).

10. Osmolovskii V.G., The free surface of a drop in a symmetric force field. Zapiski Nauchn. Sem. LOMI, 1975, v. 52, p.160-174 (in Russian)

11. Pileckas K., Solvability of a problem of a plane motion of a heavy viscous incompressible fluid with a noncompact free boundary. Diff. equations and their applications. Inst. of Math. and Cybern. Acad. Sci. Lith. SSR, Vilnius, 1981, v. 30, p. 57-96.

12. Pileckas K., Specovius-Neugebauer M., Solvability of a noncompact free boundary problem for the stationary Navier-Stokes system. Lith. Math. J., I:1989, v. 30, No 3, II:1989, v. 30, No 4 (in Russian).

13. Pileckas K., On the problem of motion of a heavy viscous incompressible fluid with noncompact free boundary. Lith. Math. J., 1988, v. 29, No 2, p. 315-333 (in Russian).

14. Pileckas K., The example of nonuniqueness of the solution to a noncompact free boundary problem for the stationary Navier-Stokes s system. Diff. equations and their applications. Inst. of Math. and Cybern. Acad. Sci. Lith. SSR, Vilnius, 1988, v. 42, p. 59-64 (in Russian).

15. Pukhnachov V.V., Plane stationary free boundary problem for the Navier-Stokes equations. Z. Prikl. Meh. i Techn. Fiz., 1972, No 3.

p. 91-102 (in Russian).
16. Pukhnachov V.V., Free boundary problems for Navier-Stokes equations. Doctoral dissertation, Novosibirsk, 1974 (in Russian).
17. Rivkind V.Ya., Theoretical justification of a method of successive approximations for stationary problems of mechanics a viscous fluid with free surfaces of separation. Zap. Nauchn. Sem. LOMI, 1982, v. 115, p. 228-235 (in Russian).
18. Rivkind V.Ya., A priori estimates and the method of successive approximations for the motion of a drop. Trudy Math. Inst. Steklov, 1983, v. 159, p. 150-166 (in Russian).
19. Socolescu D., On one free boundary problem for the stationary Navier-Stokes equations. - Methoden und Verfahren der mathematischen Physik, 1979, c. 18, p. 117-140.
20. Solonnikov V.A., Solvability of the problem of plane motion of a heavy viscous incompressible capillary liquid partially filling a container. Izv. Akad. Nauk SSSR, 1979, v. 43, p. 203-236 (in Russian), English translation: Math. USSR-Izv., 1980, v. 14, p. 193-221.
21. Solonnikov V.A., On a free boundary problem for the system of Navier-Stokes equations. Trudy Sem. S.L. Sobolev, 1978, v. 2, p. 127-140 (in Russian).
22. Solonnikov V.A., Solvability of a three-dimensional boundary value problem with a free surface for the stationary Navier-Stokes system. Partial Diff. Equations,Banach Center Publ., 1983, v. 10, p. 361-403.
23. Solonnikov V.A., Solvability of the problem of the motion of a viscous incompressible fluid bounded by a free surface. Izv. Akad. Nauk SSSR, Ser. Mat., 1977, v. 41, No 6, p. 1388-1424 (in Russian), English translation: Math. USSR-Izv., 1977, v. 11, No 6, p. 1323-1358.
24. Solonnikov V.A., On a unsteady motion of an isolated volume of a viscous incompressible fluid, Izv. Akad. Nauk SSSR, Ser. Mat., 1987, v. 51, No 5, p. 1065-1087 (in Russian).
25. Stücheli A., Flow development of a highly viscous fluid emerging from a slit onto a plate. - J. Fluid Mech., 1980, v. 97, part 2, p. 321-330.

This paper is in a final form and no similar paper has been or is being submitted elsewhere.

(received August 5, 1989.)

ON LARGE TIME BEHAVIOR OF THE TOTAL KINETIC ENERGY FOR WEAK SOLUTIONS OF THE NAVIER-STOKES EQUATIONS IN UNBOUNDED DOMAINS

Wolfgang Borchers* and Tetsuro Miyakawa**

* Fachbereich Mathematik-Informatik der Universität-Gesamthochschule Paderborn, D-4790 Paderborn, Federal Republic of Germany

** Department of Mathematics, Faculty of Science, Hiroshima University, Hiroshima 730, Japan

1. Introduction

We consider the initial boundary value problem for the Navier-Stokes equations in an unbounded domain $\Omega \subset \mathbf{R}^n$ $(n=2,3,4)$,

$$\begin{aligned} \frac{\partial u}{\partial t} + u \cdot \nabla u - \Delta u + \nabla p &= 0 \\ \nabla \cdot u &= 0 \end{aligned} \quad \text{in } (0,\infty) \times \Omega \qquad \text{(NS)}$$

$$u = 0 \text{ on } (0,\infty) \times \partial\Omega; \; u_{|t=0} = u_0 \text{ in } \Omega,$$

where $u_0 = u_0(x)$ is a prescribed initial velocity which for compatibility reasons is assumed to be solenoidal with vanishing normal component on the boundary $\partial\Omega$ of Ω. The function $u = u(t,x) = (u^1(t,x),\dots,u^n(t,x))$ is the unknown velocity and the scalar valued function $p = p(t,x)$ is the unknown pressure. The existence of a weak solution of problem (NS) satisfying the energy inequality is well known [6] while problems on uniqueness and regularity of weak solutions still remain open.
In this paper we shall study the behavior of the L^2-norm as $t \to \infty$ for weak solutions of the Navier-Stokes equations in arbitrary unbounded domains. We shall show the existence of a weak solution which decreases at a specific rate depending on the data. Moreover, as t tends to infinity, this weak solution behaves like the solution of the instationary Stokes equation which is obtained from (NS) by eliminating the nonlinear terms.

The problem of L^2 decay for weak solutions was first raised by Leray [11] in the case $\Omega = \mathbf{R}^3$. Schonbeck [15] solved this problem by an ingenious use of Fourier transform technics. Generalizations (to arbitrary space dimensions $n \geq 2$) and improvements of the decay rates were then given by Kajikiya and Miyakawa [8]. The most general results in this direction were obtained by Wiegner [16], who proved a decay rate of order $O(t^{-(n+2)/4})$. By replacing the Fourier transform by the more general spectral theory and by the fractional powers of the Stokes operator, the authors established the same results for weak solutions in halfspaces [1]. The decay rate obtained in the present work is of order $O(t^{-1/2})$. In the meantime the authors have obtained stronger results in case of smooth exterior domains $\Omega \subset \mathbf{R}^n$ $(n \geq 3)$ [3] (s. also Maremonti [13] for n = 3) and proved a decay of order $O(t^{-\beta})$ with $\beta < n/4$. However, the methods in this paper are more elementary and, moreover, apply to arbitrary domains.

We thank Professor M. Wiegner for valuable remarks and discussions.

2. Preliminaries and Main Results

We use the following notations : $L^r = L^r(\Omega)$, $1 \leq r \leq \infty$, denotes the usual Lebesgue spaces of scalar as well as vector functions defined on $\Omega \subset \mathbf{R}^n$. The norm of L^r is denoted by $\|\cdot\|_{L^r}$. If r = 2 we simply write $\|\cdot\| = \|\cdot\|_{L^2}$. $H^m = H^m(\Omega)$ denotes the usual L^2 Sobolev space of order m = 0,1,... with norm $\|\cdot\|_m$. We denote by $H_o^m = H_o^m(\Omega)$ the H^m- closure of all smooth functions with compact support in Ω. $C_{o,\sigma}^\infty(\Omega)$ is the set of all smooth solenoidal vector fields with compact support in Ω and $L_\sigma^2 = L_\sigma^2(\Omega)$ is it's L^2- closure. By P we denote the projection of L^2 onto L_σ^2. The space V is the H^1 - closure of $C_{o,\sigma}^\infty(\Omega)$.

The Stokes operator A in L_σ^2 is the nonnegative selfadjoint operator associated with the V-continuous nonnegative symmetric form

$$(\nabla u, \nabla v) = \sum_{i=1}^n (\nabla u^i, \nabla v^i), \quad u,v \in V, \tag{2.1}$$

where here and in the following $(\cdot,\cdot)$ denotes the L^2scalar product. We recall that $D(A) \subset D(A^{1/2}) = V$ and

$$\|A^{1/2}u\| = \|\nabla u\| := (\nabla u, \nabla u)^{1/2}, \quad u \in V. \tag{2.2}$$

It is also well known that -A generates a uniformly bounded holomorphic semigroup $\{e^{-tA} : t \geq 0\}$ which admits the representation

$$e^{-tA}u_o = \int_o^\infty e^{-t\lambda}\, dE_\lambda u_o, \quad t \geq o,\ u_o \in L_\sigma^2, \tag{2.3}$$

where E_λ are the spectral projections associated with A.

Let $u_o \in L_\sigma^2$. A weakly continuous function $u:[o,\infty) \to L_\sigma^2$ with $u(o)=u_o$ is called a weak solution of (NS) if

a) $u \in L^\infty(0,T;L_\sigma^2) \cap L^2(0,T;V)$ for each $o < T < \infty$, and

b) $(u(t),v(t)) + \int_s^t \{-(u,v')+(u\cdot\nabla u,v)+(\nabla u,\nabla v)\}\, d\tau = (u(s),v(s))$

for all $o \leq s \leq t \leq \infty$ and for all $v \in C^1([o,\infty);L_\sigma^2) \cap C^o([o,\infty);V)$ with compact support in $[o,\infty)$. Here v' denotes the derivative dv/dt. Due to the Sobolev inequalities

$$\|u\|_{L^4} \leq c\, \|u\|^{1-n/4} \|\nabla u\|^{n/4}, \quad n = 2,3,4, \tag{2.4}$$

with a constant $c > o$ independent of u ([4 , Th.9.3,p.24] and [10 ,p.8]), the third term on the left of b) is integrable.

We can now state our main results as follows.

<u>Theorem 1.</u> For each $u_o \in L^2_\sigma$ there is a weak solution u of (NS) such that for $t \to \infty$

(i) $\|u(t)\| = o(1)$

(ii) $\|u(t) - e^{-tA}u_o\| = o(t^{1/2-n/4})$ if $n = 2,3$; and

$\|u(t) - e^{-tA}u_o\| = O(t^{-1/2})$ if $n = 4$.

If $\|e^{-tA}u_o\| = O(t^{-\alpha_o})$ with $0 < \alpha_o < 1/2$ $(0 < \alpha_o \leq 1/2$ for $n = 4)$, then

(iii) $\|u(t)\| = O(t^{-\alpha_o})$, if $n = 3,4$,

$\|u(t)\| = O((\ln t)^{-m})$ for every $m \geq 0$, if $n = 2$,

(iv) $\|u(t) - e^{-tA}u_o\| = o(t^{-\alpha_o})$ if $n = 3$.

<u>Theorem 2.</u> Let $u_o \in L^2_\sigma$. Then the statements (i) -(iv) of Theorem 1 are valid for every weak solution u satisfying the stronger energy inequality

$$\|u(t)\|^2 + 2\int_s^t \|\nabla u(\tau)\|^2\, d\tau \leq \|u(s)\|^2 \tag{2.5}$$

for all $t \geq 0$ and almost all s in $0 \leq s \leq t$.

To see that there are sufficiently many initial data satisfying the decay requirements of (iii),(iv) , we consider initial velocities which belong to the range $R(A^\alpha)$ of the fractional power A^α $(\alpha \geq o)$, i.e. we assume $u_o = A^\alpha a$ for some $a \in D(A^\alpha)$. Then

$$\|e^{-tA}u_o\|^2 = \int_0^\infty \lambda^{2\alpha}\, e^{-2t\lambda}\, d\,\|E_\lambda a\|^2 = O(t^{-2\alpha}) \tag{2.6}$$

as $t \to \infty$. We note that $R(A^\alpha)$ is dense in L^2_σ. This is a consequence of the injectivity of A and the identity $L^2_\sigma = N(A^\alpha) \oplus \overline{R(A^\alpha)}$, where $N(A^\alpha)$ is the nullspace of A^α [9,Th. 31, p.275]. We also note that $\|e^{-tA}u_o\|$ decays exponentially if $E_\lambda u_o = 0$ in $0 \leq \lambda \leq \lambda_o$ for some $\lambda_o > 0$. The boundedness of the Stokes semigroup together with (2.6) and the denseness of $R(A)$ imply

$$e^{-tA}u_o \to 0 \text{ as } t \to \infty \text{ in } L^2_\sigma \tag{2.7}$$

for all $u_o \in L^2_\sigma$. (2.7) is needed for the proof of (i) in Theorem 1.

We shall prove Theorem 1 and 2 in Sect. 4.In Sect. 3 below we construct suitable approximate solutions to (NS) which converge to a weak solution of (NS) for which the statements of Theorem 1 are fulfilled.

Throughout this paper C, c denote positive constants which may vary from line to line.

3. Construction of Approximate Solutions

For each $\varepsilon > 0$ we consider the following regularization of (NS) ([11],[14]):

$$\frac{d}{dt} u_\varepsilon + Au_\varepsilon + B_\varepsilon(u_\varepsilon) = 0,\ t > 0,\quad u_\varepsilon(0) = u_\varepsilon \in V. \tag{RNS}$$

The nonlinear operator $u \to B_\varepsilon(u)$ is defined by $B_\varepsilon(u) = P[(g_\varepsilon * \hat{u})\cdot\nabla u]$, where g_ε is a mollifier and $g_\varepsilon * \hat{u}$ denotes the convolution in $\mathbf{R}^n$ of g_ε with the zero extension $\hat{u}$ of u. Problem (RNS) is obtained formally from the original problem (NS) by eliminating the pressure gradient via the projection P and by replacing the first factor in the convective term $u \cdot \nabla u$ by the regularization $g_\varepsilon * \hat{u}$. The following properties are elementary

$$\begin{aligned} &1)\ \|g_\varepsilon * \hat{u}\|_m \leqq \|u\|_m\ ,\quad \|g_\varepsilon * \hat{u}\|_{L^p} \leq \|u\|_{L^p}\quad (1 \leq p \leq \infty),\\ &2)\ \|g_\varepsilon * \hat{u}\|_{L^\infty} \leq C\|u\|\ , \end{aligned} \tag{3.2}$$

with $C= C(\varepsilon) > 0$ not depending on u. The first inequality follows from the corresponding identity $\|\hat{u}\|_m = \|u\|_m$ which holds for $u \in H^m_o$. Moreover, $u \in V$ implies $\operatorname{div} \hat{u} = 0$, which gives $\operatorname{div}(g_\varepsilon * \hat{u}) = 0$. This observation easily leads to

$$(B_\varepsilon(u),u) = 0 \quad \text{for all } u \in V. \tag{3.3}$$

We shall now show that for each $u_o \in V$ and $\varepsilon > 0$ problem (RNS) admits a unique solution defined on $[0,\infty)$. As in [1] this will be shown by applying the contraction mapping principle to the corresponding integral equation

$$u(t) = e^{-tA}u_o - \int_0^t e^{-(t-s)A}\, B_\varepsilon(u)(s)ds, \tag{3.4}$$

where we surpressed the dependence of u on ε to simplify the notation.

<u>Proposition 3.1.</u> For any $\varepsilon > 0$, $u_o \in V$ the problem (RNS) has a unique solution $u:[0,\infty) \to V$ such that $u \in C^o([0,T];V) \cap C^o((0,T];D(A))$ for every $T > 0$.

Proof. We denote by $\|v\|_T := \sup_{o\leq t\leq T}(\|v(t)\|+\|\nabla v(t)\|)$ the norm of the Banach space $C([0,T];V)$ and by $S = S(M,T,u_o)$ the closed subset $S=\{v:v(o)=u_o, \|v\|_T \leq M\}$. On S we consider the nonlinear operator

$$G\,v(t) = e^{-tA}u_o - \int_0^t e^{-(t-s)A}\, B_\varepsilon(u)(s)ds.$$

By (2.2) and 2) of (3.2)

$$\|B_\varepsilon(u)\| \leq \|g_\varepsilon * \hat{u}\|_{L^\infty}\|\nabla u\| \leq C\,\|u\|\,\|A^{1/2}u\|. \tag{3.5}$$

This, together with $\|A^{1/2} e^{-tA}\| \leq C\, t^{-1/2}$ (operator norm in L^2_σ), gives

$$\|G\,v\|_T \leq \|u_o\|_1 + C_1 M^2 (T+T^{1/2}),\quad v \in S.$$

Similarly, using

$$\|B_\varepsilon(u) - B_\varepsilon(v)\| \leq C_2(\|u-v\|\|\nabla u\| + \|v\|\,\|\nabla(u-v)\|),$$

we get the Lipschitz estimate

$$\| Gu - Gv \|_T \leq C_3 M (T+T^{1/2}) \| u-v \|_T , \quad u,v \in S.$$

If we choose M such that $\|u_o\|_1 \leq M/2$ and T such that $(T+T^{1/2})M\cdot\max\{C_1, C_3\} \leq 1/2$, then S is mapped into itself and G is a strict contraction. By the contraction mapping principle there is a unique $u \in S$ which solves (3.4) on some intervall $[o,T]$. By a well known bootstrap argument [5] it is then easily seen that u_t and Au are in $C^o((o,T]; L^2_\sigma)$, so that u solves (RNS) (We recall that in our general situation D(A) is not nessecarily contained in H^2). To show that u can be extended to a solution on $[0,\infty)$, it suffices to derive a priori bounds for $\|u(t)\|_1$, since (RNS) can be solved locally for initial data in V. Taking the scalar product of (RNS) with u, using (3.3) and integrating, we arrive at

$$\|u(t)\|^2 + 2\int_0^t \|\nabla u(s)\|^2\, ds = \|u_o\|^2. \tag{3.6}$$

From (3.4),(3.5) and (3.6) we conclude

$$\|A^{1/2} u(t)\| \leq \|A^{1/2} u_o\| + C \int_0^t (t-s)^{-1/2} \|A^{1/2} u(s)\| ds,$$

so that $\|A^{1/2}u(t)\|$ is dominated by the solution of the corresponding Volterra integral equation which is bounded on every compact intervall [7,p.206]. The proof is complete.

Now let $u_o \in L^2_\sigma$ and set $u_{o,k} = (1 + \frac{1}{k} A)^{-1} u_o \in V$ for $k = 1,2,\ldots$. Then $u_{o,k} \to u_o$ in L^2_σ as $k \to \infty$ and $\|e^{-tA}u_{o,k}\| \leq \|e^{-tA}u_o\|$. Let u_k be the solution of (RNS) with initial velocity $u_k(o) = u_{o,k}$ and $\varepsilon = 1/k$.

<u>Proposition 3.2.</u> There exists a subsequence of u_k which converges in $L^2_{loc}([o,\infty) \times \Omega)$ to a weak solution of (NS).

Proof. Since $u_{o,k}$ is bounded in L^2_σ , (3.6) shows that u_k is bounded in $L^2(0,T;V) \cap L^\infty(o,\infty; L^2_\sigma)$ for every finite $T > 0$. Hence, we can find a subsequence (also denoted by u_k) and a function u such that

$$u_k \to u \text{ in } L^2(0,T,V) \text{ weakly for every } T > 0,$$

and

$$u_k \to u \text{ in } L^\infty(0,\infty,L^2_\sigma) \text{ weak-star.}$$

We shall now show that this subsequence converges in $L^2_{loc}([0,\infty) \times \Omega)$. Applying the Fourier transform technic of Lions [12,pp.77-79] to (RNS) (recall du_ε/dt, $Au_\varepsilon \in L^2(0,T,L^2_\sigma)$) and using the estimates

$$\|B_\varepsilon(u)\|_{V'} \leq \|g_\varepsilon * \hat{u}\|_{L^4} \|u\|_{L^4} \leq \|u\|^2_{L^4} \leq c\, \|u\|^2_1 ,$$

we obtain that the fractional derivatives $D^\gamma_t u_k$ $(0\leq\gamma<1/4)$ are uniformly bounded in $L^2(0,T,L^2_\sigma)$. Thus, they are also uniformly bounded in $L^2(0,T,L^2(U))$ for any smooth and bounded domain U with $\bar{U} \subset \Omega$. Since $H^1(U) \subset L^2(U)$, where the injection is compact by the Rellich Kondrachow theorem, we conclude that u_k is precompact in $L^2((o,T)\times U)$ [12 Th. 5.2,p.61]. This proves $u_k \to u$ in $L^2_{loc}([o,T) \times \Omega)$. To prove that u is a weak solution of (NS), it suffices to show property b) of Sect. 2 for $s = o, t = \infty$ and for all smooth and solenoidal functions v with compact support in $[o,T) \times \Omega$. This is easily verified with the above convergence properties by taking the scalar product of (RNS) with v and integrating by parts. To pass to the limit in the nonlinear term, we have to use the convergence of u_k in $L^2_{loc}((o,T) \times \Omega)$ and the weak convergence $g_{1/k} * \hat{u}_k \to u$ in $L^2((o,T) \times \Omega)$. The latter is implied by the weak convergence of u_k. Then, a,b) of Section 2 imply the continuity of $t \to u(t)$ in the weak topology of L^2_σ as well as $u(o) = u_o$. This completes the proof.

4. Proof of the Main Results

In this section we prove Theorem 1 and 2 stated in Sect. 2 . Theorem 1 will be proved by deriving estimates on u_k of the form

$$\|u_k(t)\| \leqq f(t) \quad \text{for all } t > o,$$

where f is a continuous function independent of k. From the convergence properties obtained in Prop. 3.2 one can then conclude that $\|u(t)\| \leqq f(t)$ for all $t > o$. For details we refer to [8].

Taking the scalar product of (RNS) with u_k and integrating from s to t ($o \leqq s \leqq t$), we obtain from (3.3)

$$\|u_k(t)\|^2 + 2\int_s^t \|\nabla u_k(s)\|^2 \, ds = \|u_k(s)\|^2 , \tag{4.1}$$

for k = 1,2,... . In what follows we only make use of the corresponding inequality (2.5). All estimates on the nonlinear term which we need in the proof of Theorem 2 hold in a similar manner for the regularization $B_\varepsilon(u)$ with constants independent of ε (by (3.2) 1)). Therefore we shall verify (i) - (iv) in Theorem 1 under the assumption of Theorem 2 for simplicity in notation. We note that in case n = 2 the weak solution u is actually a strong one satisfying (4.1) with u_k replaced by u. The proofs are essentially based on the following Lemma.

Lemma 4.1. $\|E_\lambda P(u\cdot\nabla)u\| \leqq c\,\lambda^{1/2}\,\|u\|^{2-n/2}\,\|\nabla u\|^{n/2}$, $n = 2,3,4,$

for $\lambda \geqq o$ and $u \in V$. The constant is independent of u and λ.

Proof. We have

$$\|E_\lambda P(u\cdot\nabla)u\| = \sup_{\|v\|=1} |(E_\lambda P(u\cdot\nabla)u, v)|,$$

where $v \in L^2_\sigma$. Since div u = o, the scalar product can be written as

$$(E_\lambda P(u\cdot\nabla)u,v) = (u\cdot\nabla u, E_\lambda v) = -(u\cdot\nabla E_\lambda v,u).$$

It follows with (2.2),(2.4) and $\|v\| = 1$

$$|(E_\lambda P(u\cdot\nabla)u,v)| \leqq \|u\|^2_{L^4}\|\nabla E_\lambda v\| \leqq c\,\lambda^{1/2}\|u\|^{2-n/2}\,\|\nabla u\|^{n/2},$$

which proves the Lemma.

Proof of Assertions (i),(iii) and (iv).

Let u be a weak solution of (NS) satisfying (2.5) and let $\lambda = \lambda(t)$ be a smooth and positive function on (o,∞). We have

$$\|A^{1/2}u\|^2 = \int_o^\infty z \, d\,\|E_z u\|^2 \geq \int_\lambda^\infty z \, d\,\|E_z u\|^2 \geq \lambda\,(\|u\|^2 - \|E_\lambda u\|^2). \qquad (4.2)$$

Here $\|E_\lambda u\|$ can be estimated by Lemma 4.1 as follows. Since E_λ regularizes, the weak form of the equation satisfied by u allows the representation

$$E_{\lambda(t)}u(t) = E_{\lambda(t)}\mathbf{e}(t) - \int_o^t \mathbf{e}^{-(t-s)A} E_{\lambda(t)} \, P\,(u\cdot\nabla)u(s)ds\ ,$$

where $\mathbf{e}(t) = \mathbf{e}^{-tA}u_o$. Hence, by Lemma 4.1 and the Hölder inequality

$$\|E_{\lambda(t)}u(t)\| \leq \|\mathbf{e}(t)\| + c\,\lambda^{1/2}(t)(\int_o^t \|\nabla u\|^2\,ds)^{n/4}(\int_o^t \|u\|^2 ds)^{1-n/4}$$

$$\leq \|\mathbf{e}(t)\| + c\,\lambda^{1/2}(t)(\int_o^t \|u\|^2\,ds).^{1-n/4}$$

Thus, (4.2) gives

$$\|A^{1/2}u(t)\|^2 \geq \lambda(t)\,\|u(t)\|^2 - c\,\lambda^2(t)\,(\int_o^t \|u\|^2 ds)^{2-n/2} - 2\lambda(t)\|\mathbf{e}(t)\|^2 \qquad (4.3)$$

for almost all $t > o$.

Inserting (4.3) into (2.5), we obtain for $y = \|u\|^2$ an integral inequality of the form

$$y(t) - g(t,s) + \int_s^t \lambda\,(\tau)\,y\,(\tau)d\tau \leq y(s) \quad \text{a.e. in } o<s\leq t\ , \qquad (4.4)$$

with $g(t,s) = \int_s^t \{2\lambda(\tau)\|\mathbf{e}(\tau)\|^2 + c\,\lambda^2\,(\tau)\cdot(\int_o^\tau \|u(\sigma)\|^2\,d\sigma)^{2-n/2}\}\,d\tau$.

We apply the Gronwall lemma backwards, i.e. with respect to the s-variable in (4.4). Define $\phi(s) = \int_s^t \lambda\,(\tau)\,y\,(\tau)d\tau$; then ϕ is differentiable almost everywhere in $o < s < t$ with $\phi' \in L^\infty(\delta,t)$ for all small $\delta > o$. By (4.4) we have

$$\phi'(\tau) = -\lambda\,(\tau)\,y\,(\tau) \leq -\lambda(\tau)\,(\phi(\tau) + y(t) - g(t,\tau)). \qquad (4.5)$$

Now let $H \geq o$ be an integrating factor which means $H' = \lambda H$. Multiplying (4.5) by H and integrating from s to t, we arrive at the inequality

$$(H(t) - H(s))\,y\,(t) \leq H(s)\,\phi\,(s) + \int_s^t H'\,(\tau)\,g\,(t,\tau)d\tau$$

$$\leq H(s)\,(y(s)-y(t)+g(t,s)) + \int_s^t H'(\tau)g(t,\tau)d\tau \ \ \text{(by (4.4))}.$$

Since $g(t,t) = o$, partial integration yields for almost all s in $o < s \leq t$ the inequality

$$H(t)\,y\,(t) \leq H\,(s)\,y\,(s) - \int_s^t H\,(\tau)\,\frac{\partial}{\partial\tau}\,g\,(t,\tau)d\tau. \qquad (4.6)$$

If we choose $\lambda(\tau) = \alpha\,\tau^{-1}$, then $H(\tau) = \tau^\alpha$. Since y(s) is bounded, we arrive in the limit $s \to o$ at the inequality

$$\|u(t)\|^2 \leq t^{-\alpha} \int_0^t 2\alpha\, s^{\alpha-1} \|\mathbf{e}(s)\|^2\, ds + c\, t^{-\alpha} \int_0^t s^{\alpha-2} \left(\int_0^s \|u(\tau)\|^2\, d\tau\right)^{2-n/2} ds \qquad (4.7)$$

for all $t > o$, with $\alpha > o$ sufficiently large.

On account of $u \in L^\infty(o,\infty; L^2_\sigma)$, the second term in (4.7) is less than $c\, t^{1-n/2}$ and hence converges to zero as $t \to \infty$ in the cases $n = 3,4$. The first term is of order $o(1)$ since by (2.7) $\|\mathbf{e}(s)\| = o(1)$ as $s \to \infty$. This proves Assertion (i) for $n = 3,4$. We shall now prove (iii) in the case $n = 3$.

Suppose that for $j = 1,2,\ldots$

$$\|u(t)\| \leq c_j\, t^{-\beta_j}\ , \quad t > o, \quad \beta_j < 1/2\ , \qquad (4.8)$$

then (4.7) gives

$$\|u(t)\| \leq c_{j+1}\, t^{-\beta_{j+1}}\ , \quad t > o$$

with $\beta_{j+1} = \min\,(\alpha_o, 1/4 + \beta_j/2)$ and $c^2_{j+1} = 2\alpha\, \hat{c}^2 + c\, c_j/(1-2\beta_j)^{1/2}$, where $\hat{c}$ is determined by the estimate $\|\mathbf{e}(t)\| \leq \hat{c}\, t^{-\alpha_o}$. Starting from $\beta_o = o$ and $c_o = \|u_o\|$ and assuming $\alpha_o \geq 1/2$ it is easily seen that $\beta_j \to 1/2$ and $c_j \to \infty$ as $j \to \infty$. Thus, if $\alpha_o < 1/2$, then after finitely many iterations we can achieve α_o. This proves (iii)for $n = 3$. In the case $n = 4$, Assertion (iii) follows immediately from (4.7).

For $n = 2$ we choose $\lambda(\tau) = \alpha/(1+\tau)\ln(1+\tau)$ (α sufficiently large). Then we have $H(\tau) = (\ln(1+\tau))^\alpha$ and (4.6) yields

$$\|u(t)\|^2 \leq 2(\ln(1+t))^{-\alpha} \int_0^t (1+s)^{-1} (\ln(1+s))^{\alpha-1} \|\mathbf{e}(s)\|^2\ ds \qquad (4.9)$$

$$+\, c\, (\ln(1+t))^{-\alpha} \int_0^t (1+s)^{-2} (\ln(1+s))^{\alpha-2} \left(\int_0^s \|u(\tau)\|^2 d\tau\right) ds =: I_1 + I_2$$

Again, the first integral of (4.9) is of order $o(1)$ on account of $\|\mathbf{e}(s)\| = o(1)$. Assuming only the boundedness of $\|u(\tau)\|$, the second integral is less than

$$c(\ln(1+t))^{-\alpha} \int_0^t (1+s)^{-1} (\ln(1+s))^{\alpha-2} ds = c\, (\ln(1+s))^{-1}/\,(\alpha-1)$$

This proves Assertion (i) for $n = 2$.

We shall now prove (iii) for $n = 2$. If the assumptions in this case are satisfied, then for every $m > o$ we can find a constant $c = c\,(m) > 0$ such that $\|\mathbf{e}(t)\|^2 \leq c\, (\ln(1+t))^{-m}$ and we obtain

$$I_1 \leq c\, (\ln(1+t))^{-m} / (\alpha-m)\ , \quad \alpha > m\ . \qquad (4.10)$$

Now we suppose that the estimate

$$\|u(t)\|^2 \leqq c\ (\ln(1+t))^{-k}\ , \quad k = o,1,\dots \tag{4.11}$$

is already established. On account of

$$\int_o^s (\ln\ (1+\tau))^{-k}\ d\tau \leqq c\ (1+s)\ (\ln(1+s))^{-k}$$

([16], Lemma 4.1), we have

$$\begin{aligned} I_2 &\leqq c\ (\ln(1+t))^{-\alpha} \int_o^t (1+s)^{-1}\ (\ln(1+s))^{\alpha-2-k}\ ds \\ &= c\ (\ln(1+t))^{-k-1}. \end{aligned} \tag{4.12}$$

Combining (4.9) with m = k+1 and (4.10) with (4.12) we obtain (iv) by induction since (4.11) is valid for k = o.

Proof of Assertion (ii) and (iv).

Let $w = u - e$ (recall $e(t) = e^{-tA}u_o$). Taking into account the energy equality for e and (2.5) for u, we obtain

$$\begin{aligned} \|w(t)\|^2 + 2\int_s^t \|\nabla w\|^2\ d\tau &= \|u(t)\|^2 + \|\ e(t)\|^2 - 2\ (u(t),\ e(t)) \\ &\quad + 2\int_s^t \{\|\nabla u\|^2 + \|\nabla e\|^2 - 2\ (\nabla u, \nabla e)\}\ d\tau \\ &\leqq \|u(s)\|^2 + \|e(s)\|^2 - 2(u(t),\ e(t)) - 4\int_s^t (\nabla u, \nabla e)\ d\tau \\ &\leqq \|w(s)\|^2 + 2\int_s^t\ (u\cdot\nabla u, e)\ d\tau, \end{aligned} \tag{4.13}$$

where we have used property b) of Sect. 2 with $v = e$ and $s > o$. Using $(u\cdot\nabla e, e)=o$, the second term on the right of (4.13) is estimated by (n=3,4)

$$\begin{aligned} |(u\cdot\nabla u, e)| = |(u\cdot\nabla w, e)| &\leqq \|u\|_{L^n}\ \|e\|_{L^{2n/(n-2)}}\ \|\nabla w\| \\ &\leqq \|\nabla w\|^2 + c\ \|u\|^{4-n}\|\nabla u\|^{n-2}\|\nabla e\|^2 \end{aligned}$$

In view of $\|\nabla e(t)\|^2 \leqq c\ t^{-1}$, (4.13) yields

$$\|w(t)\|^2 + \int_s^t \|\nabla w\|^2\ d\tau \leqq \|w(s)\|^2 + c\int_s^t \tau^{-1}\ \|u\|^{4-n}\ \|\nabla u\|^{n-2}\ d\tau \tag{4.14}$$

for allmost all s in $o < s \leqq t$.

To estimate $\|E_\lambda w\|$, we observe that w satisfies the problem

$$w' + Aw = -P(u\cdot\nabla)u, \quad w(o) = o$$

in the weak sense. This gives

$$E_{\lambda(t)}w(t) = -\int_0^t e^{-(t-s)A} E_{\lambda(t)} P(u\cdot\nabla)u(s)\, ds$$

and with Lemma 4.1 and the Hölder inequality

$$\|E_\lambda w\| \leq c\,\lambda^{1/2} \int_0^t \|u\|^{2-n/2} \|\nabla u\|^{n/2} ds$$

$$\leq c\,\lambda^{1/2} \left(\int_0^t \|u\|^2 ds\right)^{1-n/4}.$$

From (4.2) with u replaced by w and (4.14) we obtain similar to (4.6) and (4.7) the inequality

$$\|w(t)\|^2 \leq c\, t^{-\alpha} \int_0^t \{s^{\alpha-1} \|u\|^{4-n} \|\nabla u\|^{n-2} + s^{\alpha-2} (\int_0^s \|u\|^2 d\tau)^{2-n/2}\} ds.$$

Assertion (ii) for n = 4 follows immediately. For n = 3 we have by the Hölder inequality

$$\int_0^t s^{\alpha-1} \|u\| \|\nabla u\| ds \leq c \left(\int_0^t s^{2\alpha-2} \|u\|^2 ds\right)^{1/2}$$

and

$$\int_0^t s^{\alpha-2} \left(\int_0^s \|u\|^2 d\tau\right)^{1/2} ds \leq \frac{t^{\alpha-1/2}}{\alpha-1} \left(\frac{1}{t} \int_0^t \|u\|^2 ds\right)^{1/2},$$

hence

$$\|w(t)\|^2 \leq c\, t^{-1/2} \left(\frac{1}{t} \int_0^t \|u\|^2 ds\right)^{1/2}. \tag{4.15}$$

Since $\|u(s)\| \to o$ as $s \to \infty$ by Assertion (i), it follows that

$$\frac{1}{t} \int_0^t \|u(s)\|^2 ds \to o \quad \text{as } t \to \infty .$$

Thus, Assertion (ii) follows from (4.15). To prove (iv), we assume $\|e(t)\| \leq c\, t^{-\alpha_o}$ for $t > o$ and $o < \alpha_o < 1/2$. Then (iii) gives $\|u(t)\| \leq c\, t^{-\alpha_o}$. Combining this with (4.15) yields $\|w(t)\| \leq c\, t^{-1/4-\alpha_o/2}$ and $1/4 + \alpha_o/2 > \alpha_o$. The Theorems are proved.

References

[1] Borchers,W., Miyakawa, T.: L^2 Decay for the Navier-Stokes Flow in Halfspaces. Math. Ann. 282, 139 - 155 (1988)

[2] Borchers,W., Miyakawa, T.: L^2 Decay for the Navier-Stokes Flow in Unbounded Domains. Preprint (1986)

[3] Borchers,W., Miyakawa, T.: Algebraic L^2 Decay for Navier-Stokes Flows in Exterior Domains. Preprint (1988)

[4] Friedman,A.: Partial differential equations. New York: Holt, Rinehart & Winston 1969

[5] Fujita, H., Kato, T.: On the Navier-Stokes initial value problem I Arch.Rat.Mech.Anal. 16, 269 - 315 (1964)

[6] Hopf, E.: Über die Anfangswertaufgabe für die hydrodynamischen Grundgleichungen. Math. Nachr. 4, 213 - 231 (1951)

[7] Kanwal, R.P.: Linear Integral Equations. Academic Press 1971

[8] Kajikiya, R., Miyakawa, T.: On L^2 Decay of Weak Solutions of the Navier-Stokes Equations in $\mathbf{R}^n$. Math.Z. 192, 135 - 148 (1986)

[9] Komatsu, H.: Fractional Powers of Operators. Pacific J. of Math., Vol. 19, Nr. 2, 285 - 346 (1966)

[10] Ladyzhenskaya, O.A.: The Mathematical Theory of Viscous Incompressible Flow. Gordon and Breach 1969

[11] Leray, J.: Sur le mouvement d'un liquide visqueux emplissant l'espace. Acta Math. 63, 193 - 248 (1934)

[12] Lions, J.-L.: Quelques methods de résolutions des problèmes aux limites non-linéaires. Paris: Dunod et Gauthier-Villears 1969

[13] Maremonti, P.: On the asymptotic behaviour of the L^2-norm of suitable weak solutions to the Navier-Stokes equations in three-dimensional exterior domains. Comm.Math.Phys., 118, 335-400 (1988)

[14] Rautmann, R.: Bemerkungen zur Anfangswertaufgabe einer stabilisierten Navier-Stokesschen Gleichung. ZAMM 55, T 217-221 (1975)

[15] Schonbek, M.E.: L^2 decay for weak solutions of the Navier-Stokes equations. Arch. Rat. Mech. Anal. 88, 209-222 (1985)

[16] Wiegner, M.: Decay results for weak solutions of the Navier-Stokes equations in $\mathbf{R}^n$. J. London Math. Soc. 35, 303-313 (1987)

The work of the first author is supported by Deutsche Forschungsgemeinschaft
This paper is in a final form and no similar paper has been or is being submitted elsewhere.

(received May 10, 1989).

Strong Solution for the Navier-Stokes Flow in the Half-Space

Hideo Kozono
Department of Applied Physics, Nagoya University
Nagoya 464, Japan

1. Introduction.

The purpose of this article is to report a global existence and uniqueness of strong L^n-solution and its decay properties for the Navier-Stokes equations in the n-dimensional half-space. Let $\mathbb{R}^n_+ = \{(x^1,\dots,x^n);\ x^n > 0\}$ and $Q = \mathbb{R}^n_+ \times (0,\infty)$. In Q, we consider the following Navier-Stokes equations:

$$
\left.\begin{aligned}
&\partial u/\partial t - \Delta u + u\cdot\nabla u + \nabla\pi = 0 \quad \text{in } Q, \\
&\operatorname{div} u = 0 \quad \text{in } Q, \\
&u = 0 \quad \text{on } x^n = 0, \quad t > 0, \\
&u(0) = a.
\end{aligned}\right\} \qquad \text{(N.S.)}
$$

To state our result, we need some notations. D_σ denotes the set of all C^∞-real vector functions u with compact support in $\mathbb{R}^n_+$ such that $\operatorname{div} u = 0$. X_n is the completion of D_σ in $\mathbb{L}^n(\mathbb{R}^n_+)$. By P we denote the projection operator from $\mathbb{L}^n(\mathbb{R}^n)$ onto X_n. The Stokes operator A is defined by $A = -P\Delta$ with domain $D(A) = \{u \in W^{2,n}(\mathbb{R}^n_+);\ u = 0 \text{ on } x^n = 0\} \cap X_n$.

Our definition of a strong solution is as follows:

Definition. Let $a \in X_n$. A function u defined on $[0,\infty]$ with values in X_n is called a strong solution of (N.S.) if

(i) $u \in C([0,\infty); X_n) \cap C^1((0,\infty); X_n)$;

(ii) $u(t) \in D(A)$ for $t > 0$ and $Au \in C((0,\infty); X_n)$;

(iii) $du(t)/dt + Au + Pu\cdot\nabla u = 0, \quad t > 0$

$u(0) = a$.

Now our result reads as follows:

Theorem. Let $a \in X_n$. There is a positive constant λ such that if $\|a\|_n \le \lambda$, there exists a unique strong solution u of (N.S.) satisfying the following properties:

(i) $u(t) \in \mathbb{L}^q(\mathbf{R}^n_+)$ $(t > 0)$,

$$\|u(t)\|_q = O(t^{-(n/2)(1/n - 1/q)}) \quad \text{as} \quad t \to \infty$$

for $n < q \le \infty$;

(ii) $\nabla u(t) \in \mathbb{L}^{r}(\mathbf{R}^n_+)$ $(t > 0)$,

$$\|\nabla u(t)\|_r = O(t^{-(n/2)(2/n - 1/r)}) \quad \text{as} \quad t \to \infty$$

for $n \le r < \infty$;

(iii) $\|u(t)\|_n \to 0$ as $t \to \infty$.

Here $\| \; \|_p$ denotes the norm of $\mathbb{L}^p(\mathbf{R}^n_+)$.

2. Sketch of the proof.

In this section, we shall give the outline of the proof of the theorem. We may solve the following integral equation:

$$u(t) = e^{-tA}a - \int_0^t e^{-(t-s)A} P\, u\cdot\nabla u(s)ds\,. \tag{I.E.}$$

For our purpose, we shall make use of the implicit function theorem. Let $Y = \{ u \in C([0,\infty); X_n) \cap C((0,\infty); D(A^{1/2}));\ |u|_y := \sup_{t>0} \|u(t)\|_n + \sup_{t>0} t^{1/2}\|A^{1/2}u(t)\|_n < \infty \}$. Y is a Banach space with norm $|\;|_y$. We define a map $F(\cdot,\cdot)$ on $X_n \times Y$ by

$$F(a,u)(t) = u(t) - e^{-tA}a + \int_0^t e^{-(t-s)A} Pu\cdot\nabla u(s)ds$$

for $a \in X_n$ and $u \in Y$. Then we have Proposition.

(i) F is a continuous map from $X_n \times Y$ into Y.

(ii) For each $a \in X_n$, the map $F(a,\cdot)$ is of class C^1 from Y into itself.

For the proof, we need some sharp inequalities obtained by Borchers - Miyakawa [1, Theorem 3.6].

Moreover, we see $F(o,o) = 0$ and the Fréchet derivative $D_u F(o,o)$ = identity on Y. Hence the classical implicit function theorem ensures us the existence of a unique continuous mapping u in a neighbourhood $U_\lambda = \{ a \in X_n ; \| a \|_n < \lambda \}$ of 0; $u : U_\lambda \to Y$ such that

$$u[o] = 0 \quad \text{and} \quad F(a,u[a]) = 0 \quad \text{for} \quad a \in U_\lambda .$$

This shows that $u[a]$ is the unique solution of (I.E.).

References

1. Borchers, W., Miyakawa, T.: L^2 Decay for the Navier-Stokes Flow in Halfspaces. Math. Ann. 282, 139 - 155 (1988).

2. Giga, Y., Miyakawa, T.: Solution in L_r of the Navier-Stokes initial value problem. Arch. Rational Mech. Anal. 89, 267 - 281 (1985).

(received January 18, 1989).

THE PROBLEM OF MOMENTUMLESS FLOW FOR THE NAVIER-STOKES EQUATIONS

V.V. Pukhnachov
Lavrentyev Institute of Hydrodynamics, Sibirian Division of the
USSR Academy of Sciences Novosibirsk 630090, USSR

Let Σ be a bounded closed surface in $\mathbf{R}^3$ and let Ω be the region exterior with respect to Σ . We consider the problem of determining a vector-function $\underset{\sim}{v}(x)$,(i.e. the difference between the flow velocity at the point $\underset{\sim}{x}$ and the constant velocity at infinity $\underset{\sim}{e}_1 = (1,0,0)$) and a scalar function p(x), (i.e.the pressure) satisfying a stationary system of the Navier-Stokes equations in the region Ω :

$$\Delta \underset{\sim}{v} - Re\, \partial \underset{\sim}{v} / \partial x_1 - \nabla p = Re\, \underset{\sim}{v} \cdot \nabla \underset{\sim}{v}, \qquad (1)$$

$$\nabla \cdot v = 0$$

and the following boundary conditions:

$$\underset{\sim}{v}|_{\Sigma} = \underset{\sim}{a}(x), \qquad (2)$$

$$v \to 0 \qquad \text{as } x \to \infty \qquad (3)$$

Here $\underset{\sim}{a}$ is a vector-function satisfying the condition

$$\int_{\Sigma} \underset{\sim}{a} \cdot \underset{\sim}{n}\, d\Sigma = 0 \qquad (4)$$

$\underset{\sim}{n}$ is the exterior unit normal vector to $\partial\Omega$. Equations (1) are written in a dimensionless form, so that $Re = Vl/\nu$ is the Reynolds number, ν is the kinematic viscosity coefficient, l is a characteristic length unit (for exmaple, $l = \operatorname{diam} \Sigma$).

The value V of the velocity of the flow at infinity is a natural characteristic unit of velocity, and the characteristic unit of the pressure is assumed to be equal to $\rho V\nu/l$, where ρ is the liquid density.

Furthermore , let us assume that the surface Σ belongs to the Hölder class $C^{2+\alpha}$, $0 < \alpha < 1$, and the vector components $\underset{\sim}{a}$ belong to the class $C^{2+\alpha}(\Sigma)$. Problem (1)-(4), with the function $\underset{\sim}{a}(x)$ fixed, was considered in many papers. The most substantial results in this area were obtained by R. Finn [1] and K.I. Babenko [2]. If $\underset{\sim}{a} = -\underset{\sim}{e}_1$ we arrive to a classical problem of flow past a body with a fixed impermeable boundary Σ . It is well-known that in this case the drag force acting from the liquid on the body is different from zero. Thus, to realize a stationary flow past a body, it is necessary to keep the body fixed in the liquid by external forces.

Here we will consider the problem of flow past a self-moving body, or the problem of momentumless flow. The selfmotion condition means that the full momentum, contributed into the liquid by the boundary of the body, is equal to zero. In mathematical terms this condition is expressed by the equality

$$\underset{\sim}{F} \equiv \int_{\Sigma} [-P\underset{\sim}{v} \cdot \underset{\sim}{n} + \mathrm{Re}\, \underset{\sim}{v} (\underset{\sim}{v} + \underset{\sim}{e}_1) \cdot \underset{\sim}{n}]\, d\Sigma = 0. \tag{5}$$

Here $\underset{\sim}{F}$ is the drag force, $P\underset{\sim}{v}$ is the stress tensor corresponding to the velocity field $\underset{\sim}{v}$ and the pressure p, with elements

$$(P\underset{\sim}{v})_{ij} = -p\,\delta_{ij} + \partial v_i/\partial x_j + \partial v_j/\partial x_i \quad (i,j = 1,2,3).$$

We observe that in the model under consideration the self-motion condition may hold if the surface Σ is movable or permeable. In the latter case it is natural to assume that the total flux through the surface Σ is equal to zero (condition (4)).

It is known that problem (1)-(4) with a fixed possesses at least one solution for any Reynolds number $\mathrm{Re} \geq 0$ [1-3]. At small Re, its solution is unique [2]. From here it follows that the problem of flow past a self-moving body (1)-(5) is solvable only if the additional conditions upon the function $\underset{\sim}{a}(x)$ are fullfilled. In other words, this problem may be considered as that of simultaneous determination of functions $\underset{\sim}{v}$, p and $\underset{\sim}{a}$ from (1)-(5). The self-motion condition (5), which is equivalent to three scalar relationships, admits a wide arbitrariness in the choice of function $\underset{\sim}{a}$, which may be considered as a control function. It should be noted that, by virtue of the law of conservation of momentum, condition (5) is equivalent to the same condition written for any control surface encircling Σ [4]. This justifies the name of "the problem of momentumless flow", as applied to problem (1)-(5).

Despite the natural character of the problem of flow past a self-moving body and its significance for applications, there are only a few works available, which are concerned with this problem. A plane analog of problem (1)-(5) is analysed in [5], where a potential viscous flow past a self-moving "body" is considered as an example. Its boundary consists of two joint components, a normal velocity component on each of them being equal to zero and a tangential component being constant. In [6] a two-dimensional problem of flow past a circle with a moving impermeable boundary has been investigated with the method of asymptotical matching for low Reynolds numbers. The boundary motion is prescribed so that condition (5) is fulfilled. The same method was used in the paper [7], devoted to the analysis of axis-symmetrical problem (1)-(5) with low Reynolds numbers in the case, when Σ is a sphere. Here, on the contrary, the body surface is immovable, but permeable for the liquid. A pair of indentical infinite circular cylinders in a viscous flow uniform at infinity, and rotating around their axes with opposite angular velocities is likely to be the most illustrative example. In [8,9] this problem was analyzed asymptotically for small values of the ratio between the cylinder radius and the distance between their axes.

It should be noted that the boundary condition (2) associated with the self-motion condition (5) is not the only possible model e.g. of a propeller. An alternative variant is to prescribe either self-consistent distribution of volume forces localized in a small region behind the propeller or the pressure jump over some rather small surface behind it. A numerical solution of the problem of momentumless flow past an ellipsoid of revolution in the first ot these models was realized in [10]. An asymptotical behaviour of the velocity field at big distances from self-moving body was studied in [4,11].

It should be noted that the existence of a solution of problems of momentumless flow has not been considered in any of the works mentioned above, [4-11]. This problem may be solved, even rather effectively, in the simplest case, when the Reynolds number is equal to zero. Let us designe the Stokes system (1) with Re = 0 by $(1)_0$ and relation (5) by $(5)_0$. System $(1)_0$, conditions (2)-(4) and equality $(5)_0$ form the problem, which further will be denoted by $(1)_0$ - $(5)_0$ and called the problem of flow past a self-moving body in the Stokes approximation.

Problem $(1)_0$ - $(5)_0$, in contrast to general problem (1)-(5), is a linear one, and the Stokes operator, generated by system $(1)_0$ is self-adjoint [3]. This enables us to solve problem $(1)_0$ - $(5)_0$ efficiently in terms of eigenfunctions of a certain spectral problem, which may be formulated as follows.

It is required to find the number λ and vector-function $\underset{\sim}{\phi}(x) \not\equiv 0$ determined on the surface Σ , which satisfy condition (4) and the relation

$$A \underset{\sim}{\phi} = \lambda \underset{\sim}{\phi}.$$

Here $A\phi = Pv|_{\Sigma}$, the vector Pv being calculated from the solution $\underset{\sim}{v}$, p of problem $(1)_0$ - $(5)_0$, with $\underset{\sim}{a} = \underset{\sim}{\phi}$. Operator A , initially determined on the functions $\underset{\sim}{\phi} \in \underset{\sim}{C}^{2+\alpha}(\Sigma)$, admits a self-conjugate extension up to the operator acting from $\underset{\sim}{H}^{1/2}(\Sigma)$ into $\underset{\sim}{H}^{-1/2}(\Sigma)$. The space of the traces of functions $\underset{\sim}{u} \in \underset{\sim}{H}^{1}(\Omega)$ on the surface Σ is denoted by $\underset{\sim}{H}^{1/2}(\Sigma)$ and the closure of the set of smooth solenoidal vector-function which are equal to zero for large values of $|\underset{\sim}{x}|$ in the Dirichlet norm is denoted by $H^1(\Omega)$. The space $H^{-1/2}(\Sigma)$ is the dual space of $H^{1/2}(\Sigma)$. Using the theory proposed in [12], we find the following properties of operator A : (i) it possesses the inverse A^{-1} which is completely continuous and self-adjoint; (ii) the spectrum of operator A is discrete and eigenvalues have a finite multiplicity; (iii) all the eigenvalues $\lambda_1 < \lambda_2 < \lambda_3 \ldots$ are positive; (iv) $\lambda_k \to \infty$ when $k \to \infty$; (v) eigenfunctions $\underset{\sim}{\phi}_k$ and $\underset{\sim}{\phi}_2$ corresponding to the eigenvalues λ_k and $\lambda_1 \neq \lambda_k$ are orthogonal both in the metric of $L_2(\Sigma)$ and in the metric generated by the scalar product $(A^{1/2}\underset{\sim}{\phi}\ ,\ A^{1/2}\underset{\sim}{\phi})$; (vi) the set of eigenfunctions $\{\underset{\sim}{\phi}_k\}$ forms a complete system both in space $\underset{\sim}{L}_2\ (\Sigma)$ and in $\underset{\sim}{H}^{1/2}\ (\Sigma)$.

Now it is not difficult to formulate the algorithm for solving the problem of momentumless flow in Stokes approximation. Let the Fourier coefficients of function a(x) with respect to basis $\{\underset{\sim}{\phi}_k\}$, orthonormal in $\underset{\sim}{L}_2\ (\Sigma)$ be defined by $a_k = (\underset{\sim}{a},\ \underset{\sim}{\phi}_k)$. Then condition $(5)_0$ may be written as

$$\sum_{k=1}^{\infty} (\lambda_k\ a_k\ \int_{\Sigma} \phi_{k,i}\ \ d\ \Sigma) = 0\ (i = 1,2,3)$$

where $\phi_{k,i}$ is the ith component of vector $\underset{\sim}{\phi}_k$. Condition (4) means that

$$\sum_{k=1}^{\infty} (a_k\ \int_{\Sigma} \underset{\sim}{\phi}_k \cdot \underset{\sim}{n}\ d\ \Sigma) = 0. \tag{7}$$

We choose the arbitrary element $\underset{\sim}{a} \in \underset{\sim}{H}^{1/2}\ (\Sigma)$ satisfying (6),(7) and then solve problem $(1)_0$ - $(5)_0$. Then the velocity vector $\underset{\sim}{v}$ is determined uniquely and the pressure p is determined up to an additive constant. Thus, there is a functional arbitrariness, when $(1)_0$ - $(5)_0$ is considered as the problem of determining the functions $\underset{\sim}{v}$, p and $\underset{\sim}{a}$. Therefore, it is natural to determine the function $\underset{\sim}{a}$ from the condition of the minimum of the functional

$$J = \int_{\Sigma} \underset{\sim}{v} \cdot P\underset{\sim}{v} \cdot \underset{\sim}{n} d\ \Sigma\ .$$

The value of J is equal to the work which is done by a stationary self-motion per unit time.

In the general case the problem of determining the minimum J under some natural restrictions upon the function $\underset{\sim}{a}$ is solved by the method of Lagrange multipliers. Let us give examples of such restrictions: (i) function $\underset{\sim}{a} + \underset{\sim}{e}_1$ has a support Σ' which does not concide with the whole surface Σ (the case when region $\Sigma' \subset \Sigma$ is small enough, is of interest from the standpoint of physics); (ii) function $\underset{\sim}{a} + \underset{\sim}{e}_1$ has a zero normal component, (i.e. the body boundary is impermeable); (iii) function $\underset{\sim}{a} + \underset{\sim}{e}_1$ has a zero tangential component (the body surface is immovable). If the surface Σ us a sphere, the problem of minimization of J for the second and third type of the above-mentioned restrictions is solved explicitly using the results obtained in [13]. In both these cases the minimum J turns out to be achieved on functions $\underset{\sim}{a}$ corresponding to the regime of a potential flow. It is unknown whether this property of extremals of functional J holds for the arbitrary surface Σ .

The problem of solvability of (1)-(5) for arbitrary Re $\geq$ 0 seems to be rather complicated. It is hoped, however that its solution should exist for low Reynolds numbers. The basis for such an optimism is provided by the methods proposed in [14], which gives the asymptotics of the solution of the classical problem of flow past a body ($\tilde{a} = -\tilde{e}_1$) when Re $\to 0$.

Now let us turn to the study of an asymptotical behaviour of the solution of problem (1)-(5) when $r = |\tilde{x}| \to \infty$. From now on, we do not take into account the additional condition (5) and we recall that the existence theorem of the solution to problem (1)-(4) "in the large" is valid in the class of vector-functions $\tilde{v}$ having the finite Dirichlet integral [3],

$$\int_\Omega \nabla \tilde{v} : \nabla \tilde{v} \, d x < \infty \tag{8}$$

In [2] it was established that any solution of the above-mentioned problem satisfying inequality (8) admits the estimate

$$|\tilde{v}(x)| \leqq C r^{-1/2 - \varepsilon} \quad \text{when} \quad r \to \infty \tag{9}$$

with some positive constants C and ε . Earlier, in [1] it was shown that any solution of problem (1)-(4) satisfying inequality (9) has an asymptotical behaviour

$$\tilde{v}(x) = \tilde{F} \cdot E(x) + \tilde{\zeta}(x). \tag{10}$$

Here $\tilde{F}$ is the constant vector determined by formula (5), and $\tilde{\zeta}(x)$ is the remainder, for which the following estimate has been obtained at large r :

$$|\tilde{\zeta}| \leqq C_1 \; r^{-3/2 + \varepsilon} (s + 1)^{-1 + \varepsilon} \tag{11}$$

(here $s = r - x_1$, ε is arbitrarily small, C_1 = constant > 0). Symbol $E(x)$ denotes a fundamental tensor of Oseen's system, corresponding to (1). The expressions for the elements of tensor E may be found in [1].

From (10), (11) it follows that there exists a parabolic region of the wake in the direction $\tilde{e}_1$ where $\tilde{v} = 0(r^{-1})$. Outside any circular cone with axis directed along $\tilde{e}_1$, $\tilde{v} = 0(r^{-2})$. In [15] the higher order asymptotics terms of the field $\tilde{v}(x)$ having the order $r^{-3/2}$ were obtained under the assumption that the vectors $\tilde{F}$ and $\tilde{e}_1$ are collinear. (For example, this assumption is valid in the case of axis-symmetrical flow past a rotationally-symmetric surface Σ). In [15,16] the behaviour of the vorticity at great distances from the body was studied, and it has been shown that outside the wake the vorticity decays exponentially.

Formula (10) means that far away from the body , the perturbation of the velocity field will be the same (up to higher-order small values) as that of Oseen's flow generated by the concentrated force of $\tilde{F}\delta(x-y)$. The velocity field asymptotics in the problem of momentumless flow turned out to depend on a much higher number of functionals characterizing both the body shape and the method of self-motion realization. Now let us formulate the main result.

Let v, p be the solution of problem (1)-(4) belonging to the class (8), which satisfies the additional condition (5). Then, for $r \to \infty$, the following asymptotical representation of $\underset{\sim}{v}(x)$ is valid

$$\underset{\sim}{v}(x) = R : D\, E(x) + Q : \underset{\sim}{E}(x) + \underset{\sim}{\eta}(x). \tag{12}$$

Here $R = (R_{ij})$, $Q = (Q_{ij})$ are the constant tensors $(i,j = 1,2,3)$. The elements of tensor R are expressed explicitly in terms of function $\underset{\sim}{a}(x)$. Symbols D E and E denote the third-rank tensors with elements

$$2\,(D\,E)_{ijk} = \left(\frac{\partial E_{ik}}{\partial x_j} + \frac{\partial E_{jk}}{\partial x_i}\right), \quad (\nabla E)_{ijk} = \frac{\partial E_{jk}}{\partial x_i},$$

$i,j,k = 1,2,3$. Summation in $R : D\,E$, $Q : E$ is carried out with respect to indices i and j . Function $\underset{\sim}{\eta}(x)$ admits the estimate

$$|\,\underset{\sim}{\eta}(x)\,| \leq C_2 r^{-2+\varepsilon}(s+1)^{-1/2} \tag{13}$$

with $r \to \infty$, where ε and C_2 are positive constants, ε being arbitrarily small.

Proof of formula (12) is based on the integral representation of problem (1)-(3) obtained in [1]. To estimate the volume integral

$$\underset{\sim}{N}(x) = \int_\Omega \underset{\sim}{v}(y) \cdot \underset{\sim}{v}(y) \cdot \nabla E(x-y)\, dy$$

as $r \to \infty$, the results of [15] have been used.

According to (12), (13), in the momentumless flow regime we have $\underset{\sim}{v} = 0\,(r^{-3/2})$ inside the parabolic region of the wake and $\underset{\sim}{v} = 0\,(r^{-5/2+\varepsilon})$ outside any cone having the axis $\underset{\sim}{e}_1$. Thus, a higher order decay of velocity perturbation takes place at great distances from the self-moving body, as compared to the case of a fixed one. Representation (12) also means that, at least, inside the wake the main terms of the velocity field,asymptotics in the momentumless flow problem are characterized by 18 parameters, which are the elements of tensors R and Q . In the axisymmetrical case the number ot these parameters strinks down to eight. Of significant interest is the identification of the elements of tensor Q , which are some functionals of the solution of problem (1)-(5). This problem remains open.

One of the paradoxal results associated with the classical flow problem for the Navier-Stokes equations may be formulated as follows. Suppose, under the condition (2), that $\underset{\sim}{a} = -\underset{\sim}{e}_1$, which means immobility and impermeability of the boundary.

Then, for any solution v,p of problem (1)-(3) satisfying the condition (8), we have

$$\int_{\Omega} |\underset{\sim}{v}|^2 \, dx = \infty \; .$$

Statements like this were firstly made in [17] for the problem of viscous flow past a body.

Intuitively, it is obvious that the self-moving body cannot contribute such a significate disturbance into the flow. An appropriate exact formulation may be as follows. Let v, p be the solution of problem (1)-(4), satisfying the additional conditions (5),(8). Then

$$W = \frac{1}{2} \int_{\Omega} |\underset{\sim}{v}|^2 \, dx < \infty \; . \qquad (14)$$

The characteristic feature of the problem of momentumless flow past a self-moving body expressed by inequality (14) distinguihes this solution from the solution of problem (1)-(4), if function $\underset{\sim}{a}$ entering its formulation is varied. We hope that this property my be used to investigate the existence problem of the solution of (1)-(5), if it will be considered as a certain optimization problem.

References

1. Finn R., On the exterior stationary problem for the Navier-Stokes equations, and associated perturbation problems, Arch.for Rational Mech.and Anal. 1965 - V.19, Nr. 5.
2. Babenko K.I. On stationary solutions of the problem of viscous flow past a body.- Math. Sbornik, 1973, V.9, (133), Nr. 5
3. Ladyzhenskaya O.A., Mathematical problems of a viscous incompressible flow. - Moskow, Nauka, 1970.
4. Pukhnachov V.V., On some modifications of the problem of a flow past a body.- Problems of Mathematics and Mechanics, Novosibirsk, Nauka, 1983.
5. Lugovtsov A.A., Lugovtsov B.A., Example of a viscous incompressible flow past a body with a moving boundary. - Dynamics of cont.media Nr.8, Novosibirsk, 1971.
6. Sennitsky V.L., On a flow past a self-moving body.- Journal of Appl. Mech. and Techn. Physics (PMTF) 1978, Nr. 3.
7. Sennitsky V.L., Example of an antisymmetrical flow past a self-moving body.- Journ. of Appl. Mech. and Techn. Physics (PMTF), 1984, Nr. 4.
8. Sennitsky V.L., Cylinders rotating in a viscous fluid, I - Dynamics of Cont. Media, Nr.21, Novosibirsk, 1975.
9. Sennitsky, V.L. Cylinders rotating in a viscous fluid, II - Dynamics of Cont. Media, Nr. 23, Novosibirsk, 1975.
10. Izleuov M.I., Calculation of a momentumless flow past an ellipsoid. - Problems of a viscous flow, Novosibirsk, 1975.
11. Pukhnachov V.V., Sennitsky V.L., Stationary viscous flow around self-moving bodies, IUTAM Symposium "Fluid Mechanics in the spirit of G.I. Taylor".- Cambridge, U.K., 1986.
12. Lions J.L., Magenes E. Problèmes aus limites non homogènes et applications. - Paris, Dunod, 1968.
13. Happel J., Brenner H., Low Reynolds number Hydrodynamics. - Prentice-Hall, 1965.
14. Fischer, T.M., Hsiao, G.C., Wendland W.L., Singular perturbations for the exterior three-dimensional slow viscous flow problem, J. of Math. Anal. and Applications.- 1985 - V. 110, Nr. 2.

15. Babenko K.I., Vasilev M.M., Asymptotic behavior of viscous flow past a body, Preprint Inst. of Appl. Math. of USSR Ac.Sc. Nr. 84, Moscow 1971.
16. Clark D.C., The vorticity at infinity for solutions of the stationary Navier-Stokes equations in exterior domains, Indiana Univ. Math. J. - 1971 V. 20,Nr. 7.
17. Finn R., An energy theorem for viscous fluid motions, Archive for Rational Mech. and Anal. - 1960 V.6, Nr. 5.

This paper is in its final form and no similar paper has been or is being submitted elsewhere

(received February 21, 1989).

Decay and Stability in L_p for strong solutions of the Cauchyproblem for the Navier-Stokes equations

Michael Wiegner

Mathematisches Institut der Universität

Postfach 101251, 8580 Bayreuth, FRG

We consider the Cauchyproblem for the Navier-Stokes equations on the whole space

$$(1)\qquad \begin{aligned} u_t - \Delta u + u \cdot \nabla u + \nabla \pi &= 0 \quad \text{on } \mathbb{R}^n \times \mathbb{R}^+ \\ \text{and div } u &= 0 \quad \text{on } \mathbb{R}^n \times \mathbb{R}^+ \\ u(0) &= a \in L_2, \ \text{div } a = 0. \end{aligned}$$

There are by now several techniques to prove the global existence of weak solutions $u \in L_2(\mathbb{R}^+, H_2^1) \cap L_\infty(\mathbb{R}^+, L_2)$; and some of them provide us also with solutions, having additional nice properties as e.g. an energy inequality for almost all times or solutions with decay properties for $t \to \infty$. This has been studied recently by several people, e.g. M. E. Schonbek [5], R. Kajikiya - T. Miyakawa [3] and M. Wiegner [6].

Suppose now we have a weak solution, which fulfills additionally

$$(2)\qquad a \in L_2 \cap L_p, \ u \in L_s(\mathbb{R}^+, L_p) \quad \text{with } p > n \geq 3 \quad \text{and } \tfrac{2}{s} + \tfrac{n}{p} \leq 1.$$

It is well known that the existence of such a solution for all times is assured only for small initial values (see e.g. H. Beirão da Veiga [1]), but we shall not use any kind of smallness condition in the following.

From the same paper we infer that (2) implies

$$(3)\qquad u \in L_\infty(\mathbb{R}^+, L_p) \quad \text{with } \|u(t)\|_p \leq K_1 = c(a, \exp(c\|u\|_{s,p}^s)).$$

Once this is known, one can identify this solution with the local strong one due to Serrin's uniqueness result and continue this process to get that

$$(4)\qquad u \in C_1(\mathbb{R}^+, L_p) \cap C_0(\mathbb{R}^+, H_p^2) \cap (\bar{\mathbb{R}}^+, L_p).$$

Furthermore this solution coincides with the weak solutions constructed in the papers mentioned above and therefore inherits the decay properties of the L_2-norm proved there.

It turns out that this has influence on our strong solution. We can show the following

Theorem 1: Let u be a strong solution of (1) in the sense of (2)-(4). Denote by $u_0(t) := e^{-t\triangle}a$ the solution of the heat equation and suppose $\|u_0(t)\|_2 \le C_0(1+t)^{-\alpha/2}$ (with $\alpha \neq 1$). Then

$$\|u(t) - u_0(t)\|_p \le C(1+t)^{-(\frac{d}{2}+\frac{n}{2}(\frac{1}{2}-\frac{1}{p}))}$$

with $d = \frac{n}{2} + 1 - 2\max\{1-\alpha, 0\}$.

Let us first state some consequences of this theorem.

a) We have $\|u(t)\|_p \le C(1+t)^{-(\frac{\bar{\alpha}}{2}+\frac{n}{2}(\frac{1}{2}-\frac{1}{p}))}$ with $\bar{\alpha} = \min\{\alpha, \frac{n}{2}+1\}$. This was shown for $\alpha = 0$ and $\|a\|_n$ small by B. da Veiga [1].

b) Let $a \in L_r \cap L_p$ with $1 \le r \le 2$. If $q \in [2,p]$, then $\|u(t)\|_q \le C(1+t)^{-\frac{n}{2}(\frac{1}{r}-\frac{1}{q})}$. This follows from a), as we can take $\bar{\alpha} = \alpha = n(\frac{1}{r}-\frac{1}{2})$. If $\frac{n}{2}(\frac{1}{r}-\frac{1}{q}) < 1$, this was shown by T. Kato [4] for $\|a\|_n$ small.

c) To get an impression in a concrete case: If $n = 3$, $a \in L_1 \cap L_p$, then $\|u(t) - u_0(t)\|_p \le C(1+t)^{-\frac{3}{2}(1-\frac{1}{p})-\frac{1}{2}}$. Note, that the set of initial values, such that $\|u_0(t)\|_p$ decays even exponentially fast, is dense.

Proof of Theorem 1: Let $w(t) := u(t) - u_0(t)$ and $y(t) := \|w(t)\|_p^p$. We prove first for $t \ge 1$ with positive constants c_i

$$(5) \qquad y'(t) + c_0(1+t)^{\nu\cdot\frac{d}{2}}y(t)^{1+\frac{\nu}{p}} \le c_1 y(t)^{\frac{\mu}{p}} + c_2(1+t)^{-\rho}y(t)^{\beta}$$

with $\nu = 4p/n(p-2), \mu = p(p+2-n)/(p-n), d = \frac{n}{2}+1-2\max\{1-\alpha,0\}$, and $\rho = 3(1-\frac{1}{p})+2\alpha, \beta = (p-2)/p$, if $n=3$, or $\rho = (pn-2n+4+2p\alpha)/(p+4-n), \beta = (p+2-n)/(p+4-n)$, if $n \ge 4$.

To see this multiply the equation

$$w_t - \Delta w + u \bigtriangledown u + \bigtriangledown \pi = 0 \text{ by } w|w|^{p-2}$$

and integrate by parts. With $N_p(w) = \int |\bigtriangledown w|^2 |w|^{p-2}dx$ one gets

$$\frac{1}{p}y'(t) + N_p(w) + 4(p-2)p^{-2}\int |\bigtriangledown |w^{p/2}|^2 dx \le \varepsilon N_p(w) + C_\varepsilon \int (|\pi|^2 + |u_0|^4 + |w|^4)|w|^{p-2}dx.$$

As $-\triangle\pi = \sum_{i,j}\frac{\partial^2}{\partial x_i \partial x_j}(u_i u_j)$, we have $\|\pi\|_{p*/2} \le c\|u\|_{p*}^2$ for $p* = pn/(n-2)$. Hence with $s = (p-2)p*/(p*-4)$ we get for $n \ge 4$:

$$\begin{aligned}
\int (|\pi|^2 + |u_0|^4)|w|^{p-2}dx &\le c(\|\pi\|_{p*/2}^2 + \|u_0\|_{p*}^4)\|w\|_s^{p-2} \\
&\le c(\|w\|_{p*}^4 + \|u_0\|_{p*}^4)\|w\|_p^{p+2-n}\|w\|_{p*}^{n-4} \\
&\le c\|w\|_p^{p+2-n}(N_p(w)^{n/p} + \|u_0\|_{p*}^4 N_p(w)^{(n-4)/p}) \\
&\le \varepsilon/c_\varepsilon N_p(w) + c'_\varepsilon(\|w\|_p^{\mu} + \|u_0\|_{p*}^{4p/(p+4-n)}\|w\|_p^{p\beta})
\end{aligned}$$

while for $n = 3$ this quantity is estimated by

$$c\|w\|_p^{p-2}(\|u_0\|_{2p}^4 + \|w\|_{2p}^4) \le c\|u_0\|_{2p}^4\|w\|_p^{p-2} + \varepsilon N_p(w) + c_\varepsilon\|w\|_p^\mu.$$

Last $\|w\|_{p+2}^{p+2} \le \varepsilon N_p(w) + c_\varepsilon\|w\|_p^\mu$. Here we have made use of Sobolev's inequality $\|w\|_{p*} \le cN_p(w)^{1/p}$ at several places. By the properties of the heat-kernel:

$$\begin{aligned}\|u_0(t)\|_q &\le (t/2)^{-\frac{n}{2}(\frac{1}{2}-\frac{1}{q})}\|u_0(t/2)\|_2 \\ &\le c(1+t)^{-\frac{n}{2}(\frac{1}{2}-\frac{1}{q})-\frac{\alpha}{2}} \quad \text{for } q \ge 2,\ t \ge 1.\end{aligned}$$

Collecting terms, we get

$$y'(t) + \frac{1}{2}N_p(w) \le c_1 y(t)^{\mu/p} + c_2(1+t)^{-\rho}y(t)^\beta.$$

Last $\|w\|_p \le \|w\|_2^a\|w\|_{p*}^{1-a}$ with $a = 4/(pn-2n+4)$, hence $N_p(w) \ge c_0\|w\|_p^{p+\nu}\|w\|_2^{-\nu}$. Now (5) follows, using the L_2-decay result in M. Wiegner [6].

The next step is the following calculus lemma:

$$\begin{aligned}&\text{If } \delta, \gamma \ge 0,\ r \in \mathbb{R} \text{ and} \\ &x'(t) + a(1+t)^\delta x(t)^{1+\gamma} \le b(1+t)^r \quad \text{for } t \ge 1 \\ (6)\quad &\text{then } x(t) \le C_0(1+t)^{-s_0} \\ &\text{with } s_0 = \min\{(\delta+1)/\gamma, (\delta-r)/(1+\gamma)\} \\ &\text{and } C_0 = \max\{x(1), (2b/a)^{1/(1+\gamma)}, (2s_0/a)^{1/\gamma}\}\end{aligned}$$

which follows by the standard comparison result for ODEs.

Now suppose $y(t) = C_k(1+t)^{-ps_k}$ with $s_k \ge 0, C_k \ge 1$. Then by (5) and (6)

$$y(t) \le C_{k+1}(1+t)^{-ps_{k+1}}$$

with

$$s_{k+1} = min\{\frac{1}{\nu} + \frac{d}{2}, (\frac{d}{2}\nu + min\{s_k\mu, p\beta s_k + \rho\})/(p+\nu)\}.$$

By elementary calculations (note, that $\mu > p + \nu$)

$$s_{k+1} \ge min\{\frac{1}{\nu} + \frac{d}{2}, s_k + \varepsilon_0\}$$

with some fixed $\varepsilon_0 > 0$ as long as $s_k \le \frac{1}{\nu}$. Thus after finitely many iterations $y(t) \le C(1+t)^{-p/\nu}$, which implies $y(t)^{(\mu-p)/p} \in L_1(\mathbb{R}^+)$.
Letting $z(t) := y(t)exp(-\int_1^t c_1 y(s)^{(\mu-p)/p}ds)$, we get from (5)

$$(7)\quad z'(t) + c_4(1+t)^{\nu d/2}z(t)^{1+\nu/p} \le c_5(1+t)^{-\rho}z(t)^\beta.$$

Now $z(t) \leq C_k(1+t)^{-ps_k}$ with $C_k \geq 1$, (7) and (6) imply the similar estimate with

$$s_{k+1} = min\{\frac{1}{\nu} + \frac{d}{2}, (\frac{d}{2} + p\beta s_k + \rho)/(p+\nu)\}$$

and

$$C_{k+1} \leq \gamma_0 C_k^{\beta/(1+\nu/p)}.$$

For $k \to \infty$, $s_k \to \frac{1}{\nu} + \frac{d}{2}$ and $\limsup C_k \leq \gamma_0^{(p+\nu)/(\nu+p(1-\beta))}$ which ends the proof.

Generalizing a result of H. Beirão da Veiga - P. Secchi [2], we show the following stability result.

Theorem 2: Let u_1 be a strong solution as above with initial value $a_1 \in L_2 \cap L_p$. There is an $\varepsilon > 0$, such that for $a_2 \in L_2 \cap L_p$ with $\|a_1 - a_2\|_p < \varepsilon$ a strong solution u_2 of (1) exists with $u_2 \in L_r(\mathbb{R}^+, L_p)$, if $\nu < r \leq \infty$.

Remark: In comparison to [2], we skip the assumption $u_1 \in L_\infty(\mathbb{R}^+, L_{p+2})$ and give the proof for all $n \geq 3$.

Proof: Similar as above we derive a differential inequality for $d(t) = \|u_1(t) - u_2(t)\|_p$ as long as the local strong solution $u_2(t)$ exists and get

$$\text{(8)} \qquad \begin{aligned} & d'(t) + c_0\|u_1(t) - u_2(t)\|_2^{-\nu} d(t)^{1+\nu} \leq c_1 d(t)^{\mu-p+1} + c_2 d(t)\|u_1(t)\|_{p+2}^{\gamma} \\ & \text{with } \gamma := 2(p+2)/(p+2-n). \end{aligned}$$

Suppose we have shown that $\|u_1\|_{p+2}^{\gamma} \in L_1(\mathbb{R}^+)$. Then introducing $z(t) = d(t)exp(-c_2 \int_0^t \|u_1(s)\|_{p+2}^{\gamma} ds)$ we get

$$z'(t) + C_0 z(t)^{1+\nu} \leq C_1 z(t)^{\mu-p+1}$$

which implies $z(t) \leq (C_0/C_1)^{1/(\mu-\nu-p)}$, if this inequality holds for $z(0) = d(0)$. This gives an a-priori estimate for $sup_t\|u_2(t)\|_p$, hence the theorem.
Now $\|u_1\|_{p+2} \leq c\|u_1\|_p^{1-bp} N_p(u_1)^b$ with $b = \frac{n}{p(p+2)}$. As $\gamma b < 1$ (which is equivalent to $p > n$), we get

$$\|u_1\|_{p+2}^{\gamma} \leq c N_p(u_1) + c\|u_1\|_p^{2\mu/(p+2)}.$$

Due to $\|u_1\|_p \leq c(1+t)^{-1/\nu}$ and $2\mu/(p+2) \cdot \nu > 1$, the second term belongs to $L_1(\mathbb{R}^+)$, while (with $u_0 \equiv 0$ in the differential inequality for $y(t)$)

$$\int_0^t N_p(u_1) \leq c\|a\|_p^p + c\int_0^t \|u_1\|_p^{\mu} ds \leq C$$

due to $\mu > \nu$. Thereby theorem 2 is proved.

References

[1] H. Beirão da Veiga: Existence and asymptotic behaviour for strong solutions of the Navier-Stokes equations on the whole space. Indiana Univ. Math. J., 36(1987), 149-166

[2] H. Beirão da Veiga - P. Secchi: L^p-stability for the strong solutions of the Navier-Stokes equations in the whole space. Arch. Rat. Mech. Anal. 98(1987), 65-70

[3] R. Kajikiya - T. Miyakawa: On L^2-decay of weak solutions of the Navier-Stokes equations in $\mathbb{R}^n$. Math. Z. 192(1986), 135-148

[4] T. Kato: Strong L^p-solutions of the Navier-Stokes equations in $\mathbb{R}^n$, with applications to weak solutions. Math. Z. 187(1984), 471-480

[5] M. E. Schonbek: L^2-decay for weak solutions of the Navier- Stokes equations. Arch. Rat. Mech. Anal. 88(1985), 209-222

[6] M. Wiegner: Decay results for weak solutions of the Navier-Stokes equations on $\mathbb{R}^n$. J. London Math. Soc.(2)35(1987), 303-313.

This paper is in final form and no similar paper has been or is being submitted elsewhere.

(received January 19, 1989.)

A GALERKIN APPROXIMATION FOR LINEAR EIGENVALUE PROBLEMS IN TWO AND THREE-DIMENSIONAL BOUNDARY-LAYER FLOWS

T.M. Fischer
DLR, Institute for Theoretical Fluid Mechanics
Bunsenstr. 10, D-3400 Göttingen, FR Germany

1. Introduction

The investigation of stability of small disturbances in two and three-dimensional boundary-layer flows leads to eigenvalue problems for systems of linear, ordinary differential equations on the half-line. A typical example is the *Orr-Sommerfeld equation:*

$$L\varphi := \frac{1}{Re}\left(\frac{d^4\varphi}{dz^4} - 2\alpha^2\frac{d^2\varphi}{dz^2} + \alpha^4\varphi\right) - i\alpha\left[U\left(\frac{d^2\varphi}{dz^2} - \alpha^2\varphi\right) - \frac{d^2U}{dz^2}\varphi\right] + i\omega\left(\frac{d^2\varphi}{dz^2} - \alpha^2\varphi\right) = 0 \quad \text{in} \quad 0 < z < \infty, \tag{1.1}$$

with the *boundary conditions,*

$$\varphi(0) = \frac{d\varphi}{dz}(0) = 0, \quad \varphi(z), \ \frac{d\varphi}{dz}(z) \to 0 \quad \text{as} \quad z \to \infty. \tag{1.2}$$

Here the parameters Re, $\alpha > 0$ denote the Reynolds number and the disturbance wavenumber, respectively. The boundary-layer profile U is a given, smooth function of the wall-normal coordinate z. For instance, U describes the Blasius profile in a two-dimensional flow over a flat plate [9]. The complex circular frequency ω and the amplitude function φ of the disturbance have to be determined as an eigenvalue and a corresponding eigenfunction, respectively.

The Orr-Sommerfeld eigenvalue problem, which is defined on a finite interval, has been investigated theoretically in [2] (see also the references therein). There it has been proved that the spectrum is discrete and that the eigenfunctions are complete in a certain Hilbert space. The proof is based on a compactness theorem, which is not available in the semi-infinite case. Therefore, instead of compact operators, we shall obtain bounded operators, for which only few statements can be made concerning the spectrum [6].

On the other hand, it is suggested by numerical computations (see e.g. [8]) that a *finite number of (discrete) eigenvalues* as well as a *continuous spectrum* exist. The continuous spectrum has been described analytically in [5]. Usually, shooting methods [8] or, together with coordinate transforms, spectral collocation methods [4] are used for discretizing the eigenvalue problem (1.1), (1.2). In [10] [11], a Galerkin approximation has been introduced, with an exponential mapping of the semi-infinite interval onto a finite interval and with Jacobi polynomials as basic components of the trial functions. A similar procedure, which has been proposed in [3], is the topic of the present work.

Since the velocity profile $U(z)$ approaches a constant value, say U_1, as $z \to \infty$, each eigenfunction φ decays exponentially at infinity,

$$\varphi(z) \sim e^{-\lambda z} \quad (z \to \infty), \tag{1.3}$$

with $\lambda = \alpha$ or $\lambda = \sqrt{\alpha^2 + iRe(\alpha U_1 - \omega)}$. This is the motive for taking an exponential mapping of $(0, \infty)$ onto a finite interval, though an algebraic mapping could also be chosen in the following.

One advantage of a Galerkin approximation against other numerical methods is obvious. The trial functions, which are used for the representation of the eigenfunctions, already satisfy the boundary conditions (1.2). The completeness of the trial functions in $L^2(0, \infty)$ will be shown to be essential for the convergence of the method. For instance, Laguerre orthogonal functions could be considered. However, to avoid an overflow in the computations, the Laguerre quadrature formulas may only be evaluated for a limited number of polynomials [12]. Therefore, in a first step, we shall prove that the system of trial functions, which has been introduced in [3], is complete. Furthermore, a weak formulation of the Orr-Sommerfeld eigenvalue problem (1.1), (1.2) will be defined and analysed. The resulting stability of the Galerkin approximation can then be used to establish a qualitative convergence result. Moreover, the high accuracy of the method will be demonstrated by means of a numerical example.

2. The Jacobi Orthogonal Functions

In this section, we shall introduce a system of trial functions which is based on the Jacobi polynomials and which is complete in

$$L^2 := L^2(0, \infty), \quad (\varphi, \psi)_{L^2} := \int_0^\infty \varphi(z)\overline{\psi(z)}dz \quad (\varphi, \psi \in L^2), \tag{2.1}$$

the Hilbert space of all measurable, square-integrable functions on the semi-infinite interval $(0, \infty)$.

Let $a, \kappa > 0$. Then the trial functions $\varphi^{(n)}$, $n = 1,2, \ldots$, are defined by

$$\varphi^{(n)}(z) := \frac{\sqrt{a}\, 2^{2+\kappa/2}}{\sqrt{h_{n-1}^{(4,\kappa-1)}}} \cdot (1 - \tilde{z})^2 \cdot P_{n-1}^{(4,\kappa-1)}(2\tilde{z} - 1) \cdot \tilde{z}^{\kappa/2}, \quad \tilde{z} = e^{-az} \quad (z \geq 0), \tag{2.2}$$

where $P_m^{(\beta,\gamma)}(s)$ $(-1 \leq s \leq 1)$, for $\beta, \gamma > -1$, are the *Jacobi polynomials* (see e.g. [12]),

$$\begin{aligned} P_m^{(\beta,\gamma)}(s) &:= [s - b_m]\, P_{m-1}^{(\beta,\gamma)}(s) - c_m P_{m-2}^{(\beta,\gamma)}(s), \quad m = 2,3, \ldots, \\ P_0^{(\beta,\gamma)}(s) &:= 1, \quad P_1^{(\beta,\gamma)}(s) := s + \frac{(\beta - \gamma)}{(\beta + \gamma + 2)}, \end{aligned} \tag{2.3}$$

$$\begin{aligned} b_m &:= \frac{(\beta + \gamma)(\gamma - \beta)}{(2m + \beta + \gamma)(2m + \beta + \gamma - 2)}, \quad m = 2,3, \ldots, \\ c_m &:= \frac{4(m-1)(m+\beta-1)(m+\gamma-1)(m+\beta+\gamma-1)}{(2m+\beta+\gamma-1)(2m+\beta+\gamma-2)^2(2m+\beta+\gamma-3)}, \quad m = 3,4, \ldots, \\ c_2 &:= \frac{4(\beta+1)(\gamma+1)}{(\beta+\gamma+3)(\beta+\gamma+2)^2}, \end{aligned} \tag{2.4}$$

and

$$h_m^{(\beta,\gamma)} := \frac{2^{\beta+\gamma+2m+1} m!\,\Gamma(\beta+m+1)\Gamma(\gamma+m+1)\Gamma(\beta+\gamma+m+1)}{(\beta+\gamma+2m+1)[\Gamma(\beta+\gamma+2m+1)]^2}, \quad m = 0,1, \ldots. \tag{2.5}$$

We note that the boundary conditions (1.2) are satisfied,

$$\varphi^{(n)}(0) = \frac{d\varphi^{(n)}}{dz}(0) = 0\,, \quad \varphi^{(n)}(z) \sim e^{-a\frac{\kappa}{2}z} \quad (z \to \infty)\,.$$

For this system of trial functions, the following *orthogonality relation* holds,

$$\int_0^\infty \varphi^{(n)}(z)\varphi^{(m)}(z)dz = \delta_{nm} := \begin{cases} 1\,, & \text{if } n = m, \\ 0\,, & \text{if } n \neq m, \end{cases} \tag{2.6}$$

which can be obtained directly from the orthogonality relation [12] which is valid for the Jacobi polynomials,

$$\int_{-1}^1 (1-s)^\beta (1+s)^\gamma P_n^{(\beta,\gamma)}(s) P_m^{(\beta,\gamma)}(s)ds = \begin{cases} h_m^{(\beta,\gamma)}\,, & \text{if } n = m, \\ 0\,, & \text{if } n \neq m. \end{cases} \tag{2.7}$$

Let $J_N \subset L^2$ denote the finite-dimensional space which is spanned by the trial functions $\varphi^{(1)}, \ldots, \varphi^{(N)}$. We can prove that the union of all spaces J_N is dense in L^2.

Lemma 1. *Let* $\varphi \in L^2$. *Then*

$$\| \varphi - \sum_{n=1}^N (\varphi, \varphi^{(n)})_{L^2} \varphi^{(n)} \|_{L^2} \to 0 \quad (N \to \infty)\,, \tag{2.8}$$

i.e. the system (2.2) *of the orthonormal functions* $\varphi^{(n)}$, $n = 1,2, \ldots$, *is complete in* L^2.

P r o o f . First, we observe that the function

$$f(\tilde z) := \frac{\varphi(-\frac{1}{a}\ln \tilde z)}{\sqrt{a}\,\tilde z^{1/2}} \quad (0 < \tilde z < 1)$$

is square-integrable on (0,1),

$$f \in L^2(0,1)\colon \quad \int_0^1 |f(\tilde z)|^2 d\tilde z = \int_0^\infty |\varphi(z)|^2 dz < \infty\,.$$

Furthermore, the system of functions

$$f^{(n)}(\tilde z) := \frac{\varphi^{(n)}(-\frac{1}{a}\ln \tilde z)}{\sqrt{a}\,\tilde z^{1/2}} \quad (0 < \tilde z < 1)\,, \ n = 1,2, \ldots,$$

is orthonormal in $L^2(0,1)$, where

$$(f, f^{(n)})_{L^2(0,1)} := \int_0^1 f(\tilde z)\overline{f^{(n)}(\tilde z)}d\tilde z = (\varphi, \varphi^{(n)})_{L^2}\,.$$

Now let $\varepsilon > 0$ be given. Since $C_0^\infty(0,1)$, the space of all functions, which are differentiable of infinite order and have compact supports in (0,1), is dense in $L^2(0,1)$ [1], there exists a function $f_\varepsilon \in C_0^\infty(0,1)$ such that

$$\|f - f_\varepsilon\|_{L^2(0,1)} < \frac{\varepsilon}{2}\,.$$

Then we define

$$g_\varepsilon(\tilde z) := \begin{cases} \dfrac{f_\varepsilon(\tilde z)}{(1-\tilde z)^2 \tilde z^{(\kappa-1)/2}}\,, & \text{if } 0 < \tilde z < 1, \\ 0\,, & \text{if } \tilde z = 0 \text{ or } \tilde z = 1. \end{cases}$$

We have $g_\varepsilon \in C_0^\infty(0,1)$, because f_ε has a compact support in (0,1), and, moreover, the following expansion of g_ε with respect to the Jacobi polynomials $P_m^{(4,\,\kappa-1)}$, $m = 0,1, \ldots$, holds [15]. There is an integer N_0 such that

$$\max_{0 \le \tilde{z} \le 1} \left| g_\varepsilon(\tilde{z}) - \sum_{n=1}^{N} \hat{g}_{\varepsilon,n-1} P_{n-1}^{(4,\,\kappa-1)}(2\tilde{z}-1) \right| < \frac{\varepsilon}{2}\sqrt{\kappa}$$

whenever $N \ge N_0$, where

$$\hat{g}_{\varepsilon,n-1} := \frac{1}{h_{n-1}^{(4,\,\kappa-1)}} \int_{-1}^{1} (1-s)^4 (1+s)^{\kappa-1} P_{n-1}^{(4,\,\kappa-1)}(s)\, g_\varepsilon((1+s)/2)\, ds\,.$$

Using the best approximation in $L^2(0,1)$, we obtain:

$$\begin{aligned}
&\| \varphi - \sum_{n=1}^{N} (\varphi, \varphi^{(n)})_{L^2} \varphi^{(n)} \|_{L^2} \\
&= \| f - \sum_{n=1}^{N} (f, f^{(n)})_{L^2(0,1)} f^{(n)} \|_{L^2(0,1)} \\
&\le \| f - \sum_{n=1}^{N} \frac{\sqrt{h_{n-1}^{(4,\,\kappa-1)}}}{2^{2+\kappa/2}} \hat{g}_{\varepsilon,n-1} \cdot f^{(n)} \|_{L^2(0,1)} \\
&\le \|f - f_\varepsilon\|_{L^2(0,1)} + \| (1-\tilde{z})^2 \tilde{z}^{(\kappa-1)/2} \left(g_\varepsilon - \sum_{n=1}^{N} \hat{g}_{\varepsilon,n-1} P_{n-1}^{(4,\,\kappa-1)} \right) \|_{L^2(0,1)} \\
&< \frac{\varepsilon}{2} + \frac{\varepsilon}{2}\sqrt{\kappa} \left(\int_0^1 \tilde{z}^{\kappa-1} d\tilde{z} \right)^{1/2} = \varepsilon\,. \quad \square
\end{aligned}$$

Finally, let us consider subspaces of the *Sobolev space* [1] [16]

$$H^2(0,\infty) := \left\{ \varphi \in L^2 \,\middle|\, \frac{d\varphi}{dz}, \frac{d^2\varphi}{dz^2} \in L^2 \right\}, \tag{2.9}$$

which is a Hilbert space, where the inner product is given by

$$(\varphi, \psi)_{H^2} := \int_0^\infty \left(\frac{d^2\varphi}{dz^2}(z) \overline{\frac{d^2\psi}{dz^2}(z)} + \frac{d\varphi}{dz}(z) \overline{\frac{d\psi}{dz}(z)} + \varphi(z)\overline{\psi(z)} \right) dz \tag{2.10}$$

$$(\varphi, \psi \in H^2(0,\infty))\,.$$

The finite-dimensional space which is spanned by the trial functions $\varphi^{(1)}, \ldots, \varphi^{(N)}$, together with the H^2-norm, will be denoted by H_N, and the closure of the union of all spaces H_N with respect to the H^2-norm by H. From Lemma 1, it follows immediately that H is dense in L^2.

3. The Weak Orr-Sommerfeld Eigenvalue Problem

For every classical solution φ of the Orr-Sommerfeld eigenvalue problem (1.1), (1.2) and for arbitrary, linear combinations ψ of the trial functions $\varphi^{(n)}$, $n = 1,2, \ldots$, the bilinear form $(L\varphi, \psi)_{L^2}$ associated with the Orr-Sommerfeld operator L can be written in the form,

$$(L\varphi, \psi)_{L^2} = \frac{1}{Re} \int_0^\infty \left(\frac{d^2\varphi}{dz^2} \frac{d^2\bar{\psi}}{dz^2} + 2\alpha^2 \frac{d\varphi}{dz} \frac{d\bar{\psi}}{dz} + \alpha^4 \varphi\bar{\psi} \right) dz$$

$$+ i\alpha \left[\int_0^\infty U \left(\frac{d\varphi}{dz} \frac{d\bar{\psi}}{dz} + \alpha^2 \varphi\bar{\psi} \right) dz - \int_0^\infty \frac{dU}{dz} \varphi \frac{d\bar{\psi}}{dz} dz \right] \tag{3.1}$$

$$- i\omega \int_0^\infty \left(\frac{d\varphi}{dz} \frac{d\bar{\psi}}{dz} + \alpha^2 \varphi\bar{\psi} \right) dz ,$$

by using integration by parts.

Now let $\varphi \in H^2(0, \infty)$. Then the Sobolev imbedding theorem [1] implies that both φ and $d\varphi/dz$ are continuous on the closed interval $0 \leq z < \infty$ and that

$$\varphi(z) , \quad \frac{d\varphi}{dz}(z) \to 0 \quad \text{as} \quad z \to \infty . \tag{3.2}$$

Moreover, for $\varphi \in H$, we obtain

$$\varphi(0) = \frac{d\varphi}{dz}(0) = 0 . \tag{3.3}$$

Therefore, it is meaningful to consider the following weak formulation of the Orr-Sommerfeld eigenvalue problem:

Find $\omega \in \mathbb{C}$ and $\varphi \in H$, $\varphi \neq 0$, such that

$$(L\varphi, \psi)_{L^2} = 0 \quad (\psi \in H) . \tag{3.4}$$

If, in addition,

$$\frac{d^3\varphi}{dz^3} , \quad \frac{d^4\varphi}{dz^4} \in L^2 ,$$

then $L\varphi \in L^2$ and we conclude $L\varphi = 0$, since H is dense in L^2.

Lemma 2. *Let*

$$\max_{0 \leq z < \infty} |U(z)| < \infty \quad \text{and} \quad \mu := \frac{1}{2} \max_{0 \leq z < \infty} \left| \frac{dU}{dz}(z) \right| < \infty .$$

Then, for $\text{imag}(\omega) \geq \mu$, *there exists a uniquely determined bijective, in both directions continuous, linear operator* $S\colon H \to H$, *with*

$$(L\varphi, \psi)_{L^2} = (S^{-1}\varphi, \psi)_{H^2} \quad (\varphi, \psi \in H) \tag{3.5}$$

and with the norm,

$$\|S\| \leq \frac{Re}{\min(1, \alpha^4)} . \tag{3.6}$$

P r o o f . To begin with, using the Cauchy-Schwarz inequality, we have

$$\left| \int_0^\infty \frac{dU}{dz} \varphi \frac{d\bar{\varphi}}{dz} dz \right| \leq \mu \left(\frac{1}{\alpha} \left\| \frac{d\varphi}{dz} \right\|_{L^2}^2 + \alpha \|\varphi\|_{L^2}^2 \right) .$$

Obviously, the bilinear form $(L\varphi, \psi)_{L^2}$, $\varphi, \psi \in H$, is continuous. Moreover, we can now prove that, for $\text{imag}(\omega) \geq \mu \geq 0$, it is strictly positive:

$$|\operatorname{real}(L\varphi,\varphi)_{L^2}| \geq \frac{1}{Re}\int_0^\infty\left(\left|\frac{d^2\varphi}{dz^2}\right|^2 + 2\alpha^2\left|\frac{d\varphi}{dz}\right|^2 + \alpha^4|\varphi|^2\right)dz$$

$$+ \operatorname{imag}(\omega)\int_0^\infty\left(\left|\frac{d\varphi}{dz}\right|^2 + \alpha^2|\varphi|^2\right)dz$$

$$-\alpha\left|\int_0^\infty \frac{dU}{dz}\varphi\frac{d\overline{\varphi}}{dz}dz\right|$$

$$\geq \frac{1}{Re}\int_0^\infty\left(\left|\frac{d^2\varphi}{dz^2}\right|^2 + 2\alpha^2\left|\frac{d\varphi}{dz}\right|^2 + \alpha^4|\varphi|^2\right)dz$$

$$\geq \frac{\min(1,\alpha^4)}{Re}\;\|\varphi\|^2_{H^2}.$$

Lemma 2 then follows from the Lax-Milgram theorem [16]. □

Next, we consider the operator $M := \frac{d^2}{dz^2} - \alpha^2$. Let $\varphi, \psi \in H$. Then we obtain

$$(M\varphi,\psi)_{L^2} = -\int_0^\infty\left(\frac{d\varphi}{dz}\frac{d\overline{\psi}}{dz} + \alpha^2\varphi\overline{\psi}\right)dz = (\varphi, M\psi)_{L^2}, \tag{3.7}$$

by using integration by parts, and the following estimate holds:

$$|(M\varphi,\psi)_{L^2}| \leq \sqrt{2}\,\max(1,\alpha^2)\cdot\|\varphi\|_{L^2}\cdot\|\psi\|_{H^2}. \tag{3.8}$$

Applying the Riesz representation theorem (cf. [16]) and taking into account that H is dense in L^2, we find that *there exists a uniquely determined continuous, linear operator* $T\colon L^2 \to H$, *with*

$$(M\varphi,\psi)_{L^2} = (T\varphi,\psi)_{H^2} \quad (\varphi,\psi\in H). \tag{3.9}$$

Furthermore, let ω_0 be an arbitrary but fixed complex number, for which $\operatorname{imag}(\omega_0) \geq \mu$, and let $L(\omega_0)$ be the Orr-Sommerfeld operator at $\omega = \omega_0$. Then, from the relations (3.5) and (3.9), it follows that

$$(L\varphi,\psi)_{L^2} = (L(\omega_0)\varphi,\psi)_{L^2} + i(\omega-\omega_0)\,(M\varphi,\psi)_{L^2}$$

$$= (S^{-1}\varphi,\psi)_{H^2} + i(\omega-\omega_0)\,(T\varphi,\psi)_{H^2} \quad (\varphi,\psi\in H),$$

which implies that the weak Orr-Sommerfeld eigenvalue problem (3.4) is equivalent to the following problem:

Find $\omega\in\mathbb{C}$ *and* $\varphi\in L^2$, $\varphi\neq 0$, *such that*

$$\varphi + i(\omega-\omega_0)\,S\circ T\varphi = 0. \tag{3.10}$$

Since the operator $S\circ T$ maps L^2 into $H\subset L^2$, each solution of this equation satisfies $\varphi\in H$.

4. The Galerkin Method

The Galerkin approximation of the eigenvalue problem (3.4) is given by the following discrete problem:

Find $\omega\in\mathbb{C}$ *and* $\varphi\in H_N$, $\varphi\neq 0$, *such that*

$$(L\varphi,\psi)_{L^2} = 0 \quad (\psi\in H_N). \tag{4.1}$$

Similar as in the infinite-dimensional case, this problem can be written in the form:

Find $\omega \in \mathbb{C}$ and $\varphi \in J_N$, $\varphi \neq 0$, such that

$$\varphi + i(\omega - \omega_0)\, S_N \circ T_N \varphi = 0\,, \tag{4.2}$$

with uniquely determined continuous, linear operators $S_N: H_N \to H_N$ and $T_N: J_N \to H_N$.

The estimates (3.6) and (3.8) can also be applied to the operators S_N and T_N, respectively:

$$\|S_N\| \leq \frac{Re}{\min(1, \alpha^4)}\,, \quad \|T_N\| \leq \sqrt{2}\, \max(1, \alpha^2)\,. \tag{4.3}$$

Consequently, the sequence of operators $S_N \circ T_N$, $N = 1,2, \dots$, is *stable*. Moreover, we can prove the convergence of $S_N \circ T_N \to S \circ T$ $(N \to \infty)$ in the sense of [13].

Lemma 3. *For every $\varphi \in L^2$ and every sequence of elements $\varphi_N \in J_N$, $N = 1,2, \dots$, there holds*

$$\|\varphi - \varphi_N\|_{L^2} \to 0 \;\Rightarrow\; \|(S \circ T)\varphi - (S_N \circ T_N)\varphi_N\|_{H^2} \to 0 \quad (N \to \infty)\,. \tag{4.4}$$

P r o o f . Let Q_N be the orthogonal projection $H \to H_N$. For $\psi \in H$, we have

$$\|\psi - Q_N \psi\|_{H^2} \to 0 \quad (N \to \infty)\,,$$

since the union of all spaces H_N is dense in H. Furthermore, the operators S_N and T_N admit the representations:

$$S_N = (Q_N S^{-1}|_{H_N})^{-1}\,, \quad T_N = Q_N T|_{J_N}\,.$$

Now let $\varphi \in L^2$ and $\varphi_N \in J_N$, $N = 1,2, \dots$, with $\varphi_N \to \varphi$ $(N \to \infty)$. Let $I: H \to H$ denote the identity operator. We then obtain

$$\begin{aligned}
&\|(S \circ T)\varphi - (S_N \circ T_N)\varphi_N\|_{H^2} \\
&\quad \leq \|(I - S_N Q_N S^{-1}) S \circ T \varphi\|_{H^2} + \|S_N \circ Q_N T(\varphi - \varphi_N)\|_{H^2} \\
&\quad \leq \|(I - S_N Q_N S^{-1})(I - Q_N) S \circ T \varphi\|_{H^2} + \|S_N \circ Q_N T(\varphi - \varphi_N)\|_{H^2} \\
&\quad \leq (1 + \|S_N\|\, \|S^{-1}\|) \cdot \|(S \circ T)\varphi - Q_N (S \circ T)\varphi\|_{H^2} + \|S_N\| \cdot \|T\| \cdot \|\varphi - \varphi_N\|_{L^2}\,,
\end{aligned}$$

which tends to zero as $N \to \infty$. □

We remark that the existence of a sequence of elements $\varphi_N \in J_N$, $N = 1,2, \dots$, which converges to $\varphi \in L^2$, is ensured by Lemma 1 and, thus, the operator $S \circ T$ and the sequence of operators $S_N \circ T_N$, $N = 1,2, \dots$, are *consistent*.

As an immediate consequence of Lemma 3, we have the following result.

Theorem 1. *Let $\omega_N \in \mathbb{C}$ and $\varphi_N \in H_N$, $\|\varphi_N\|_{L^2} = 1$, $N = 1,2, \dots$, be respectively eigenvalues and corresponding eigenfunctions of the Galerkin equations* (4.1). *Assume that $\omega \in \mathbb{C}$ is a number and $\varphi \in L^2$ is a function, for which*

$$|\omega - \omega_N| \to 0\,, \quad \|\varphi - \varphi_N\|_{L^2} \to 0 \quad (N \to \infty)\,. \tag{4.5}$$

Then ω is an eigenvalue and $\varphi \in H$, $\|\varphi\|_{L^2} = 1$, is a corresponding eigenfunction of the weak Orr-Sommerfeld eigenvalue problem (3.4).

Moreover, under the assumptions of Theorem 1, we can prove $\|\varphi - \varphi_N\|_{H^2} \to 0$ $(N \to \infty)$. For convergence results on the basis of compact operators, see e.g. [7] [14].

Finally, in order to demonstrate the accuracy of the Galerkin approximation (4.1), let us consider the case $Re = 580$, $\alpha = 0.179$ for Blasius flow, a case, for which the first eigenvalue $\omega^{(1)}/\alpha$ has been given to eight decimal places in [4]. The relative error, $|\omega^{(1)} - \omega_N| / |\omega_N|$, which is obtained for the Galerkin approximation ω_N of this eigenvalue is shown in dependence upon the number N of Jacobi polynomials in figure 1. Here the numerical integration has been performed by using the computer program JACOBI of [12]. A more detailed study of the parameter dependence of the Galerkin solution is contained in [3]. Note that, according to Theorem 1, the convergence of the calculated eigenfunctions must also be tested.

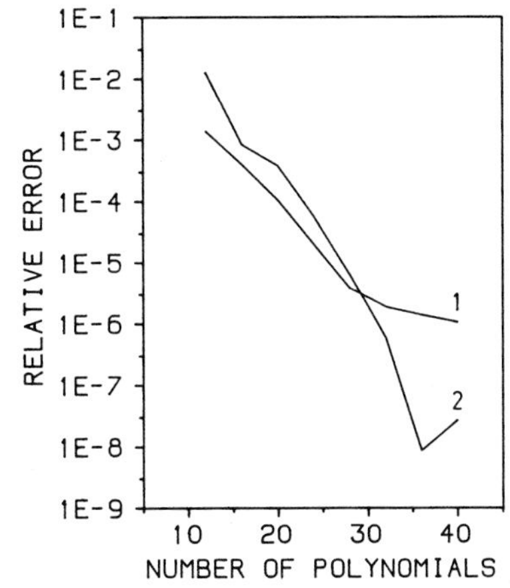

Fig. 1. Relative error of the Galerkin approximation, with (1) $a = 1/4$, $\kappa = 1$ and (2) $a = 1/8$, $\kappa = 1$.

In a further work, the results concerning the perturbation of the spectrum should be refined by using the actual decay properties of $U - U_1$ and dU/dz at infinity and, with that, the compactness of the corresponding operators in (3.1) (cf. [17]).

Acknowledgment

The author is indebted to Prof. Dr. K. Kirchgässner and Prof. Dr. W.L. Wendland for helpful discussions and suggestions.

References

[1] Adams, R.A.: Sobolev Spaces. New York: Academic Press 1975
[2] Di Prima, R.C.; Habetler, G.J.: A completeness theorem for non-selfadjoint eigenvalue problems in hydrodynamic stability. Arch. Rat. Mech. Anal. **34** (1969) 218 - 227
[3] Fischer, T.M.: Ein Spektral-Galerkin-Verfahren zur Untersuchung linearer Instabilität in Grenzschichten. DFVLR-IB 221-88 A 10 (1988)
[4] Grosch, C.E.; Orszag, S.A.: Numerical solution of problems in unbounded regions: coordinate transforms. J. Comp. Phys. **25** (1977) 273 - 296
[5] Grosch, C.E.; Salwen, H.: The continuous spectrum of the Orr-Sommerfeld equation. Part 1. The spectrum and the eigenfunctions. J. Fluid Mech. **87** (1978) 33 - 54
[6] Kato, T.: Perturbation Theory for Linear Operators. Berlin - Heidelberg - New York: Springer 1976
[7] Krasnosel'skii, M.A.; Vainikko, G.M.; Zabreiko, P.P.; Rutitskii, Y.B.; Stetsenko, V.Y.: Approximate Solution of Operator Equations. Groningen: Wolters-Noordhoff 1972

[8] Mack, L.M.: A numerical study of the temporal eigenvalue spectrum of the Blasius boundary layer. J. Fluid Mech. **73** (1976) 497 - 520

[9] Schlichting, H.: Grenzschicht-Theorie. Karlsruhe: Braun 1982

[10] Spalart, P.R.: A spectral method for external viscous flows. Contemp. Math. **28** (1984) 315 - 335

[11] Spalart, P.R.: Numerical simulation of boundary layers: part 1. Weak formulation and numerical method. NASA TM-88222 (1986)

[12] Stroud, A.H.; Secrest, D.: Gaussian Quadrature Formulas. Englewood Cliffs: Prentice-Hall 1966

[13] Stummel, F.: Diskrete Konvergenz linearer Operatoren. I. Math. Ann. **190** (1970) 45 - 92

[14] Stummel, F.: Diskrete Konvergenz linearer Operatoren. II. Math. Z. **120** (1971) 231 - 264

[15] Tricomi, F.G.: Vorlesungen über Orthogonalreihen. Berlin - Göttingen - Heidelberg: Springer 1955

[16] Wloka, J.: Partielle Differentialgleichungen. Stuttgart: Teubner 1982

[17] Rannacher, R.: Zur asymptotischen Störungstheorie für Eigenwertaufgaben mit diskreten Teilspektren. Math. Z. **141** (1975) 219 - 233

This paper is in final form and no similar paper has been or is being submitted elsewhere.

(received January 8, 1989).

SYMMETRY-BREAKING EFFECTS OF DISTANT SIDEWALLS IN RAYLEIGH-BÉNARD CONVECTION

Wayne Nagata
University of British Columbia, Vancouver, Canada

1. Introduction and linear analysis

Convection in a fluid heated from below is a widely studied phenomenom, since it is accessible to both detailed mathematical treatments and accurate experimental observations. Much theory has been developed for spatially periodic convection in an infinite layer with no sidewalls. In this case, for an Oberbeck-Boussinesq fluid with free boundaries, the minimum value of the Rayleigh number R (a parameter that measures the temperature difference between the top and bottom of the fluid layer), for which there is a spatially periodic instability, is $R_0 = 27\pi^4/4$. This value corresponds to a marginally stable mode with spatial periodicity $2\pi/\alpha_0$, where $\alpha_0 = \pi/\sqrt{2}$. It is well-known that a supercritical bifurcation of stationary convective solutions occurs at $R = R_0$.

For a finite fluid layer with distant sidewalls, Segel [9] used a multiple-scale perturbation analysis to study the onset of convection in a three-dimensional fluid layer with large typical aspect ratio L, and showed that the critical Rayleigh number, at which bifurcation occurs, is raised by an amount of order L^{-2}. Daniels [2] made a careful linearized stability analysis for a two-dimensional fluid layer, and showed that in fact, two pairs of convective solutions bifurcate at values of R that are separated by an amount of order L^{-3}. One pair of solutions, called even, have stream functions that are even about the vertical centreline of the fluid layer, and the other, odd, pair of solutions have stream functions that are odd. Both pairs of stationary solutions bifurcate supercritically, but only the pair that bifurcate at the lower value of R are stable.

In this paper we study the two-dimensional model of convection in a layer with sidewalls considered in [1]-[5], and investigate the non-linear interaction between even and odd convective modes as they bifurcate at almost the same value of $R = R_0 + O(L^{-2})$. We reduce the problem to a bifurcation equation that accounts for the effects of distant sidewalls as a small perturbation of the classical, infinite layer problem. By applying results of Golubitsky and Schaeffer [8], on bifurcation in the presence of symmetry, we show that generically there exist "mixed-mode" patterns of steady convection, neither even nor odd, that were not considered in the previous studies. These solutions arise from secondary bifurcations, and can be involved in the adjustment in the number of convection rolls in the layer, very near the onset of convection.

We consider two-dimensional Rayleigh-Bénard convection, in the Oberbeck-Boussinesq approximation, described by the equations

$$
(1.1)\qquad
\begin{aligned}
u_t &= P(u_{xx} + u_{zz} - p_x) - (uu_x + wu_z),\\
w_t &= P(w_{xx} + w_{zz} - p_z + \theta) - (uw_x + ww_z),\\
u_x &+ w_z = 0,\\
\theta_t &= \theta_{xx} + \theta_{zz} + Rw - (u\theta_x + w\theta_z),
\end{aligned}
$$

in the spatial region $-L \le x \le L$, $0 \le z \le 1$. Here u and w are the horizontal and vertical components of the fluid velocity, p is the reduced pressure, and θ is the temperature deviation from the constant-gradient profile $T = T_0 - Rz$. All variables have been non-dimensionalized, and the parameters R, L and P are all positive constants. The Rayleigh number R is proportional to the difference between the constant temperatures of the lower (warmer) and upper (cooler) surfaces $z = 0$ and $z = 1$; the Prandtl number P is the ratio of kinematic viscosity to thermal diffusivity, considered fixed at some value. The boundary conditions are

$$(1.2)\qquad u_z = w = \theta = 0 \quad \text{on} \quad z = 0, 1,$$

corresponding to free, isothermal upper and lower surfaces, and

$$(1.3)\qquad u = w = \theta_x = 0 \quad \text{on} \quad x = \pm L,$$

corresponding to rigid, insulated sidewalls. As usual, the reduced pressure terms in (1.1) may be eliminated by introducing a stream function, or by a projection onto solenoidal fluid motions.

For any value of L, we have two discrete symmetries. The problem (1.1)-(1.3) is covariant with respect to the horizontal reflection symmetry

$$(1.4a)\qquad \kappa:\ x \to -x,\ u \to -u,$$

and the symmetry

$$(1.4b)\qquad \rho:\ x \to -x,\ z \to 1-z,\ u \to -u,\ w \to -w,\ \theta \to -\theta$$

that corresponds to inversion through the midpoint of the layer. We note that if the Oberbeck-Boussinesq approximation is not used (for example, if the viscosity of the fluid is temperature-dependent), then we cannot assume covariance under (1.4b). The symmetries (1.4a,b) generate the group of symmetries $\mathbf{Z}_2 \oplus \mathbf{Z}_2$, and the trivial solution $u = w = \theta = 0$ is $(\mathbf{Z}_2 \oplus \mathbf{Z}_2)$- invariant.

The linearized stability of the trivial (motionless) solution is determined by the eigenvalue problem for the linearization, which may be solved by separation of variables [2], [6]. One obtains, for all $P > 0$, two interlaced curves $R = R^E(L)$ and $R = R^O(L)$, along which 0 is an eigenvalue, with each of these curves corresponding to a one-dimensional subspace of eigenfunctions (i.e., marginally stable modes). See [2, Fig. 1].

The critical, even and odd, eigenfunctions along the curves $R = R^E(L)$ and $R = R^O(L)$ have the form $\phi^E = (u^E, w^E, \Theta^E)$ and $\phi^O = (u^O, w^O, \Theta^O)$, where the components u^E, u^O, etc. can be explicitly expressed in terms of elementary functions [2]. Furthermore, we note that the critical eigenfunctions transform under the symmetries (1.4) as

$$\kappa\phi^E = -\phi^E, \quad \rho\phi^E = \phi^E, \quad \kappa\phi^O = \phi^O, \quad \rho\phi^O = -\phi^O \tag{1.5}$$

Daniels [2] studied the eigenvalue problem for the linearization as $L \to \infty$, and obtained asymptotic expressions for $R^E(L)$ and $R^O(L)$. His results imply that there is an infinite sequence of aspect ratios $2L$, with $L = L_\ell$, $\ell = 1,2,3,\ldots$, for which the curves $R = R^E(L)$ and $R = R^O(L)$ cross transversally:

$$L_\ell = (\sqrt{2})^{-1}\ell - (\sqrt{2}\pi)^{-1} \operatorname{Arctan} (8\sqrt{2}/13) + O(\ell^{-1}) \tag{1.6}$$

2. Reduction to planar bifurcation equations

To study the nonlinear interactions between even and odd convective modes for (1.1)-(1.3), we apply Liapunov-Schmidt reduction at each intersection point (L_ℓ, R_ℓ), $\ell = 1,2,\ldots$, of the two curves $R = R^E(L)$ and $R = R^O(L)$, and obtain a two-dimensional system of bifurcation equations for each ℓ. Since both the functional-analytic reformulation of the problem (e.g., see [10]), and the reduction to bifurcation equations (e.g., see [8]), are standard, we omit the details.

By the Liapunov-Schmidt reduction, all small-amplitude solutions of (1.1)-(1.3) may be expressed as $\phi = (u, w, \Theta)$, with

$$\phi = \phi(x,z) = \xi\phi^E(x,z) + \eta\phi^O(x,z) + \psi(\xi,\eta,\lambda,\sigma)(x,z), \tag{2.1}$$

where $\lambda = R - R_\ell$, $\sigma = L - L_\ell$ and $\psi = O(|\xi,\eta|^2 + |\xi,\eta||\lambda,\sigma|)$.

The small amplitudes ξ, η satisfy the $(\mathbf{Z}_2 \oplus \mathbf{Z}_2)$-symmetric pair of analytic bifurcation equations

$$\begin{aligned} 0 &= \xi[\alpha\lambda + \gamma\sigma + A\xi^2 + B\eta^2 + f(\xi^2,\eta^2,\lambda,\sigma)], \\ 0 &= \eta[\beta\lambda + \delta\sigma + C\xi^2 + D\eta^2 + g(\xi^2,\eta^2,\lambda,\sigma)], \end{aligned} \tag{2.2}$$

where $f,g = O(|\xi^2,\eta^2|^2 + |\xi^2,\eta^2||\lambda,\sigma| + |\lambda,\sigma|^2)$.

The bifurcation equations (2.2) are covariant with respect to

(2.3) $\kappa: (\xi,\eta) \to (\xi,-\eta),\ \rho: (\xi,\eta) \to (-\xi,\eta),$

which are the co-ordinate representations of the $(\mathbf{Z}_2 \oplus \mathbf{Z}_2)$-actions (1.4), acting on the subspace spanned by the critical eigenvectors ϕ^E and ϕ^O.

Using asymptotic approximations of the critical eigenfunctions, we carry out the Liapunov-Schmidt reduction and obtain the coefficients in (2.3):

$$\alpha,\beta = \frac{2P}{9\pi^2(1+P)} + O(\ell^{-2}), \qquad \gamma,\delta = O(\ell^{-2}),$$

(2.4) $$A,\ B,\ C,\ D = -\frac{P}{8(1+P)^2}\,[1+\Delta\ell^{-1}+O(\ell^{-2})],$$

$$\Delta = (\sqrt{2}/9\pi)\sin(\pi L_\ell/\sqrt{2})\,[13\cos(\pi L_\ell/\sqrt{2})-8\ 2\sin(\pi L_\ell/\sqrt{2})].$$

3. Results and discussion

Using (2.4), the bifurcation equations (2.2) may be written as

(3.1)
$$0 = \xi h_\ell(\xi^2+\eta^2,\lambda) + \ell^{-2}\xi p_\ell(\xi^2,\eta^2,\lambda,\sigma),$$
$$0 = \eta h_\ell(\xi^2+\eta^2,\lambda) + \ell^{-2}\eta q_\ell(\xi^2,\eta^2,\lambda,\sigma)$$

where

$$h_\ell(\xi^2+\eta^2,\lambda) = \frac{2P}{9\pi^2(1+P)}\lambda - \frac{P}{8(1+P)^2}(1+\Delta\ell^{-1})(\xi^2+\eta^2).$$

For large ℓ, (3.1) may be considered as a small perturbation of the classical bifurcation equation that describes the onset of convection in an infinite layer with periodicity $2\pi/\alpha_0$ in the x-direction:

(3.2) $$0 = \zeta\left[\frac{2P}{9\pi^2(1+P)}\lambda - \frac{P}{8(1+P)^2}|\zeta|^2\right], \qquad \zeta = \xi+i\eta \in \mathbf{C}.$$

Corresponding to the reflection symmetry $\kappa: x \to -x$ and the continuous family of translation symmetries $x \to x + \tau \pmod{2\pi/\alpha_0}$, for the convection problem (1.1)-(1.2) in the infinite layer $(-\infty < x < \infty)$, the bifurcation equation (3.2) possesses the group $O(2)$ of symmetries of the circle:

(3.3a,b) $\zeta \to \bar{\zeta},\quad \zeta \to \exp(i\alpha_0\tau)\zeta$

In (3.1), the effect of the distant sidewalls at $x = \pm L$, $L = O(\ell)$, is to "break" the continuous symmetry (3.3b) by introducing a small perturbation of order $O(\ell^{-2})$ that possesses only the discrete symmetries (2.3). For fixed $\lambda > 0$, the "unperturbed" problem (3.2) has an infinite family of bifurcating solutions (the phase of ζ is arbitrary), while the "perturbed" problem (3.1) generically has only a discrete set of bifurcating solutions (e.g., even or odd modes). However, in the limit of (3.1) as $\ell \to \infty$, the domain of validity of the bifurcation equations shrinks to zero. The linear analysis of Daniels [2] implies that, when λ is increased to an amount $O(\ell^{-2})$, another two pairs of even and odd modes, distinct from those considered above, become marginally stable. These additional pairs may interact nonlinearly with the previously considered modes via secondary bifurcations, so we must restrict $\lambda \le O(\ell^{-2})$ and ξ, $\eta \le O(\ell^{-1})$ in (2.2) and (3.1). On the other hand, the results of Daniels [2] imply that γ, $\delta = O(\ell^{-3})$ in (2.2) and we can take $\sigma \le O(1)$ as $\ell \to \infty$, i.e., the bifurcation diagrams for (2.2) correspond to those for the original system (1.1)-(1.3) even for values of L not necessarily very near some L_ℓ.

To study the existence and stability of solutions of (1.1)-(1.3), we apply results of Golubitsky and Schaeffer [8, Chapter X] (see also [7]) to our bifurcation equations (2.2). Suppose in (2.2) that the following, generically satisfied, nondegeneracy conditions hold:

(3.4a) $\alpha\delta - \beta\gamma \neq 0,$

(3.4b) $\alpha \neq 0,\ \beta \neq 0,\ A \neq 0,\ D \neq 0,$

(3.4c) $A\beta - C\alpha \neq 0,\ B\beta - D\alpha \neq 0,\ AD - BC \neq 0.$

In fact, the results of Daniels [2] imply that (3.4a) holds at order ℓ^{-3}, and hence (at least) for all sufficiently large ℓ. Furthermore, (2.4) implies (3.4b) for all sufficiently large ℓ. However, we cannot use (2.4) to verify (3.4c) at order ℓ^{-1}. We assume that (3.4c), like (3.4a), is satisfied at some higher order. In this case, there is a transformation of the variables ξ,η,λ,σ that preserves the $(\mathbf{Z}_2\oplus\mathbf{Z}_2)$ -symmetry and local bifurcations and takes (2.2) into

(3.5)
$$0 = \lambda\xi - \xi^3 - m\xi\eta^2,$$
$$0 = (\lambda - \sigma)\eta - n\xi^2\eta - \eta^3,$$

where

$$m = \left|\frac{\beta}{D\alpha}\right|B, \quad n = \left|\frac{\alpha}{A\beta}\right|C.$$

In other words, (2.2) and (3.5) are both universal unfoldings of $(\mathbf{Z}_2\oplus\mathbf{Z}_2)$-equivalent bifurcation problems. Moreover, the stability of a bifurcating solution of (3.5) can be shown to be the same as the stability of the corresponding solution of (2.2). We note that in [8], the convention is that eigenvalues with positive real parts correspond to stability. Throughout this paper, however, eigenvalues with negative real parts corresponds to stability, and thus (3.5) corresponds to [8, equation (X.4.6)].

From (2.4) we obtain

(3.6a) $m = 1 + O(\ell^{-2}),\quad n = 1 + O(\ell^{-2}),$

while (3.4) implies

(3.6b) $m \neq 1,\ n \neq 1,\ mn \neq 1.$

Thus for large ℓ, the different bifurcations for (2.2) correspond to the regions in the (m,n)-plane labelled 1, 2, 2', 3, 3' and 4a in [8, p. 432, Figure X.4.1]. The corresponding bifurcation diagrams for $\sigma = 0$ are shown in [8, p. 433, Figure X.4.2], where the "x-mode" is our even mode ($\eta = 0$, $\xi \neq 0$), and the "y-mode" is our odd mode ($\xi = 0$, $\eta \neq 0$). Due to the symmetries (2.3), even modes occur in pairs $(\pm\xi,0)$, as do odd modes $(0,\pm\eta)$, while mixed modes occur in fours $(\pm\xi,\pm\eta)$.

The bifurcation diagrams for $\sigma \neq 0$ are given in [8, p. 434, Figure X.4.3]. Note that σ in (2.3) may correspond to either σ or $-\sigma$ in (3.5), depending on the sign of $\alpha\delta - \beta\gamma$. In all cases, there is (at least) an open right- or left-neighborhood of $\sigma = 0$ for which there exist mixed-mode solutions. Correspondingly, there exist right- or left-neighborhoods of each $L = L_\ell$, ℓ large, such that (1.1)-(1.3) have (nonlinear) convective solutions with corresponding stream functions having the form

$$\left\{ \frac{2\sqrt{2}}{3\pi(1+P)} [\xi \cos(\pi x/\sqrt{2}) - \eta \sin(\pi x/\sqrt{2})]\cos(\pi x/2L_\ell)+O(\ell^{-1})\right\} \cos \pi z$$
$$+ O(|\xi,\eta|^2+|\xi,\eta||\lambda,\sigma|),$$

with $\xi\neq0$, $\eta\neq0$. In the central region of the fluid layer, the mixed-mode convection pattern is roll-like, but is positioned laterally so that it is neither even nor odd about $x = 0$. To leading order, the rolls are "slowly" modulated by an envelope, or "amplitude" $A_0(X) \approx \cos(\pi X/2)$, $X = x/L_\ell$. As λ increases, the higher-order terms $O(|\xi,\eta|^2+|\xi,\eta||\lambda,\sigma|)$ contribute more significantly to the shape of $A_0(X)$. Indeed, the work of Daniels [3] implies that $A_0(X)$ develops boundary layers near $X = \pm1$.

Cases 2 and 3 [2' and 3'] imply that for values of L belonging to an open right- or left-neighborhood of L_ℓ (ℓ large), convection appears at onset as an even [odd] mode, but at slightly higher Rayleigh numbers (still in the range $R_0 + O(L^{-2})$) the even [odd] mode is unstable and there is a stable odd [even] mode. The transition from even to odd [odd to even] modes may be "hard" due to unstable mixed-mode solutions, or "soft" due to stable mixed mode solutions, depending on the sign of AD-BC. See Figure 1. In both cases the number of rolls in the convection pattern changes as the Rayleigh number increases. In this sense, such mixed-mode solutions would have a similar effect as the "phase-winding" solutions considered in [1],[4],[5] at $R = R_0 + O(L^{-1})$, but are not of the same form.

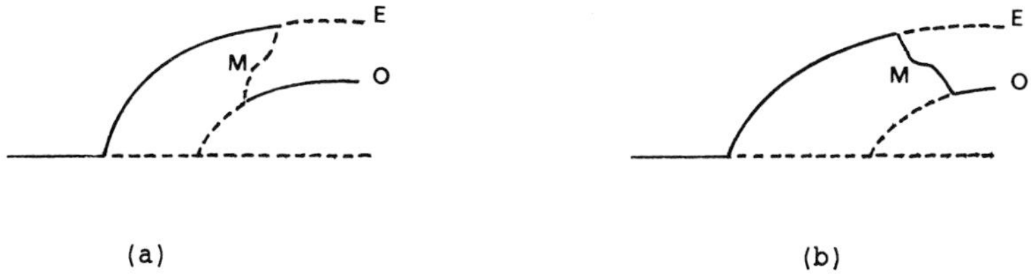

Figure 1: Bifurcation diagrams representing (a) hard, and (b) soft transitions from even (E) to odd (O) modes of convection, due to secondary bifurcations of mixed (M) modes. Solid lines represent stable solutions, and broken lines represent unstable solutions.

References

[1] M.C. Cross, P.G. Daniels, P.C. Hohenberg and E.D. Siggia, "Phase-winding solutions in a finite container above the convective threshold," J. Fluid Mech. 127, 155-183 (1983).

[2] P.G. Daniels, "Asymptotic sidewall effects in rotating Bénard convection," Z. Angew. Math. Phys. 28, 575-584 (1977).

[3] P.G. Daniels, "The effect of distant sidewalls on the transition to finite amplitude Bénard convection," Roc. Roy. Soc. London A358, 173-197 (1977).

[4] P.G. Daniels, "The effect of distant sidewalls on the evolution and stability of finite-amplitude Rayleigh-Bénard convection," Proc. Roy. Soc. London A378, 539-566 (1981).

[5] P.G. Daniels, "Roll-pattern evolution in finite-amplitude Rayleigh-Bénard convection in a two-dimensional fluid layer bounded by distant sidewalls," J. Fluid Mech. 143, 125-152 (1984).

[6] P.G. Drazin, "On the effects of sidewalls on Bénard convection," Z. Agnew. Math. Phys. 27, 239-245 (1975).

[7] M. Golubitsky and D. Schaeffer, "Imperfect bifurcation in the presence of symmetry," Commun. Math. Phys. 67, 205-232 (1979).

[8] M. Golubitsky and D.G. Schaeffer, Singularities and Groups in Bifurcation Theory, Volume 1, Springer-Verlag, New York, 1985.

[9] L.A. Segel, "Distant side-walls cause slow amplitude modulation of cellular convection," J. Fluid Mech. 38, 203-224 (1969).

[10] R. Temam, Navier-Stokes Equations, Theory and Numerical Analysis, North-Holland, Amsterdam, 1979.

Acknowledgment: This research was supported in part by NSERC, Canada.

This paper is in its final form and no similar paper has been or is being submitted elsewhere.

(received February 21, 1989.)

Applications of Degenerate Bifurcation Equations to the Taylor Problem and the water wave problem

HISASHI OKAMOTO

Department of Pure and Applied Sciences,
University of Tokyo, Meguro-ku, Tokyo 153, Japan

§1. Introduction. Our purpose in this paper is to explain usefulness of certain bifurcation equations. Those equations arise in many bifurcation problems. In this paper we consider applications to two bifurcation problems in fluid dynamics. One is the Taylor-Couette problem of the viscous fluid motion between two concentric cylinders. The other is Stokes' problem of progressive water waves of two dimensional irrotational flow. We consider two sets of algebraic equations. They are:

$$(1.1)\qquad \begin{cases} x(\epsilon\lambda + \alpha + ax^2 + bz^2 + cx^2 z) + (\beta + ez^2)xz = 0, \\ z(\delta\lambda + \hat{a}x^2 + \hat{b}z^2) \pm x^2 = 0, \end{cases}$$

and

$$(1.2)\qquad \begin{cases} x(\epsilon\lambda + \alpha + ax^2 + bz^2) \pm xz = 0, \\ z(\delta\lambda + \hat{a}x^2 + \hat{b}z^2 + \hat{c}x^2 z) + (\beta + dx^2)x^2 = 0, \end{cases}$$

where λ, x and z are real variables, $\epsilon, \delta, a, b, c, d, e, \hat{a}, \hat{b}$ and $\hat{c}$ are real constants. We explain in the subsequent sections how the solutions (λ, x, z) to (1.1,2) fit the bifurcation diagrams given in [**9,11**]. Tavener and Cliffe [**9**] numerically computed Taylor vortices of new type bifurcating from the Couette flow. Shōji [**11**] gave new bifurcation diagrams by plotting numerical capillary-gravity waves. Our objective is to prove that some of the numerical results in [**9,11**] are obtainable by (1.1,2).

The equations (1.1,2) are derived from a certain degeneration of the equations given in Fujii, Mimura and Nishiura [**1**]. The meaning of the degeneration, which we will assume later, is that certain coefficients of the mappings in (I-2) of [**1**] are very small. We interpret this as follows: we can describe the phenomena by some unfoldings of the degenerate mappings. The equation (1.1) are considered in Fujii, Nishiura and Hosono

[2], but (1.2) seems to be new. Although they consider in [1,2] a reaction diffusion system which has nothing to do with the Taylor-Couette problem or the water wave problem, the local structure of the bifurcation is of the same category. The reason is because the orthogonal group O(2) acts on all the problems in [1,2,9,11].

In this paper we announce a results in [6,7,12] in which we employed the singularity theoretic approach by Golubitsky, Stewart and Schaeffer ([3,4,5]) to see the structure of the set of the solutions in a systematic way. In §2 we state a precise formulation of the Taylor-Couette problem to be considered here. In §3 the relation between (1.1,2) and the Taylor-Couette problem is given. In §4 we compare the results in [9] with the zeros of (1.1,2). In §5 we formulate the water wave problem considered in [11]. §6 contains a discussion on the water wave problem.

§2. The Taylor-Couette problem. In this section we recall some newly developed analysis in [9] for the mechanism of vortex number exchange. The problem considered by them is to determine a fluid velocity field (u, v, w) and the pressure p which satisfy the following stationary Navier-Stokes equation (2.1-4):

$$\Delta u - \frac{u}{r^2} - R\left[u\frac{\partial u}{\partial r} + w\frac{\partial u}{\partial z} - \frac{v^2}{r} + \frac{\partial p}{\partial r}\right] = 0, \tag{2.1}$$

$$\Delta v - \frac{v}{r^2} - R\left[u\frac{\partial v}{\partial r} + w\frac{\partial v}{\partial z} + \frac{uv}{r}\right] = 0, \tag{2.2}$$

$$\Delta w - R\left[u\frac{\partial w}{\partial r} + w\frac{\partial w}{\partial z} + \frac{\partial p}{\partial z}\right] = 0, \tag{2.3}$$

$$\frac{1}{r}\frac{\partial}{\partial r}(ru) + \frac{\partial w}{\partial z} = 0, \tag{2.4}$$

where the cylindrical coordinates are adopted, u, v, w are r, θ, z-component, respectively and R is the Reynolds number. In this paper, as in [9], we consider only velocity fields which are independent of the azimuthal coordinate θ and the time t. We have suitably nondimensionalized the variables so that (2.1-4) are satisfied in

$$\{(r, z); \quad \eta < r < 1, \quad 0 < z < \Gamma\}.$$

Here, η and Γ are positive constants called a radius ratio and an aspect ratio, respectively. The following boundary conditions are imposed :

$$(u, v, w) = (0, 1, 0) \qquad (r = \eta) \tag{2.5}$$

$$(u, v, w) = (0, 0, 0) \qquad (r = 1) \tag{2.6}$$

$$\left(\frac{\partial u}{\partial z}, \frac{\partial v}{\partial z}, w\right) = (0, 0, 0) \qquad (z = 0, \Gamma) \tag{2.7}$$

We can easily prove that all the requirements are satisfied by the Couette flow $(u, v, w, p) = (0, v_0(r), 0, p_0(r))$ with

$$v_0(r) = Ar + \frac{B}{r}, \qquad p_0(r) = \int_\eta^r \frac{v_0(s)^2}{s} ds.$$

if $A = -\eta/(1-\eta^2)$ and $B = \eta/(1-\eta^2)$ hold true. The problem is to study the solutions bifurcating from this Couette flow. Notice that the condition (2.7) makes it difficult to carry on a laboratory experiment but there is no difficulty in performing a computer simulation. In fact, [**9**] presents a penetrating description of the stability exchange of the solutions. Let us take a brief survey of the results in [**9**]. They fix the parameter $\eta = 0.615$ and let the aspect ratio Γ vary. It is known that the primary bifurcation branch consists of the two-cell Taylor vortices for small value of Γ ($\Gamma \approx 2(1-\eta)$) and that it consists of the four-cell Taylor vortices for larger value of Γ ($\Gamma \approx 4(1-\eta)$) ([**9**]). The number of the vortices, however, is integer and Γ can vary continuously. Therefore, there exist flows of " mixed type " when Γ is in a intermediate range. They appear near the double critical point at which the curve of the critical Reynolds number for two cell flow intersects with that for four xell flow. The bifurcation diagrams in [**9**] describe qualitatively how this exchange from two-cell to four cell occurs.

Although they consider the exchange mechanism between two-cell and six-cell or four-cell and six-cell, we consider only the transition from two to four, which requires the least mathematical technique to theorize. No normal form seems to be available for the remaining two cases in [**9**].

§3. **Degenerate bifurcation equations.** In this section we show how (1.1,2) are derived. Our starting point is to observe a hidden symmetry in (2.1-7). Let us consider another problem to seek (u^*, v^*, w^*, p^*) which satisfies (2.1-4) in $\eta < r < 1$, $-\Gamma < z < \Gamma$, the boundary condition (2.5,6) and the periodic boundary condition on $z = -\Gamma, \Gamma$. (Note that the height of the cylinder is double.) We call this problem $[2\Gamma]$ and the original problem $[\Gamma]$. This new problem $[2\Gamma]$ has an advantage that it is O(2)-equivariant in the following sense: Let us define an action of the orthogonal group O(2) by

$$(3.2) \quad \gamma(u(r,z), v(r,z), w(r,z), p(r,z)) = (u(r,z+\alpha), v(r,z+\alpha), w(r,z+\alpha), p(r,z+\alpha))$$

if $\gamma \in O(2)$ is a rotation with angle α,

$$(3.3) \qquad \gamma(u(r,z), v(r,z), w(r,z), p(r,z)) = (u(r,-z), v(r,-z), -w(r,-z), p(r,-z))$$

if $\gamma \in O(2)$ is a reflection. Then we can easily verify that the problem $[2\Gamma]$ is O(2)-equivariant in the sense of [**3**]. Since $[\Gamma]$ is equivariant with respect to only a discrete subgroup of $O(2)$, the use of $[2\Gamma]$ is advantageous. The relation between $[2\Gamma]$ and $[\Gamma]$ is given by

PROPOSITION 3.1. *If (u^*, v^*, w^*, p^*) is a solution to $[2\Gamma]$ and if (u^*, v^*, w^*, p^*) is invariant with respect to (3.3), then it satisfies $[\Gamma]$ in $0 < z < \Gamma$. Conversely, if (u, v, w, p) satisfies $[\Gamma]$ and if we extend it to (u^*, v^*, w^*, p^*) in such a way that u^*, v^*, p^* are even extension of u, v, p, respectively, and w^* is an odd extension of w, then (u^*, v^*, w^*, p^*) is a solution to $[2\Gamma]$.*

In short, finding solutions to $[\Gamma]$ is equivalent to finding " symmetric " solutions to $[2\Gamma]$. From now on we consider $[2\Gamma]$. In the remaining part of this section, we explain how the analysis in [**6**] is applied to $[2\Gamma]$.

For a fixed (Γ, η) we consider the linearized problem of $[2\Gamma]$. This is a linear eigenvalue problem with a parameter R. For each positive integer n, this eigenvalue problem has a nontrivial solution of the following form

$$g_n(r,z) = \Big(U_n(r)\cos(n\psi), V_n(r)\cos(n\psi), W_n(r)\sin(n\psi), P_n(r)\cos(n\psi)\Big),$$

$$h_n(r,z) = \Big(U_n(r)\sin(n\psi), V_n(r)\sin(n\psi), W_n(r)\cos(n\psi), P_n(r)\sin(n\psi)\Big),$$

where $\psi = 2\pi z/\Gamma$. In these expressions $U_n(r), V_n(r), W_n(r), P_n(r)$ are functions of r only and are determined through a certain ordinary differential equation (see [**7**]). Let $R_n^*(\Gamma, \eta)$ denote the critical Reynolds number associated with the above (g_n, h_n)

There is a numerical evidence of the existence of the following function Γ_0 : for an arbitrarily fixed η, there is a $\Gamma_0(\eta)$ such that the linearlized operator of (2.1-4) with (2.5,6) and the periodic boundary condition has a four dimensional null space spanned by g_2, h_2, g_4 and h_4. This shows that in the (Γ, η) plane there is a 1-dimensional variety where the critical Reynolds number of 2-cell flow and that of 4-cell flow coincide. Let R^* denote this critical Reynolds number at $(\Gamma, \eta) = (\Gamma_0(\eta), \eta)$. We assume the existence of Γ_0.

We follow the Lyapunov-Schmidt procedure at $\Gamma = \Gamma_0(\eta)$. To this end, we rewrite $[2\Gamma]$ as an abstract operator equation in some Banach space. Several way of rewriting is possible. We can find one in Temam [**10**], which we used in [**7**]. But the following analysis does not depend on the specific form of the operator equation. Let $F = F(R, \Gamma, \eta; u, v, w, p)$ be a nonlinear functional for $[2\Gamma]$ realized in some Banach space, and let P be a projection onto the 4-dimensional space spanned by g_2, h_2, g_4, h_4. Then the bifurcation of the Taylor vortices is governed in a neighborhood of $(R^*, \Gamma_0(\eta), \eta; 0, v_0, 0, p_0)$ by

$$G(R, \Gamma, \eta; x, y, z, w) = PF(R, \Gamma, \eta; (0, v_0, 0, p_0) + xg_1 + yg_2 + zg_3 + wg_4 + \phi),$$

where ϕ is in a complement of the range of P. ϕ is a function of x, y, z and w, determined by an implicit function theorem. Since O(2)-equivariance of the original equation is inherited to this bifurcation equation, G must be of a special form given in [**1,6**].

In order to explain the form in [**1**,**6**], we identify real 4-dimensional space of (x, y, z, w) with $\mathbb{C}^2$ by $\xi = x + iy$ and $\zeta = z + iw$. Then G must be of the following form: $G = (G_1, G_2)$,

$$G_1 = f_1\xi + f_2\overline{\xi}\zeta, \tag{3.4}$$

$$G_2 = f_3\zeta + f_4\xi^2, \tag{3.5}$$

where $f_j (j = 1, 2, 3, 4)$ are functions of $R, \Gamma, \eta, |\xi|^2, |\zeta|^2$ and $\mathrm{Re}(\xi^2\overline{\zeta})$ only. For the proof, see [**1**] (see also [**5**,**6**]). Since G is a bifurcation equation, it holds that $f_1(\sharp; 0, 0, 0) = f_3(\sharp; 0, 0, 0) = 0$, where $\sharp = (R^*, \Gamma_0(\eta), \eta)$. Under this conditions, the most general case is

$$\text{(A)} \qquad f_2(\sharp; 0, 0, 0) \neq 0, \qquad f_4(\sharp; 0, 0, 0) \neq 0$$

which is considered in [**1**]. The degeneration which we mentioned at the beginning of the present paper is

$$\text{(B)} \qquad f_2(\sharp; 0, 0, 0) = 0, \qquad f_4(\sharp; 0, 0, 0) \neq 0,$$

and

$$\text{(C)} \qquad f_2(\sharp; 0, 0, 0) \neq 0, \qquad f_4(\sharp; 0, 0, 0) = 0.$$

The case (B) is considered in [**2**] but (C) seems to be new.

We regard R as a bifurcation parameter and Γ, η as splitting parameters. Thus there is a good possibility that the following assumption holds true:

*In the (Γ, η) plane there is a 1-dimensional variety where the critical Reynolds number of 2-cell flow coincide with that of 4-cell flow. On this variety there are points at which (B) or (C) holds. If we assume that these points exist and are not far from the values chosen in [**9**], then it is natural that the figures in [**9**] are captured by some perturbation of (3.4,5) in the case of (B) or (C).*

Since (3.4,5) is applicable to a number of problems, we think it is useful to study (3.4,5) systematically. To this end, the machineries by Golubitsky and Schaeffer [**3**,**4**] and Golubitsky, Stewart and Schaeffer [**5**] are easy to handle for application-oriented mathematicians. We analyzed (3.4,5) with (A), (B) or (C) in [**6**] via the method of [**3**,**4**,**5**]. There we obtained normal forms and universal unfoldings. Below we summarize the results in [**6**]. In the case of (A), the bifurcation equation is, when slightly perturbed, O(2)-equivalent to

$$\begin{cases} \xi(\epsilon\lambda + \alpha + b|\zeta|^2) \pm \overline{\xi}\zeta = 0, \\ \zeta(\delta\lambda + \hat{b}|\zeta|^2) \pm \xi^2 = 0, \end{cases} \tag{3.6}$$

where $\lambda = R - R^*$ and α is a perturbation (unfolding) parameter. Here we fix Γ and η. Note that we only consider an O(2)-equivariant perturbation. By the O(2)-equivalence we mean that the bifurcation equation is transformed to one of this form by a suitable coordinates change which preserves the $O(2)$-equivariance. Roughly, we can say that (3.6) is a universal unfolding in the case of (A). Of course, a certain generic condition is necessary for the O(2)-equivalence. This is given in [**6**]. In the case of (B),

$$(3.7) \qquad \begin{cases} \xi[\epsilon\lambda + \alpha + a'|\xi|^2 + b'|\zeta|^2 + c'\mathrm{Re}(\xi^2\overline{\zeta})] + (\beta + e'|\zeta|^2)\overline{\xi}\zeta = 0, \\ \zeta(\delta\lambda + \hat{a}|\xi|^2 + \hat{b}|\zeta|^2) \pm \xi^2 = 0, \end{cases}$$

is a universal unfolding. Here $\alpha, \beta, |a - a'|, |b - b'|, |c - c'|$ and $|e - e'|$ are small. In the case of (C),

$$(3.8) \qquad \begin{cases} \xi(\epsilon\lambda + \alpha + a|\xi|^2 + b|\zeta|^2) \pm \overline{\xi}\zeta = 0, \\ \zeta[\delta\lambda + \hat{a}'|\xi|^2 + \hat{b}'|\zeta|^2 + \hat{c}'\mathrm{Re}(\xi^2\overline{\zeta})] + (\beta + d'|\xi|^2)\xi^2 = 0, \end{cases}$$

is a universal unfolding. When we consider the problem $[\Gamma]$, the solutions are invariant with respect to the reflection (3.3). Therefore we obtain the bifurcation equations for $[\Gamma]$ by restricting complex variables ξ, ζ to real ones x, z. This restriction reduces (3.7,8) to (1.1,2), respectively.

§4. **Discussion on the Taylor-Couette problem.** We now show how the phenomena in [**9**] is explained by (1.1,2). We choose α and β suitably and draw the set of zeros of (1.1). Fig. 1-5 show the set of zeros of (1.1), which we drew by a computer and an X-Y plotter. The horizontal axis represents the Couette flow. The parabolic curves represent four cell flows. Others correspond to flows of mixed nature, though the solutions on the loop near the Couette flow are almost two cell flow. The difference between Fig. 2 and Fig 3 is that the loop in Fig. 3 is bent and there are two turning points on the loop, while no turning point is present in Fig. 2 . In Fig. 4 two transcritical branches intersect with that of the four cell flow. These explain the bifurcation diagrams in [**9**] for large Γ (Fig. 4.2 (d,f,g,h,i)). In Fig. 1-5, we have chosen a $\beta > 0$. α changes its sign from $-$ to $+$ as we change the figures from 1 to 5. For the details about the choice of the coefficients, see [**7**]. From (1.2) we obtain Fig. 6, which explains diagrams for smaller Γ (Fig. 4.2c of [**9**]). Although there is a pitchfork bifurcation in Fig. 4.2 (a,b,c) of [**9**], this cannot be explained by (1.2). This, however, may be a consequence of more global bifurcation. Except for this, our pictures fit the diagrams in [**9**] quite well. For the complete set of the bifurcation diagrams of (1.1,2), see [**7**].

§5. Bifurcation of water waves. In this section we would like to introduce the result of Shōji [**11**]. The problem considered in [**11**] is the water waves of two dimensional irrotational flow of inviscid incompressible fluid. The problem is to find a function $\theta = \theta(\sigma)$ which is 2π-periodic in σ and satisfies the following

$$(5.1) \qquad e^{2K\theta}\frac{dK\theta}{d\sigma} - pe^{-K\theta}\sin\theta + q\frac{d}{d\sigma}\left(\frac{d\theta}{d\sigma}e^{K\theta}\right) = 0,$$

where K is an operator defined by:

$$K\left(\sum_{n=1}^{\infty}\left(a_n\sin(n\sigma) + b_n\cos(n\sigma)\right)\right) = \sum_{n=1}^{\infty}\frac{1+r^n}{1-r^n}\left(-a_n\sin(n\sigma) + b_n\cos(n\sigma)\right).$$

The equation (5.1) is originally due to Nekrasov and Levi-Civita. For the detail of derivation of (5.1), see [**8**]. We observe that there are three parameters p, q and r. They satisfy $0 \le p < \infty, 0 \le q < \infty, 0 \le r < 1$. We may call p, q, r, nondimensional gravity, nondimensional surface tension and aspect ratio, respectively. Whatever these parameters may be, $\theta \equiv 0$ is a solution to (5.1). The problem is to see the structure of the bifurcation from this trivial solution.

We consider the following action of O(2) on θ:

$$\theta(\sigma) \mapsto \theta(\sigma + \alpha)$$

for a rotation of angle α, and

$$\theta(\sigma) \mapsto -\theta(-\sigma)$$

for the reflection. In [**8**] we give an abstract operator equation in some Banach space so that the operator is equivariant with respect to the above action of O(2). Accordingly we can apply the method and the bifurcation equation given in the preceding sections.

The linearized equation of (5.1) is:

$$(5.2) \qquad \frac{dK\theta}{d\sigma} - p\theta + q\frac{d^2\theta}{d\sigma^2} = 0.$$

In what follows, we assume that $r = 0$, which implies that the depth of the fluid is infinite. In this case, (5.2) has a nontrivial solution if and only if p and q satisfy $1 = p/n + nq$ for some positive integer n. If this is the case, the eigenspace is spanned by $\sin(n\sigma)$ and $\cos(n\sigma)$. At $(p, q) = (2/3, 1/3)$, the equation $1 = p/n + nq$ is satisfied both for $n = 1$ and for $n = 2$. Therefore the eigenspace is spanned by $\sin(\sigma), \cos(\sigma), \sin(2\sigma), \cos(2\sigma)$ at $(p, q) = (2/3, 1/3)$. Consequently we are in the same situation as in §3.

The formulation of Shōji [**11**] is different from the above one but the existence of the solutions can be seen more easily in the present formulation. We only consider in this paper the bifurcation near $(p, q) = (2/3, 1/3)$.

§6. Discussion on the water wave problem. Fig. 7 through Fig 10 are borrowed from [**11**]. The difference between Fig. 7 and Fig. 8 is that the secondary bifurcation branch of Fig. 8 has two turning points (limit points) while that of Fig. 7 does not. In Fig. 9, the secondary bifurcation point coalesces the pitchfork bifurcation point and makes two transcritical branches. In Fig. 10 the loop intersects with the horizontal axis, while it intersects with the parabolic curve in Fig. 7-8. In these figures the horizontal axis represents the trivial solution: $\theta \equiv 0$. The bifurcation parameter is $1/p$. Each figure is drawn for different q. The dotted curve indicates that the solution is unphysical (see [**11**]). In Fig. 11 through 14 we depicted diagrams obtained by (1.1). Here we changed α from $-$ to $+$. The values of other coefficients differ from those in §4. Fig. 11-14 fit Fig. 7-10, respectively, very well. In [**11**], she obtained a lot of other new bifurcation diagrams. The relation between her diagrams and abstract bifurcation equations will be studied in more detail in a forthcoming paper [**12**].

REFERENCES

1. H. Fujii, M. Mimura and Y. Nishiura, *A picture of the global bifurcation diagram in ecological interacting and diffusing systems*, Physica 5D (1982), 1–42.
2. H. Fujii, Y. Nishiura and Y. Hosono, *On the structure of multiple existence of stable stationary solutions in systems of reaction-diffusion equations*, Patterns and Waves- Qualitative analysis of nonlinear differential equations, eds. T. Nishida, M. Mimura and H. Fujii , North-Holland (1986), 157–219.
3. M. Golubitsky and D. Schaeffer, *Imperfect bifurcation in the presence of symmetry*, Comm. Math. Phys. **67** (1979), 205–232.
4. M. Golubitsky and D. Schaeffer, "Singularities and Groups in bifurcation theory, vol 1," Springer Verlag, New York, 1985.
5. M. Golubitsky, I. Stewart and D. Schaeffer, "Singularities and Groups in bifurcation theory, vol 2," Springer Verlag, New York, 1988.
6. H. Okamoto, *O(2)-equivariant bifurcation equations with mode (1,2)*, Sci. Papers College of Arts and Sci., Univ. of Tokyo **39** (1989), 1–43.
7. H. Okamoto and S.J. Tavener, *Degenerate O(2)-equivariant bifurcation equations and their application to the Taylor problem*, preprint.
8. H. Okamoto, *On the problem of water waves of permanent configuration*, Nonlinear Anal. Theory and Appl. (in press).
9. S.J. Tavener and K.A. Cliffe, *Primary flow exchange mechanisms in the Taylor apparatus applying impermeable stress-free boundary conditions.*
10. R. Temam, "Navier-Stokes Equations," North-Holland, Amsterdam, New York, Oxford, 1984.
11. M. Shōji, *New bifurcation diagrams in the problem of permanent progressive waves*, J. Fac. Sci., Univ. Tokyo, Sec. IA **36** (1989), 571–613.
12. H. Okamoto and M. Shōji, *Normal forms of the bifurcation equations in the problem of capillary-gravity waves*, preprint.

This paper is in its final form and no similar paper is being or has been submitted elsewhere

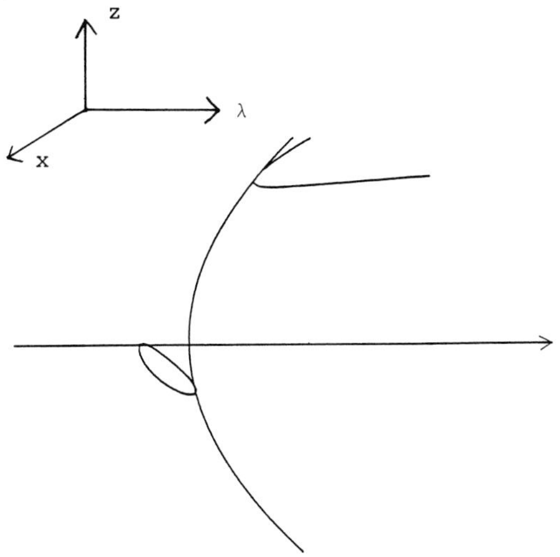

Fig. 1

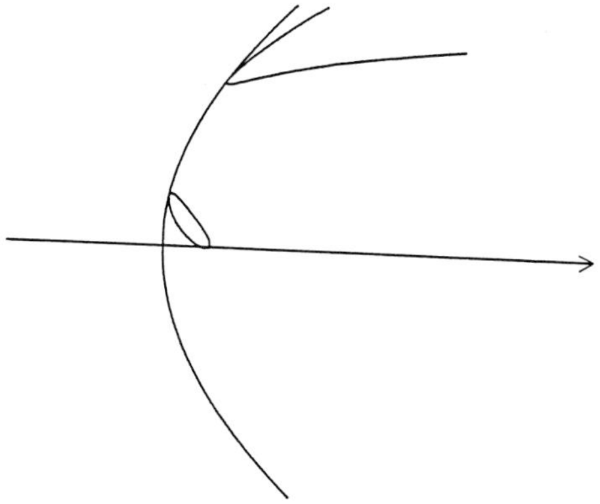

Fig. 2

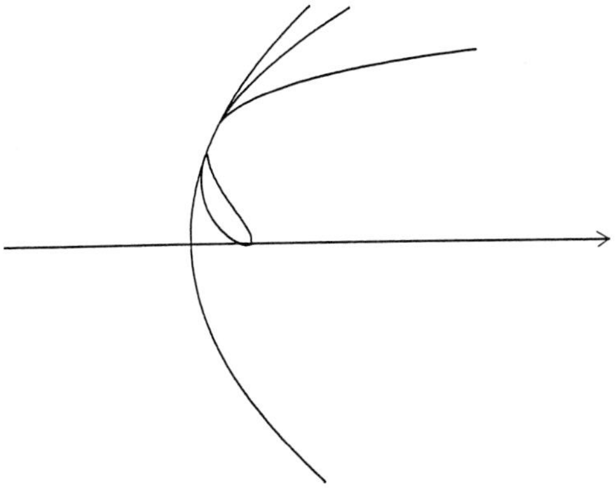

Fig. 3

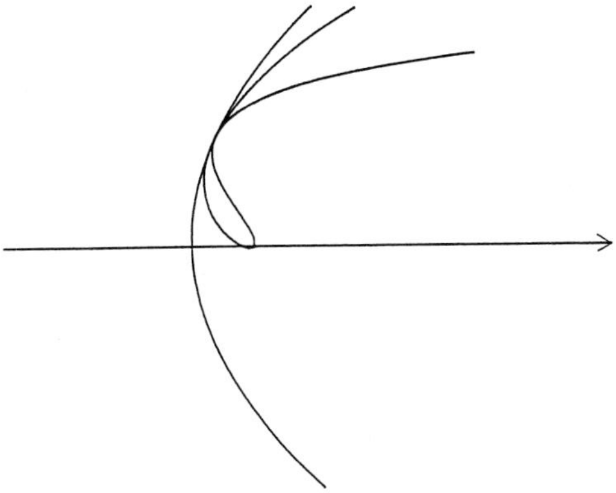

Fig. 4

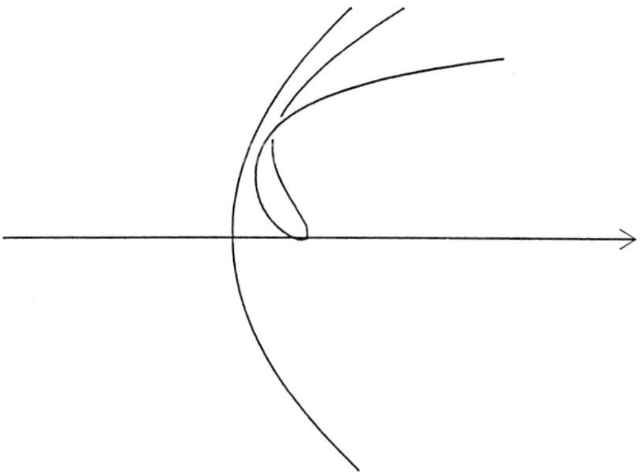

Fig. 5

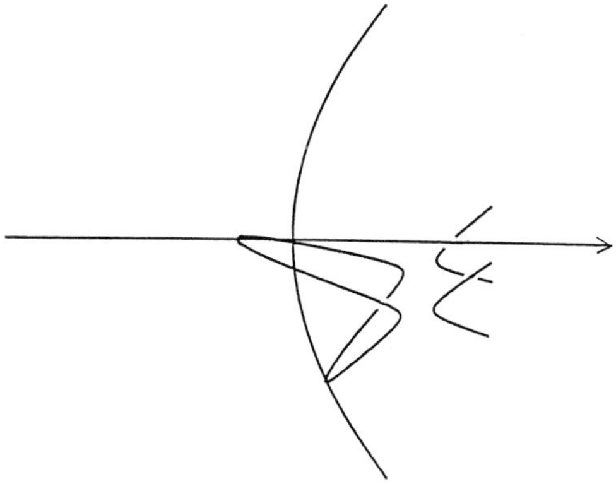

Fig. 6

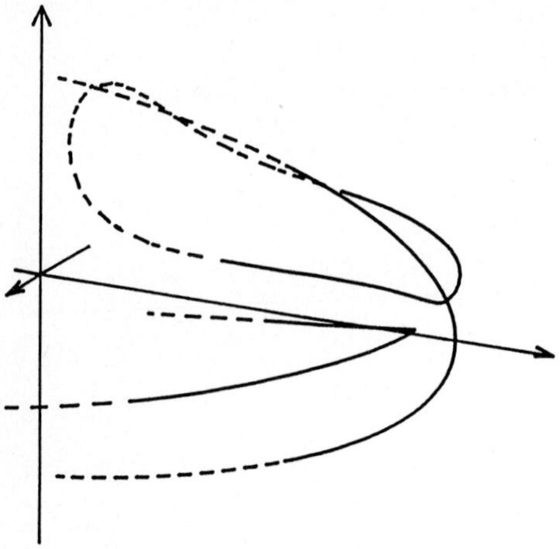

Fig. 7

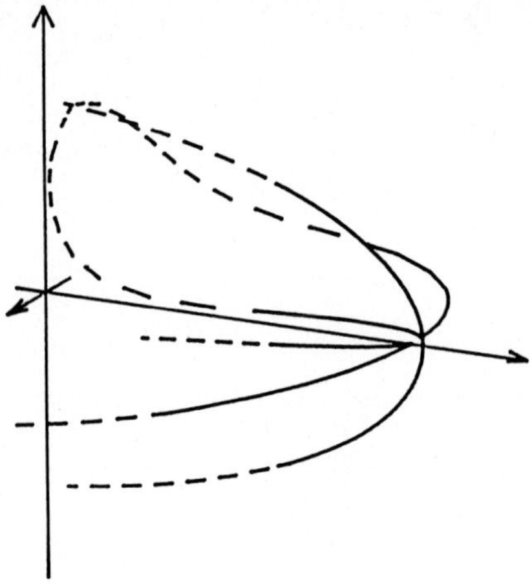

Fig. 8

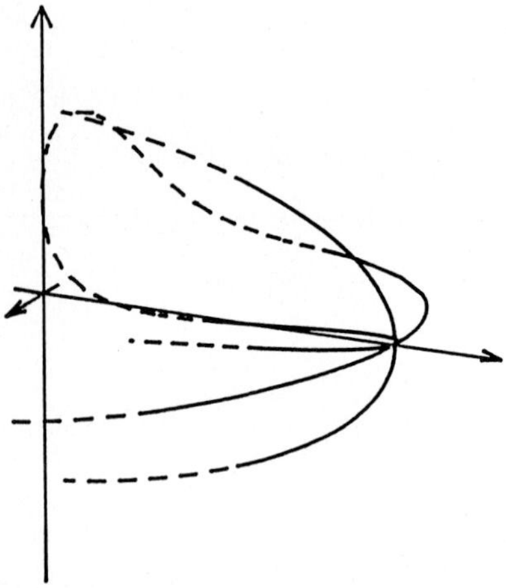

Fig. 9

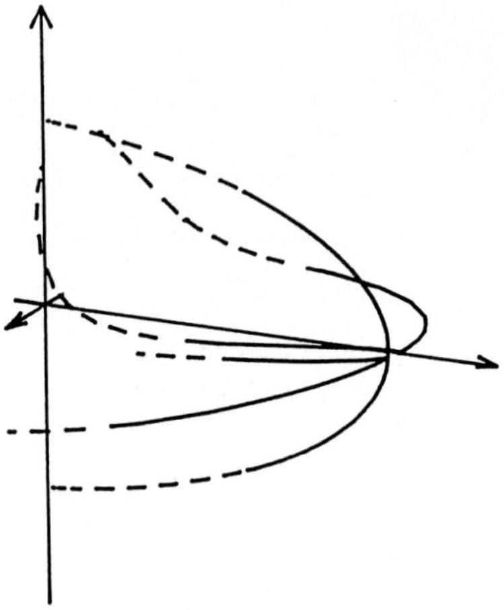

Fig. 10

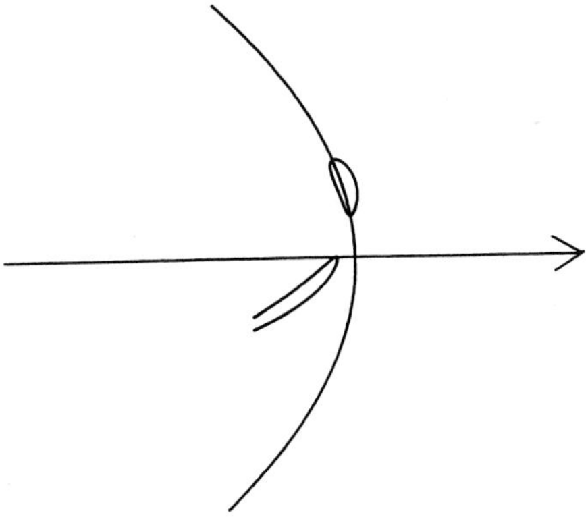

Fig. 11

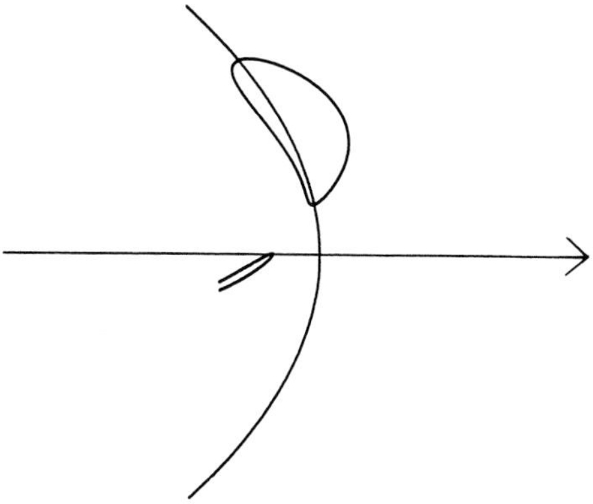

Fig. 12

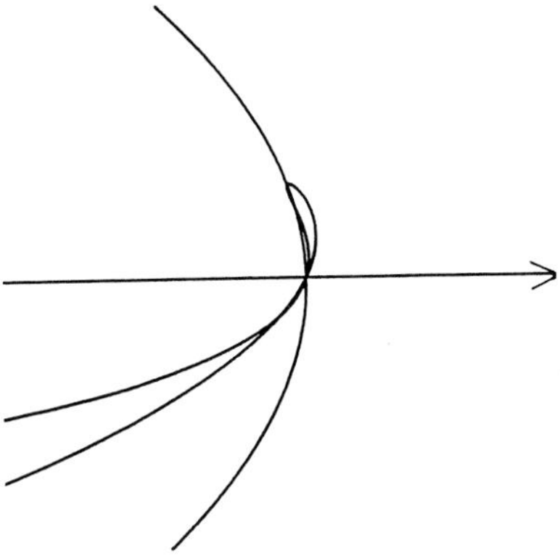

Fig. 13

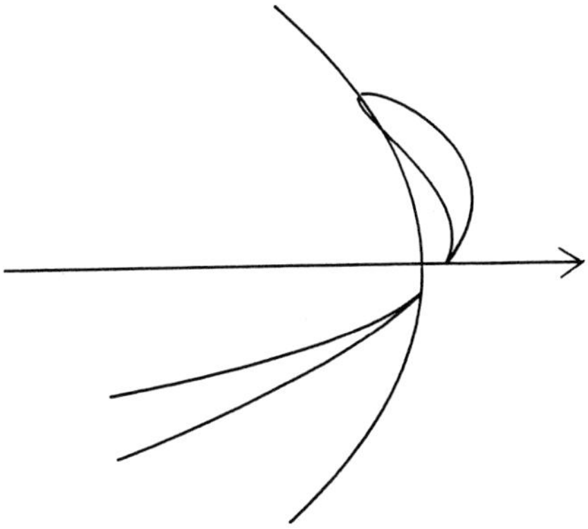

Fig. 14

A UNIQUENESS CRITERION FOR THE SOLUTION OF THE STATIONARY NAVIER-STOKES EQUATIONS

Giovanni Prouse
Dipartimento di Matematica, Politecnico di Milano
Piazza Leonardo da Vinci 32, 20133 Milano

1 - Introduction. Consider, in an open bounded 2- or 3-dimensional set Ω, the homogeneous Dirichlet problem for the stationary Navier-Stokes equations associated to the motion of a viscous, incompressible fluid. Denoting by $\mathcal{N}$ the set of divergence-free vectors, indefinitely differentiable and with compact support in Ω and by N^1 the closure of in $H^1(\Omega)$, such a problem can be formulated as follows:

Given $\vec{f} \in L^2$, find $\vec{u} \in N^1$ such that

$$\nu(\vec{u},\vec{\varphi})_{H_0^1} + (\vec{u}\cdot\nabla)\vec{u},\vec{\varphi})_{L^2} = (\vec{f},\vec{\varphi})_{L^2} \tag{1.1}$$

$\forall\ \vec{\varphi} \in N^1$, $\vec{f}$ representing the external force and ν the viscosity of the fluid, whose density is assumed to be equal to 1.

It is well known (see for instance, [1] , [2]) that this problem admits a solution, in the sense outlined above, but the solution may be not unique.

The aim of the present paper is to give a criterion by which it is possible to single out, among all the solutions of (1.1), a "privileged" one, which has some "special" characteristics.

One of the simplest and most natural physically meaningful conditions which could be imposed is that the functional $L_0(\vec{u}) = (\vec{f},\vec{u})_{L^2}$, which represents the mechanical power of the flow described by $\vec{u}$, be maximum; this condition alone is however not sufficient to identify a single solution. It will, however, be shown that, adding to this an appropriate "stability" condition, a uniqueness property is achieved.

More precisely, denoting by U the set of solutions of (1.1) corresponding to $\vec{f}$ and setting

$$L_\vartheta(\vec{u}) = (\vec{f}, (-\Delta)^\vartheta \vec{u})_{L^2} \qquad (0 \leq \vartheta \leq \frac{1}{2}), \tag{1.2}$$

we shall say that a <u>solution</u> $\tilde{\vec{u}}$ <u>is maximal</u> θ-<u>stable</u> if, $\forall\ \vec{u} \in U$,

$$L_o(\tilde{\vec{u}}) \geq L_o(\vec{u}) \tag{1.3}$$

<u>and there exists a neighbourhood</u> $(0, \vartheta_u)$ <u>of</u> $\vartheta = 0$ <u>such that</u>

$$L_\vartheta(\tilde{\vec{u}}) > L_\vartheta(\vec{u}) \quad \forall\ \vartheta \in (0, \vartheta_u). \tag{1.4}$$

The uniqueness result we shall prove can be thus stated.
<u>If</u> $\vec{f}$ <u>is a generic vector</u> $\in L^2$, <u>there exists one, and only one solution</u> $\tilde{\vec{u}}$ <u>of</u> (1.1) <u>which is maximal</u> ϑ-<u>stable</u>.
By generic $\vec{f}$ we mean the following: let $\{\vec{g}_j\}$, $\{\lambda_j\}$ denote the eigenfunctions and eigenvalues of the operator $-\Delta$ from H_o^1 to H^{-1} $((\vec{g}_j, \vec{g}_k)_{L^2} = \delta_{jk},\ 0<\lambda_1\leq\lambda_2\leq \dots \leq\lambda_n \leq \dots,\ \lim_{n\to\infty} \lambda_n = \infty)$. $\vec{f}$ is generic if $(\vec{f}, \vec{g}_j)_{L^2} \neq 0,\ \forall\ j$.
With the notations just introduced, bearing in mind that $\{\vec{g}_j\}$ is a basis in L^2 and H_o^1, we can set

$$\begin{aligned} \vec{u} &= \sum_{j=1}^{\infty} \alpha_j \vec{g}_j & \alpha_j &= (\vec{u}, \vec{g}_j)_{L^2}, \\ \vec{f} &= \sum_{j=1}^{\infty} \varphi_j \vec{g}_j & \varphi_j &= (\vec{f}, \vec{g}_j)_{L^2} \end{aligned} \tag{1.5}$$

Hence, by (1.2),

$$L_\vartheta(\vec{u}) = \sum_{j=1}^{\infty} \lambda_j^\vartheta \alpha_j \varphi_j . \tag{1.6}$$

<u>Remark 1.</u> The condition imposed on $\vec{f}$ is obviously necessary. In fact if, for instance, it were $\varphi_p=0$, the p-th term in the expansions (1.6) would drop out, leaving the coefficient α_p undetermined.

<u>Remark 2</u>. Assume that $\vec{f} \in H^{\bar{\vartheta}}$ $(\bar{\vartheta} > 0)$; by (1.2) we have then, $\forall\ \vartheta \in [0, \bar{\vartheta}]$,

$$L_\vartheta(\vec{u}) = ((-\Delta)^\vartheta \vec{f}, \vec{u})_{L^2} . \tag{1.7}$$

The condition that $\vec{u}$ is maximal ϑ-stable can then also be interpreted in the following way:

a) The mechanical power of the flow is maximum (condition (1.3));

b) This maximum is stable with respect to the set $\vec{f}_\vartheta = (-\Delta)^{\vartheta}\vec{f}$ of functions approximating $\vec{f}$.

2 - Some auxiliary lemmas. We now state some auxiliary lemmas which are essential for the proof of the uniqueness theorem. Complete proofs of these lemmas are given in [3] .

Lemma 1 : Let U denote the set of solutions corresponding to $\vec{f}$ (in the sense defined at § 1). U is compact in H_o^s, $\forall$ s < 1.

Lemma 2 : The function $\vartheta \to L_\vartheta(\vec{u})$ can be expanded in the power series

$$L_\vartheta(\vec{u}) = \sum_{j=1}^{\infty} \lambda_j^{\vartheta}\alpha_j\varphi_j = \sum_{k=0}^{\infty} \frac{\vartheta^k}{k!} \sum_{j=1}^{\infty} (\log\lambda_j)^k \alpha_j\varphi_j \tag{2.1}$$

in the neighbourhood $|\vartheta| < \frac{1}{2}$ of $\vartheta = 0$.

Lemma 3 : The functionals

$$\vec{u} \to \sum_{j=1}^{\infty} (\log\lambda_j)^k \alpha_j\varphi_j = \left.\frac{\partial^k L_\vartheta(\vec{u})}{\partial\vartheta^k}\right|_{\vartheta=0} \quad (\vec{u} \in U,\ k=0,1,\dots) \tag{2.2}$$

are H_o^s - continuous, $\forall$ s $\in [0,1)$.

Lemma 4 : Assume that $(\vec{f},\vec{g}_j)_{L^2} = \varphi_j \neq 0$ $\forall j$; the set

$$\vec{y}_k = \sum_{j=1}^{\infty} (\log\lambda_j)^k \varphi_j\vec{g}_j \tag{2.3}$$

is then a basis in $H^{-\varepsilon}$, $\forall$ $\varepsilon > 0$.

3 - Proof of the uniqueness criterion. We can now prove the uniqueness theorem stated in section 1.

Consider, at first, the functional $\vec{u} = L_o(\vec{u}) = (\vec{f},\vec{u})_{L^2}$, $\vec{f} \in L^2$, $\vec{u} \in U$; this functional is obviously H_o^s-continuous, $\forall$ s $\in [0,1]$. Since, by lemma 1, the set U of solutions is H_o^s- compact (s<1), there exists

$\vec{u}_1 \in U$ such that

$$L_o(\vec{u}_1) \geq L_o(\vec{u}) \qquad \forall\, \vec{u} \in U\,. \tag{3.1}$$

Setting

$$\sigma_1 = L_o(\vec{u}_1)\ , \tag{3.2}$$

we denote by U_1 the subset of U constituted by the solutions $\vec{u}$ such that

$$L_o(\vec{u}) = \sigma_1. \tag{3.3}$$

The set U_1 is obviously $\neq \emptyset$ and is H_o^s-compact, $\forall\, s < 1$. In fact, by lemma 1, if $\{\vec{u}_n\}$ is a sequence $\in U_1$, then $\vec{u}_n, \to \vec{u}$ weakly in H_o^1, with $\vec{u} \in U$; it is obvious, moreover, that $L_o(\vec{u}_n) = (\vec{f},\vec{u}_n)_{L^2}\ (\vec{f},\vec{u})_{L^2} = L_o(\vec{u}) = \sigma_1$. Consider now the functional

$$\vec{u} \to \left.\frac{\partial L_\vartheta(\vec{u})}{\partial\vartheta}\right|_{\vartheta=0} = \sum_{j=1}^{\infty} (\log\lambda_j)\,\alpha_j\varphi_j. \tag{3.4}$$

We have, obviously,

$$\begin{aligned} \Big|\sum_{j=1}^{\infty} (\log\lambda_j)\varphi_j\alpha_j\Big| &\leq \sum_{j=1}^{\infty} (M_\varepsilon + \lambda_j^\varepsilon)\,|\varphi_j|\,|\alpha_j| \leq M_\varepsilon \|\vec{f}\|_{L^2}\|\vec{u}\|_{L^2} + \\ &+ \|\vec{f}\|_{L^2}\|\vec{u}\|_{H_o^{2\varepsilon}} \leq C_\varepsilon \|\vec{f}\|_{L^2}\|\vec{u}\|_{H_o^{2\varepsilon}}\ , \end{aligned} \tag{3.5}$$

and, consequently, $\left.\frac{\partial L_\vartheta(\vec{u})}{\partial\vartheta}\right|_{\vartheta=0}$ is H_o^s-continuous $\forall\, s \in [0,1]$ and $\forall\, \vec{u} \in U$.

Since U_1 is H_o^s-compact $\forall\, s < 1$, there exists $\vec{u}_2 \in U_1$ such that

$$\sigma_2 = \left.\frac{\partial L_\vartheta(\vec{u}_2)}{\partial\vartheta}\right|_{\vartheta=0} \geq \left.\frac{\partial L_\vartheta(\vec{u})}{\partial\vartheta}\right|_{\vartheta=0} \qquad \forall\, \vec{u} \in U_1\,. \tag{3.6}$$

We then denote by U_2 the H_o^s-compact set $\subset U_1$ constituted by the solutions $\vec{u}$ such that

$$\left.\frac{\partial L_\vartheta(\vec{u})}{\partial\vartheta}\right|_{\vartheta=0} = \sigma_2 \tag{3.7}$$

and repeat this procedure for the functionals

$$\vec{u} \to \left.\frac{\partial^k L_\vartheta(\vec{u})}{\partial\vartheta^k}\right|_{\vartheta=0} = \sum_{j=1}^{\infty} (\log\lambda_j)^k \varphi_j \alpha_j \quad (k=2,3\ldots)\,. \tag{3.8}$$

We obtain in this way a monotonic decreasing sequence of H_o^s-compact sets $\neq \emptyset$, defined by

$$U_{k+1} = \{\vec{u} \in U_k;\ \left.\frac{\partial^k L_\vartheta(\vec{u})}{\partial\vartheta^k}\right|_{\vartheta=0} = \sigma_{k+1}\} \quad U_k\ , \tag{3.9}$$

with

$$\sigma_{k+1} = \max_{\vec{u}\in U_k} \left.\frac{\partial^k L_\vartheta(\vec{u})}{\partial\vartheta^k}\right|_{\vartheta=0}\ , \qquad U_o = U. \tag{3.10}$$

Let

$$\tilde{U} = \lim_{k\to\infty} U_k\ ; \tag{3.11}$$

it is evident, by definition, that $\tilde{U} \neq \emptyset$; moreover, if $\vec{u} \in \tilde{U}$, $\vec{u}$ is a solution such that

$$\left.\frac{\partial^k L_\vartheta(\vec{u})}{\partial\vartheta^k}\right|_{\vartheta=0} = \sigma_{k+1}, \quad \forall\ k = 0,1,\ldots \tag{3.12}$$

Let us show that $\tilde{U}$ is constituted by a single element, $\tilde{\vec{u}}$, i.e. if $\vec{u}$, $\vec{v} \in \tilde{U}$, then $\vec{u}=\vec{v}$. We have, in fact, in this case

$$\left.\frac{\partial^k L_\vartheta(\vec{u})}{\partial\vartheta^k}\right|_{\vartheta=0} = \left.\frac{\partial^k L_\vartheta(\vec{v})}{\partial\vartheta^k}\right|_{\vartheta=0} = \sigma_{k+1} \quad (k=0,1,\ldots) \tag{3.13}$$

and, by (3.8), setting $\alpha_j=(\vec{u},\vec{g}_j)_{L^2}$, $\beta_j=(\vec{v},\vec{g}_j)_{L^2}$, $\eta_j=\alpha_j-\beta_j$,

$$\sum_{j=1}^{\infty} (\log\lambda_j)^k \varphi_j \eta_j = 0 \qquad (k = 0,1,\ldots)\,. \tag{3.14}$$

Equations (3.14) can be written in the form

$$< \vec{y}_k, \vec{w} > = 0 \qquad (k = 1,2,\ldots) \tag{3.15}$$

where $\vec{y}_k$ is given by (2.3) and $\vec{w} = \vec{u} - \vec{v} = \sum_{j=1}^{\infty} \gamma_j \vec{g}_j$.

Since, by lemma 4, $\{\vec{y}_k\}$ is a basis in H^{-1} and $\vec{w} \in H_o^1$, it follows that $\vec{w} = 0$.

Finally, we prove that $\vec{u}$ is the only maximal ϑ-stable solution. Denoting, in fact, by $\vec{u}$ any other solution (necessarily $\notin \tilde{U}$), there exists $p \geq 0$ such that

$$\left.\frac{\partial^p L_\vartheta(\vec{u})}{\partial \vartheta^p}\right|_{\vartheta=0} > \left.\frac{\partial^p L_\vartheta(\vec{u})}{\partial \vartheta^p}\right|_{\vartheta=0}, \tag{3.16}$$

while, if $p \geq 1$,

$$\left.\frac{\partial^k L_\vartheta(\vec{u})}{\partial \vartheta^k}\right|_{\vartheta=0} = \left.\frac{\partial^k L_\vartheta(\vec{u})}{\partial \vartheta^k}\right|_{\vartheta=0} \quad \text{for } k=0,\ldots,p-1. \tag{3.17}$$

Bearing in mind the expansion (2.1), it follows then that

$$L_\vartheta(\vec{\tilde{u}}) > L_\vartheta(\vec{u}) \tag{3.18}$$

in an appropriate neighbourhood $(0, \vartheta'_u)$ of $\vartheta = 0$.

The theorem is thus proved.

References

[1] O.A. Ladyzhenskaja: The mathematical theory of viscous, incompressible flow. Gordon and Breach, 1969.

[2] R. Temam: Navier-Stokes equations. North Holland, 1984.

[3] G. Prouse: A uniqueness criterion for the solution of the stationary Navier-Stokes equations. To appear on Rend. Acc. Naz. Lincei, 1988.

Note: This paper is in final form and no similar paper has been or is being submitted elsewhere.

(received January 23, 1989.)

ON DECAY PROPERTIES OF THE STOKES EQUATIONS IN EXTERIOR DOMAINS

Hermann Sohr

Fachbereich Mathematik-Informatik, Universität-GH-Paderborn

Warburger Straße 100, D-4790 Paderborn

Werner Varnhorn

Fachbereich Mathematik, Technische Hochschule Darmstadt,

Schloßgartenstraße 7, D-6100 Darmstadt

1. Introduction

Let $G \subset \mathbb{R}^3$ be an exterior domain (i. e. a domain which is the complement of the closure of some nonempty bounded domain) with smooth boundary ∂G of class $C^{2+\mu}$ $(0<\mu<1)$. In G we consider the equations of Stokes in the general form

$$\begin{gathered} -\Delta u + \nabla p = f, \quad \operatorname{div} u = g, \quad u_{|\partial G} = \Phi, \\ \lim\, (u(x) - (a+Ax)) = 0 \text{ for } |x| \longrightarrow \infty. \end{gathered} \tag{1.1}$$

Here $u = (u_1,u_2,u_3)$, $f = (f_1,f_2,f_3)$ are vector fields, p and g are scalar functions; $a \in \mathbb{R}^3$ and $A = (a_{ij})_{i,j=1,2,3}$ prescribe the behaviour of u at infinity; $\Phi = (\Phi_1,\Phi_2,\Phi_3)$ is the given boundary value of u on ∂G, and f, g are given in G. As usual, Δ denotes the Laplacian, $\nabla = (D_1,D_2,D_3)$ the gradient, and $\operatorname{div} u = \nabla\cdot u = D_1u_1 + D_2u_2 + D_3u_3$ the divergence of u.

Besides the potential theoretical method there are essentially two possibilities to investigate decay properties of the solution u, p for large $|x|$. The first method solves (1.1) in weighted Sobolev spaces (Choquet-Bruhat [6], Giroire [12], Sequeira [15], Specovius-Neugebauer [17]). An alternative way is to multiply the system (1.1) by certain weight functions obtaining modified equations and then to solve these new equations in the framework of standard Sobolev spaces. Up to now, the second approach does not seem to be used for (1.1). The aim of the present paper is to develop this method and applying it to obtain decay properties for the solution of (1.1).

To outline the notation set $\mathbb{N}_0 = \{0,1,2,\dots\}$ and, for $m \in \mathbb{N}_0$ and $1 \le r < \infty$ let $H^{m,r}(G)$ denote the usual Sobolev space ($H^{0,r}(G) = L^r(G)$). With $\overline{G}$ as the closure of G, the space $C^m(\overline{G})$ is the space of restrictions $v_{|\overline{G}}$ of functions $v \in C^m(\mathbb{R}^3)$. Using $D_i = \partial/\partial x_i$ $(i = 1,2,3;\ x = (x_1,x_2,x_3) \in G)$, $\alpha = (\alpha_1,\alpha_2,\alpha_3) \in \mathbb{N}_0^{\,3}$, $|\alpha| = \alpha_1 + \alpha_2 + \alpha_3$, $D^\alpha = D_1^{\alpha_1} D_2^{\alpha_2} D_3^{\alpha_3}$, $\|v\|_r = (\int_G |v(x)|^r dx)^{1/r}$ where in this case $|\cdot|$ is the Euclidian norm, and $\|\nabla^m v\|_r = (\sum_{|\alpha|=m} \|D^\alpha v\|_r^r)^{1/r}$, the norm in $H^{m,r}(G)$ is defined by $\|v\|_{m,r} = (\|v\|_r^r + \|\nabla v\|_r^r + \cdots + \|\nabla^m v\|_r^r)^{1/r}$. By $H_0^{m,r}(G)$ we denote the closure of $C_0^m(G)$ with respect to $\|\cdot\|_{m,r}$, where $C_0^m(G)$ is the space of all $v \in C^m(\overline{G})$ having compact support in G. The space $H_{loc}^{m,r}(G)$ contains all functions v such that $v_{|G\cap B} \in H^{m,r}(G\cap B)$ holds for every open ball $B \subset \mathbb{R}^3$. Moreover, we need the usual fractional order Sobolev space $W^{2-1/r,r}(\partial G)$, which contains the traces $u_{|\partial G}$ of all $u \in H^{2,r}(G)$. Let $L^\infty(G)$ denote the usual space of essentially bounded functions with norm $\|\cdot\|_\infty$. By $L^r(G)^3$, $H^{m,r}(G)^3$, ... we mean the corresponding spaces for vector fields $v = (v_1,v_2,v_3)$, whose norms are denoted as in the scalar case above. Finally, concerning the decay of functions, we sometimes use the notation $u(x) = o(g(x))$ for $|x| \longrightarrow \infty$ to express that $(u(x)/g(x)) \longrightarrow 0$ for $|x| \longrightarrow \infty$. Now our main result is stated in

<u>1.2</u> <u>Theorem</u> : Let $1 < r < 3/2$ and let s,q be defined by $s = (1/r - 1/3)^{-1}$ and $q = (1/r - 2/3)^{-1}$. Let $f \in L^r(G)^3 \cap L^q(G)^3$, $g \in L^s(G)$ with $\nabla g \in L^r(G)^3 \cap L^q(G)^3$, $\Phi \in W^{2-1/q,q}(\partial G)^3$, $a \in \mathbb{R}^3$, $A = (a_{ij})_{i,j=1,2,3}$ with $a_{ij} \in \mathbb{R}$ and $a_{11} + a_{22} + a_{33} = 0$, and let $u_\infty(x) = a + Ax$ for $x \in \mathbb{R}^3$.

Then there is a unique solution u,p of (1.1) with $u - u_\infty \in H^{2,q}(G)^3$,

$p \in H^{1,q}(G)$, and for $|x| \longrightarrow \infty$ it holds

$$(u(x) - (a + Ax)) \longrightarrow 0, \quad \nabla u(x) \longrightarrow A, \quad p(x) \longrightarrow 0. \tag{1.3}$$

In addition, let $1 < r* < r$, $s* = (1/r* - 1/3)^{-1}$, $q* = (1/r* - 2/3)^{-1}$, $f \in L^{r*}(G)^3$, $g \in L^{s*}(G)$ with $\nabla g \in L^{r*}(G)^3$. Consider a function $M \in C^2(\overline{G})$ with $Mf \in L^r(G)^3 \cap L^q(G)^3$, $Mg \in L^s(G)$, $M\nabla g \in L^r(G)^3 \cap L^q(G)^3$, and $\|\nabla M\|_\rho < \infty$, $\|\nabla^2 M\|_\tau < \infty$ $(\rho = (1/r - 1/s*)^{-1}$, $\tau = (1/r - 1/q*)^{-1})$ as well as $\|\nabla M\|_\infty < \infty$, and $\|\nabla^2 M\|_\infty < \infty$.

Then we have $M(u - u_\infty) \in H^{2,q}(G)^3$, $Mp \in H^{1,q}(G)$, and for $|x| \longrightarrow \infty$

$$\begin{gathered} u(x) = a + Ax + o(1/M(x)), \\ \nabla u(x) = A + o(1/M(x)), \qquad p(x) = o(1/M(x)). \end{gathered} \tag{1.4}$$

<u>Remark</u> : Setting $a_{ij} = 0$ we obtain a solution u, which tends to a prescribed constant vector a at infinity. The above result admits very general weight functions, which even may depend on the direction of tending to infinity.

From 1.2 we obtain immediately

<u>1.5</u> <u>Corollary</u> : Let $g \in L_1(G) \cap L_\infty(G)$, $\nabla g, f \in L_1(G)^3 \cap L_\infty(G)^3$, and let Φ, a, A, u_∞ as in Theorem 1.2. Then there is a unique solution u,p of (1.1) with $u - u_\infty \in H^{2,q}(G)^3$, $p \in H^{1,q}(G)$ for all $q > 3$. For $|x| \longrightarrow \infty$ u, p satisfy (1.3).

In addition, let $M \in C^2(\overline{G})$ be given with $|\nabla^k M(x)| \le c\,|x|^{\alpha - k}$ (k=0,1,2) for sufficiently large $|x|$ and some constants $c > 0$ and $0 < \alpha < 1$, and let $Mg \in L_1(G) \cap L_\infty(G)$ and $M\nabla g$, $Mf \in L_1(G)^3 \cap L_\infty(G)^3$. Then, for all $q > 3/(1-\alpha)$ we have $M(u - u_\infty) \in H^{2,q}(G)^3$, $Mp \in H^{1,q}(G)$, and (1.4) for $|x| \longrightarrow \infty$.

The following corollary yields a decay property of the Stokes operator A_s which is defined as follows: For $1 < s < \infty$ let $L_\sigma^s(G)^3$ denote the closure of

$$C_{o,\sigma}^\infty(G)^3 = \{v \in C_o^\infty(G)^3 \mid \operatorname{div} v = 0\}$$

in $L^s(G)^3$. Then A_s is defined by

$$A_s u = - P_s \Delta u, \qquad u \in D(A_s) = H^{2,s}(G)^3 \cap H_0^{1,s}(G)^3 \cap L_\sigma^s(G)^3,$$

where

$$P_s \ : \ L^s(G)^3 \longrightarrow L_\sigma^s(G)^3$$

is the projection associated with the Helmholtz decomposition([9], [13], [16]).

<u>1.6 Corollary</u> : Let $1 < s < \infty$, $0 < \alpha < \beta \leq 1$, and $u \in D(A_s)$. If $A_s u \in L^\infty(G)^3$ and $|A_s u(x)| \leq c|x|^{-2-\beta}$ holds for all sufficiently large $|x|$ and some constant $c > 0$, then, after redefinition on a set of measure zero, u is continuous and $|u(x)||x|^\alpha \longrightarrow 0$ for $|x| \longrightarrow \infty$.

2. Preliminaries

Besides the spaces already defined we need the homogeneous Sobolev spaces $\hat{H}_0^{m,r}(G)$, which are the completions of $C_0^\infty(G)$ with respect to the norm $\|\nabla^m \cdot \|_r$ ($m \in \mathbb{N}_0$, $1 < r < \infty$). Using the Poincaré inequality on neighbourhoods of ∂G we see that $\hat{H}_0^{m,r}(G) \subset H_{loc}^{m,r}(G)$. The corresponding spaces of vector fields are denoted by $\hat{H}_0^{m,r}(G)^3$.

In the following, $G_b \subset \mathbb{R}^3$ defines a bounded domain with smooth boundary at least of class C^1. The first proposition concerns special statements of Bogovski's theory [2] about the equation $\operatorname{div} u = f$ in bounded domains. A more general and more detailed investigation of this theory is given in [4].

<u>2.1 Proposition</u> : Let $1 < r < \infty$ and $1 < s < \infty$. Then for every $f \in L^r(G_b) \cap L^s(G_b)$ with $\int_{G_b} f(x)\,dx = 0$ there exists some $u \in H_0^{1,r}(G_b)^3 \cap H_0^{1,s}(G_b)^3$ satisfying $\operatorname{div} u = f$ and

$$\|\nabla u\|_r \le c\,\|f\|_r \quad , \|\nabla u\|_s \le c\,\|f\|_s$$

with some constant $c = c(G_b, r, s) > 0$.

Moreover, if $f \in H_0^{1,r}(G_b) \cap H_0^{1,s}(G_b)$ with $\int_{G_b} f(x)\,dx = 0$, then there exists some $u \in H_0^{2,r}(G_b)^3 \cap H_0^{2,s}(G_b)^3$ with $\operatorname{div} u = f$ and

$$\|\nabla^2 u\|_r \le c\,\|\nabla f\|_r, \qquad \|\nabla^2 u\|_s \le c\,\|\nabla f\|_s$$

with some constant $c = c(G_b, r, s) > 0$.

<u>Proof</u> : Following [2] we first assume that G_b is starlike with respect to some open ball B with $\overline{B} \subset G_b$. Then u can be constructed explicitly using Bogovski's formula

$$u(x) = \int_{G_b} K(x,y)\,f(y)\,dy, \quad K(x,y) = \frac{x-y}{|x-y|^3} \int_{|x-y|}^{\infty} h\Big(y + t\frac{x-y}{|x-y|}\Big)\,t^2\,dt,$$

where here $h \in C_0^\infty(B)$ can be chosen arbitrarily with $\int_B h(x)\,dx = 1$. With help of Calderon-Zygmund estimates the assertion in 2.1 follows by straight-forward calculations in this case. For the more general domain G_b presented above, the assertions in 2.1 can be shown by an additional localization procedure [4].

The following more special statements about the equation $\operatorname{div} u = f$ in exterior domains can be developed from Bogovski's theory. In a more general approach similar results have been proved in [4].

<u>2.2</u> <u>Proposition</u> : Let $1 < q < 3$. Then for every $f \in L^q(G)$ there exists some $u \in \hat{H}_0^{1,q}(G)^3$ such that $\operatorname{div} u = f$ and

$$\|\nabla u\|_q \le c\,\|f\|_q$$

with some constant $c = c(G,q) > 0$.

Let $1 < r < 3/2$ and $s = (1/r - 2/3)^{-1}$. Then for every $f \in \hat{H}_0^{1,r}(G) \cap \hat{H}_0^{1,s}(G)$ there exists some $u \in \hat{H}_0^{2,r}(G)^3 \cap H_0^{2,s}(G)^3$ such that $\operatorname{div} u = f$ and

$$\|\nabla^2 u\|_r \leq c\,\|\nabla f\|_r, \qquad \|\nabla^2 u\|_s \leq c\,\|\nabla f\|_s$$

with some constant $c = c(G,r,s) > 0$.

<u>Proof</u> : Let $F(x) = 1/(4\pi|x|)$ denote the fundamental solution of the Laplacian, i. e. $-\Delta F(x) = \delta(x)$ in the sense of distributions. Extending f by zero to the whole space, let $f_j \in C_0^\infty(\mathbb{R}^3)$ be a sequence with $\|f - f_j\|_q \longrightarrow 0$, and define $h_j(x) = -\int_{\mathbb{R}^3} F(x-y)\,f_j(y)\,dy$. Then we have $h_j \in C^\infty(\mathbb{R}^3)$, $\operatorname{div}\nabla h_j = f_j$, and $|\nabla^n h_j(x)| \leq c\,|x|^{-(n+1)}$ for $|x| \longrightarrow \infty$ and $n \in \mathbb{N}_0$. Moreover, by well-known estimates for the Riesz potential and by the Calderon - Zygmund theorem we obtain

$$\|\nabla h_j\|_p \leq c_1\,\|\nabla^2 h_j\|_q \leq c_2\,\|f_j\|_q$$

with $p = (1/q-1/3)^{-1}$. Hence we can find some $w \in L^p(\mathbb{R}^3)^3$ with $\|w-\nabla h_j\|_p \longrightarrow 0$ and $\|\nabla w - \nabla^2 h_j\|_q \longrightarrow 0$ as $j \longrightarrow \infty$, with $\operatorname{div} w = f$, and with

$$\|w\|_p \leq c_1\,\|\nabla w\|_q \leq c_2\,\|f\|_q.$$

Next we show that w can be approximated by C_0^∞-functions with respect to the norm $\|\nabla\cdot\|_q$. This follows by proving that for every fixed j there is a sequence $h_{jk} \in C_0^\infty(\mathbb{R}^3)$ with $\|\nabla^2(h_j - h_{jk})\|_q \longrightarrow 0$ as $k \longrightarrow \infty$. To construct such a sequence, let $\varphi \in C_0^\infty(\mathbb{R}^3)$ denote a cut off function such that $0 \leq \varphi \leq 1$, $\varphi(x)=1$ for $|x| \leq 1$, $\varphi(x)=0$ for $|x| \geq 2$, and define $\varphi_k(x) = \varphi(x/k)$. Then for $n \in \mathbb{N}$ we have

$$\nabla^n(\varphi_k(x)) = k^{-n}(\nabla^n\varphi)_k(x) = k^{-n}(\nabla^n\varphi)(x/k),$$

and $\nabla^n(\varphi_k(x)) = 0$ for all x with $|x| \leq k$ or $|x| \geq 2k$. Let us define $R_k = \{x \in \mathbb{R}^3 \mid k < |x| < 2k\}$. Then the functions $h_{jk} = h_j\varphi_k \in C_0^\infty(\mathbb{R}^3)$ satisfy the asserted convergence, because for $k \longrightarrow \infty$ we have

$$\|(\nabla^2 h_j)\,(\varphi_k - 1)\|_q \longrightarrow 0,$$

$$\|(\nabla h_j)\cdot(\nabla\varphi_k)\|_q = k^{-1}\Big(\int_{R_k} |\nabla h_j(x)\cdot(\nabla\varphi(x/k))|^q dx\Big)^{1/q} \leq$$

$$c_3\,k^{-3}\Big(\int_{R_1} |\nabla\varphi(y)|^q k^3\,dy\Big)^{1/q} \leq c_4\,k^{3/q-3} \longrightarrow 0 \quad \text{since } q>1,$$

and

$$\| h_j(\nabla^2\varphi_k) \|_q \leq c_5 k^{3/q-3} \longrightarrow 0.$$

Next let us choose some open ball B with $\partial G \subset B$ and some function $\Psi \in C_0^\infty(\mathbb{R}^3)$ with $\Psi = 1$ in a neighbourhood of ∂G and $\Psi = 0$ in a neighbourhood of ∂B. Setting $G_b = G \cap B$, we construct a solution v of the equation

$$\operatorname{div} v = \operatorname{div}(\Psi w)$$

as in 2.1, which is possible because due to $\operatorname{div} w = f = 0$ in $\mathbb{R}^3 \setminus G$ the condition $\int_{G_b} \operatorname{div}(\Psi w)\, dx = 0$ is satisfied. Now for

$$u = w - \Psi w + v$$

we obtain $\operatorname{div} u = f$ and

$$\begin{aligned} \|\nabla u\|_q &\leq \|\nabla w\|_q + \|\nabla(\Psi w)\|_q + \|\nabla v\|_q \leq c_5 (\|f\|_q + \|(\nabla\Psi)\cdot w\|_q \\ &\leq c_6 \|f\|_q + \|(\nabla\Psi)\cdot w\|_p) \leq c_7 \|f\|_q . \end{aligned}$$

Here we used that $\nabla\Psi$ has a compact support. Thus it follows $u \in \hat{H}_0^{1,q}(G)^3$ from the construction above and the first assertion is proved.

To prove the second assertion let $f \in \hat{H}_0^{1,r}(G) \cap \hat{H}_0^{1,s}(G)$. By means of the usual mollification and cut off techniques as above we can find a sequence $f_j \in C_0^\infty(G)$ such that for $j \longrightarrow \infty$ we have $\nabla f_j \longrightarrow \nabla f$ with respect to $\|\cdot\|_r$ as well as to $\|\cdot\|_s$. Extending f and f_j by zero to the whole space, we construct the functions w, v, u as above, using r instead of q. With Calderon-Zygmund estimates and well-known Sobolev embeddings we obtain

$$\begin{gathered} \|\nabla^2 w\|_r \leq c_1 \|\nabla f\|_r, \qquad \|\nabla^2 w\|_s \leq c_2 \|\nabla f\|_s, \\ \|w\|_s \leq c_3 \|\nabla w\|_p \leq c_4 \|\nabla^2 w\|_r, \end{gathered}$$

where p is defined by $p=(1/r-1/3)^{-1}$. Thus we get $w \in H^{2,s}(\mathbb{R}^3)^3$ and, noting that $\nabla\Psi$ and $\nabla^2\Psi$ have compact supports, it follows

$$\|\nabla^2(\Psi w)\|_r \leq c_5 \|\nabla f\|_r, \qquad \|\nabla^2(\Psi w)\|_s \leq c_6 \|\nabla f\|_s .$$

Finally, applying 2.1 we obtain

$$\|\nabla^2 v\|_r \le c_7 \|\nabla \operatorname{div}(\Psi w)\|_r \le c_8 \|\nabla f\|_r,$$
$$\|\nabla^2 v\|_s \le c_9 \|\nabla \operatorname{div}(\Psi w)\|_s \le c_{10} \|\nabla f\|_s,$$

and this yields the assertion.

The next proposition follows from 2.2 by a duality argument. For $m \in \mathbb{N}_0$ and $1 < r < \infty$ let $\hat{H}^{-m,r}(G) = (\hat{H}_0^{m,r'})^*$ denote the dual space of $\hat{H}_0^{m,r'}(G)$ where $1/r + 1/r' = 1$. Its norm is defined by

$$\|u\|_{-m,r} = \sup_{0 \neq \xi \in \hat{H}_0^{m,r'}(G)} \frac{|\langle u,\xi\rangle|}{\|\nabla^m \xi\|_{r'}}$$

with $\langle u,v\rangle = \int_G uv\, dx$. The corresponding spaces of vector fields are given by $\hat{H}^{-m,r}(G)^3$, where in this case we have $\langle u,v\rangle = \int_G (u_1v_1+u_2v_2+u_3v_3)dx$. Moreover, we sometimes use the term $\langle u,v\rangle$ also to denote the value of a distribution u at v. Recently, a result similar to 2.3 has been proved already by other methods ([10], [14]).

<u>2.3</u> <u>Proposition</u> : Let $1 < r < 3$ and let $p = (1/r-1/3)^{-1}$. Then for every distribution u with $\nabla u \in L^r(G)^3$ there is a constant K_u such that $u + K_u \in L^p(G)$ and

$$\|u+K_u\|_p \le c \|\nabla u\|_r$$

with some $c = c(G,r) > 0$.

If additionally $u \in L^q(G)$ for some q with $1 < q < \infty$, then we have

$$\|u\|_p \le c \|\nabla u\|_r.$$

<u>Proof</u> : Let r' and p' be defined by $1/r + 1/r' = 1$ and $1/p + 1/p' = 1$, hence $1 < p' < 3$ and $r' = (1/p'-1/3)^{-1}$. From 2.2 it follows that the operator

$$\mathrm{div} \quad : \quad \hat{H}_0^{1,p'}(G)^3 \longrightarrow L^{p'}(G)$$

has the closed range $R(\mathrm{div}) = L^{p'}(G)$. By the closed range theorem [18] the transposed operator (with respect to $\langle\ ,\ \rangle$)

$$-\nabla \quad : \quad L^p(G) \longrightarrow \hat{H}^{-1,p}(G)^3$$

has also a closed range $R(-\nabla)$, and from $w \in \hat{H}^{-1,p}(G)^3$, $\langle w,v\rangle = 0$ for all $v \in N(\mathrm{div}) = \{v \in \hat{H}_0^{1,p'}(G)^3 \,|\, \mathrm{div}\, v = 0\}$, we obtain $w \in R(-\nabla)$, hence $w = \nabla h$ for some $h \in L^p(G)$. Because of

$$|\langle\nabla u, v\rangle| \leq \|\nabla u\|_r \, \|v\|_{r'} \leq c \, \|\nabla u\|_r \, \|\nabla v\|_{p'}$$

by Sobolev's imbedding theorem, we have $\nabla u \in \hat{H}^{-1,p}(G)^3$ with $\|\nabla u\|_{-1,p} \leq c \, \|\nabla u\|_r$, and it also holds $\langle\nabla u, v\rangle = 0$ for all $v \in N(\mathrm{div})$. To see the latter, we use that $C_{0,\sigma}^{\infty}(G)^3$ is dense in $L_{\sigma}^{r'}(G)^3$ by definition and that $\hat{H}_0^{1,p'}(G)^3 \subset L^{r'}(G)^3$, which implies $N(\mathrm{div}) \subset L_{\sigma}^{r'}(G)^3$. Thus every $v \in N(\mathrm{div})$ can be approximated by a sequence $v_k \in C_{0,\sigma}^{\infty}(G)^3$ with $\|v_k - v\|_{r'} \longrightarrow 0$ as $k \longrightarrow \infty$ and, since $\mathrm{div}\, v_k = 0$, for $k \longrightarrow \infty$ it follows

$$\langle\nabla u, v\rangle = \lim \langle\nabla u, v_k\rangle = -\lim \langle u, \mathrm{div}\, v_k\rangle = 0.$$

Now having shown that $\nabla u \in R(-\nabla)$, there exists some $u^* \in L^p(G)$ with $\nabla u = \nabla u^*$ and we obtain some constant K_u such that $u + K_u = u^*$. Because G is an exterior domain, the null space $N(-\nabla)$ is equal to $\{0\}$, and, since $R(-\nabla)$ is closed, we conclude that

$$\|u + K_u\|_p = \|u^*\|_p \leq c_1 \|\nabla u^*\|_{-1,p} = c_1 \|\nabla u\|_{-1,p} \leq c_2 \|\nabla u\|_r ,$$

and the first assertion is proved.

The second assertion follows observing that $K_u = 0$ if $u \in L^q(G)$ and $u + K_u \in L^p(G)$.

Applying 2.3 twice we get the following result:

2.4 Corollary : Let $1<r<3/2$ and p, q be defined by $p=(1/r-1/3)^{-1}$, $q=(1/r-2/3)^{-1}$. If $u \in L^s(G)$ with $\nabla u \in L^t(G)^3$ for some s,t $(1<s,t<\infty)$ and if $\|\nabla^2 u\|_r < \infty$, then $u \in L^q(G)$, $\nabla u \in L^p(G)^3$ and

$$\|u\|_q \leq c_1 \|\nabla u\|_p \leq c_2 \|\nabla^2 u\|_r$$

with some constants $c_1 = c_1(G,r) > 0$, $c_2 = c_2(G,r) > 0$.

In the next section, the above statements about the divergence equation $\operatorname{div} u = f$ are used to reduce the general Stokes equations (1.1) to the case $\operatorname{div} u = 0$.

3. The Stokes Equations

Applying 2.4 to a result of Solonnikov [16] we obtain the following

3.1 Proposition : Let $1<r<3/2$ and let s,q be defined by $s=(1/r-1/3)^{-1}$ and $q=(1/r-2/3)^{-1}$. Then for every $f \in L^r(G)^3$ there exists a unique $u \in L^q(G)$ with $\|\nabla u\|_s < \infty$, $\|\nabla^2 u\|_r < \infty$ and a unique $p \in L^s(G)$ with $\|\nabla p\|_r < \infty$ such that

$$-\Delta u + \nabla p = f, \quad \operatorname{div} u = 0, \quad u_{|\partial G} = 0 \tag{3.2}$$

is satisfied in G. The estimate

$$\|\nabla^2 u\|_r + \|\nabla p\|_r \leq c \|f\|_r \tag{3.3}$$

holds with some $c = c(G,r) > 0$.

Proof : It is known that for every $f \in C_0^\infty(G)^3$ there exist distributions u,p with $\|\nabla^2 u\|_r < \infty$ and $\|\nabla p\|_r < \infty$ satifying (3.2) and the estimate (3.3) with a constant $c=c(G,r)>0$ not depending on the support of the given f [16, p.476].

Moreover, the potential representation given in [16] yields that u, ∇u, p belong to certain L^p- spaces, which by 2.4 implies $u \in L^q(G)^3$, $\|\nabla u\|_s < \infty$, and $p \in L^s(G)$ as well as $\|u\|_q \le c_1 \|\nabla u\|_s \le c_2 \|\nabla^2 u\|_r$ and $\|p\|_s \le c_3 \|\nabla p\|_r$ with some constants independent of u and p. Now assuming $f \in L^r(G)^3$ we take a sequence $f_k \in C_0^\infty(G)^3$ with $\|f - f_k\|_r \longrightarrow 0$ for $k \longrightarrow \infty$. Since the constants c, c_1, c_2, c_3 are independent of k, we obtain a solution u,p with the asserted properties if k tends to infinity. The uniqueness of u and p follows easily observing that $\|u\|_q = 0$ and $\|p\|_s = 0$ if u and p satisfy the homogeneous equations $-\Delta u + \nabla p = 0$, $\operatorname{div} u = 0$, $u_{|\partial G} = 0$.

3.4 Proposition : Let $1 < r < 3/2$ and q be defined by $q = (1/r - 2/3)^{-1}$. Then for every $f \in L^r(G)^3 \cap L^q(G)^3$ there exist a unique $u \in H^{2,q}(G)^3$ with $\|\nabla^2 u\|_r < \infty$, and a unique $p \in H^{1,q}(G)$ with $\|\nabla p\|_r < \infty$ satisfying (3.2) and

$$\|u\|_{2,q} + \|\nabla^2 u\|_r + \|p\|_{1,q} + \|\nabla p\|_r \le c\,(\|f\|_r + \|f\|_q) \qquad (3.5)$$

with a constant $c = c(G,r) > 0$.

Proof : Using $f \in L^r(G)^3$, let u,p denote the solution of (3.2) constructed in 3.1. Then from $u \in L^q(G)^3$ and $-\Delta u + \nabla p = f \in L^q(G)^3$, $\operatorname{div} u = 0$, $u_{|\partial G} = 0$ we conclude that $u \in D(A_q)$. This follows from Miyakawa's duality result [13,p.122] and the continuity of the form

$$v \longrightarrow \langle f,v\rangle = \langle u, A_{q'} v\rangle \qquad (v \in D(A_{q'}) \cap D(A_{r'}))$$

with respect to $\|v\|_{q'}$. It follows $u \in H^{2,q}(G)^3$, and from $\nabla p = f + \Delta u$ we quote $\nabla p \in L^q(G)^3$. Using $u - \Delta u + \nabla p = f + u$, hence $(I + A_q) u = P_q f + u$, we obtain

$$\begin{aligned} \|u\|_{2,q} &\le c_1 \|(I+A_q)u\|_q \le c_2(\|f\|_q + \|u\|_q) \\ &\le c_3(\|f\|_q + \|\nabla^2 u\|_r) \le c_4(\|f\|_q + \|f\|_r), \\ \|\nabla p\|_q &\le \|f + \Delta u\|_q \le c_5(\|f\|_q + \|f\|_r). \end{aligned}$$

Due to $\nabla p \in L^r(G)^3 \cap L^q(G)^3$ we have $\nabla p \in L^s(G)^3$ for $s = (1/r - 1/3)^{-1}$ with $\|\nabla p\|_s \le c_6 (\|\nabla p\|_q + \|\nabla p\|_r)$. Since $p \in L^s(G)$ by 3.1, using 2.4 this implies

$p \in L^q(G)$ with

$$\|p\|_q \le c_7 \|\nabla p\|_s \le c_8(\|\nabla p\|_q + \|\nabla p\|_r) \le c_9(\|f\|_q + \|f\|_r),$$

where here for the last estimate also (3.3) is used. The uniqueness assertion follows as in 3.1.

In the next proposition we consider the more general case $\operatorname{div} u = g \neq 0$.

<u>3.6 Proposition</u> : Let $1 < r < 3/2$ and let s,q be defined by $s = (1/r - 1/3)^{-1}$ and $q = (1/r - 2/3)^{-1}$. Then for every $f \in L^r(G)^3 \cap L^q(G)^3$ and every $g \in \hat{H}_0^{1,r}(G) \cap \hat{H}_0^{1,q}(G)$ there exists a unique $u \in H^{2,q}(G)^3$ with $\|\nabla u\|_s < \infty$, $\|\nabla^2 u\|_r < \infty$, and a unique $p \in H^{1,q}(G)$ with $\|p\|_s < \infty$, $\|\nabla p\|_r < \infty$ such that

$$-\Delta u + \nabla p = f, \quad \operatorname{div} u = g, \quad u_{|\partial G} = 0 \tag{3.7}$$

is satisfied in G. The estimate

$$\|u\|_{2,q} + \|\nabla^2 u\|_r + \|p\|_{1,q} + \|\nabla p\|_r \le c\,(\|f\|_r + \|f\|_q + \|\nabla g\|_r + \|\nabla g\|_q) \tag{3.8}$$

holds with a constant $c = c(G,r) > 0$, and it follows

$$u(x) \longrightarrow 0, \quad \nabla u(x) \longrightarrow 0, \quad p(x) \longrightarrow 0,$$

if $|x|$ tends to infinity.

<u>Proof</u> : Applying 2.2 we choose $v \in \hat{H}_0^{2,r}(G)^3 \cap \hat{H}_0^{2,q}(G)^3$ with $\operatorname{div} v = g$ and

$$\|\nabla^2 v\|_r \le c_1 \|\nabla g\|_r, \qquad \|\nabla^2 v\|_q \le c_1 \|\nabla g\|_q.$$

Then we solve the equations

$$-\Delta(u-v) + \nabla p = f + \Delta v, \quad \operatorname{div}(u-v) = 0, \quad (u-v)_{|\partial G} = 0$$

using 3.4. Setting $w = u - v$ the estimate (3.5) yields

$$\|w\|_{2,q} + \|\nabla^2 w\|_r + \|p\|_{1,q} + \|\nabla p\|_r \le c_2(\|f + \Delta v\|_q + \|f + \Delta v\|_r),$$

and this leads to (3.8) with w instead of u. Noting that the Sobolev imbedding theorem [8, p.24] implies $\hat{H}_0^{2,r}(G) \subset \hat{H}_0^{1,s}(G) \subset L^q(G)$ with continuous injection, r,s,q as above, we obtain $\|v\|_q \le c_3\|\nabla v\|_s \le c_4\|\nabla^2 v\|_r$, and, due to $\|\nabla^2 v\|_s \le c_5(\|\nabla^2 v\|_q + \|\nabla^2 v\|_r)$ by interpolation, 2.4 implies $\|\nabla v\|_q \le c_6\|\nabla^2 v\|_s$. Collecting all estimates, we have $v \in H^{2,q}(G)^3$ satisfying (3.8) with v instead of u, thus the same is valid for $u = w + v \in H^{2,q}(G)^3$. The uniqueness of u,p follows as in 3.4. To show the asserted decay we take balls B(x) with centre at x and fixed radius. Due to $u \in H^{2,q}(G)^3$ and $p \in H^{1,q}(G)$ with $q > 3$, we obtain by Sobolev's imbedding theorem for $|x| \longrightarrow \infty$

$$|u(x)| \le c_7\|u\|_{1,q,B(x)} \longrightarrow 0,$$
$$|\nabla u(x)| \le c_8\|u\|_{2,q,B(x)} \longrightarrow 0,$$
$$|p(x)| \le c_9\|p\|_{1,q,B(x)} \longrightarrow 0,$$

since the constants c_7, c_8, c_9 do not depend on x. Thus 3.6 is proved.

<u>Proof of Theorem 1.2</u> : To prove the first assertion we reduce the general equations (1.1) to the case $u_{|\partial G} = 0$, $g_{|\partial G} = 0$, $u_\infty(x) = 0$ as in 3.6 (note that functions belonging to $\hat{H}_0^{1,p}(G)$ vanish on the boundary ∂G). For this purpose we take some open ball $B \subset \mathbb{R}^3$ with $\partial G \subset B$, define $G_b = G \cap B$, and choose a function $\Phi_0 \in W^{2-1/q,q}(\partial B)^3$ such that

$$\int_{G_b} g(x)\,dx = \int_{\partial G} \Phi \cdot N\,do + \int_{\partial B} \Phi_0 \cdot N\,do, \qquad (*)$$

where here $\Phi \in W^{2-1/q,q}(\partial G)^3$ is the given boundary value and N denotes the exterior normal vector on $\partial G_b = \partial G \cup \partial B$. Due to (*), in the bounded domain G_b we can apply Cattabriga's theorem [5] to obtain a function v with

$$v \in H^{2,q}(G_b)^3 \cap H^{2,r}(G_b)^3,$$
$$\operatorname{div} v = g, \quad v = \Phi \text{ on } \partial G, \quad v = \Phi_0 \text{ on } \partial B,$$

where we used that the assumptions made on g imply $g \in H^{1,q}(G_b) \cap H^{1,r}(G_b)$. Now we take a function $\Psi \in C_0^\infty(G_b)$ such that $\Psi = 1$ in a neighbourhood of ∂G and extend it by zero into the whole G. Setting

$$w = \Psi v + (1-\Psi) u_\infty$$

let us consider in G the equations

$$\begin{aligned} -\Delta(u-w) + \nabla p &= f + \Delta w, \\ \operatorname{div}(u-w) &= g - \operatorname{div} w, \\ (u-w)_{|\partial G} &= 0. \end{aligned} \tag{3.9}$$

It follows from the construction of w that in $G \setminus G_b$ we have $\Delta w = 0$ and $\operatorname{div} w = \operatorname{div} u_\infty = 0$, and that on ∂G we have $g - \operatorname{div} w = g - \operatorname{div} v = 0$, hence $f + \Delta w \in L^r(G)^3 \cap L^q(G)^3$ and $g - \operatorname{div} w \in \hat{H}_0^{1,r}(G) \cap \hat{H}_0^{1,q}(G)$. Applying 3.6 to (3.9) we obtain $u - w \in H^{2,q}(G)^3$, $p \in H^{1,q}(G)$, thus $u - u_\infty \in H^{2,q}(G)^3$, and u, p have all the properties asserted in the first part of the theorem. Again, the uniqueness assertion follows from the uniqueness proved in 3.6.

To prove the second assertion we set

$$w^* = u - v, \qquad f^* = f + \Delta w, \qquad g^* = g - \operatorname{div} w,$$

and we consider (3.9) in the form

$$-\Delta w^* + \nabla p = f^*, \quad \operatorname{div} w^* = g^*, \quad w^*_{|\partial G} = 0.$$

Let $M \in C^2(\overline{G})$ be a weight function as in Theorem 1.2. Multiplying the above equations by M we obtain

$$-\Delta(Mw^*) + \nabla(Mp) = f^\circ, \quad \operatorname{div}(Mw^*) = g^\circ, \quad Mw^*_{|\partial G} = 0, \tag{3.10}$$

where

$$\begin{aligned} f^\circ &= Mf^* - 2\nabla w^* \nabla M - w^* \Delta M + p \nabla M, \\ g^\circ &= Mg^* + (\nabla M) \cdot w^*. \end{aligned}$$

In the following we only have to prove that f°, g° satisfy the assumptions made on f, g in the first part of the theorem and that the resulting solution of (3.10) may be identified with $Mw*$ and Mp, respectively. Then we have $Mw* \in H^{2,q}(G)^3$ and $p \in H^{1,q}(G)$, and the asserted decay properties follow as above. To do so we apply the first part of the theorem twice, using the indices $r < s < q$ as well as $r* < s* < q*$ $(q* = (1/s* - 1/3)^{-1})$, which, roughly speaking, gives us some margin to estimate the additional terms of f^o and g^o.

First we proof that $f^o \in L^r(G)^3$. Due to $Mf \in L^r(G)^3$ we have $Mf* \in L^r(G)^3$ and because of $1/r = 1/s* + 1/\rho = 1/q* + 1/\tau$ it holds

$$\begin{aligned} \|\nabla w* \nabla M\|_r &\leq \|\nabla w*\|_{s*} \|\nabla M\|_\rho , \\ \|w* \triangle M\|_r &\leq \|w*\|_{q*} \|\nabla^2 M\|_\tau , \\ \|p \nabla M\|_r &\leq \|p\|_{s*} \|\nabla M\|_\rho . \end{aligned}$$

Next, since $Mf \in L^q(G)^3$ and

$$\begin{aligned} \|\nabla w* \nabla M\|_q &\leq \|\nabla w*\|_q \|\nabla M\|_\infty , \\ \|w* \triangle M\|_q &\leq \|w*\|_q \|\nabla^2 M\|_\infty , \\ \|p \nabla M\|_q &\leq \|p\|_q \|\nabla M\|_\infty , \end{aligned}$$

we have $f^o \in L^q(G)^3$. The fact that $g^o \in L^s(G)$ follows from the assumption $Mg \in L^s(G)$ and, noting that $1/s = 1/\rho + 1/q*$, from the estimate

$$\|(\nabla M)\cdot w*\|_s \leq \|\nabla M\|_\rho \|w*\|_{q*} .$$

Moreover, since $g \in L^{s*}(G)$ implies $g* \in L^{s*}(G)$, and since by 2.4 and $\nabla g \in L^s(G)^3$ it follows $g \in L^q(G)$, hence also $g* \in L^q(G)$, and since finally $M\nabla g* \in L^r(G)^3 \cap L^q(G)^3$ due to $M\nabla g \in L^r(G)^3 \cap L^q(G)^3$ by assumption, we obtain $\nabla g^o = \nabla M g* + M\nabla g* + \nabla^2 M w* + \nabla M \nabla w* \in L^r(G)^3 \cap L^q(G)^3$ from the estimates

$$\begin{aligned} \|(\nabla M) g*\|_r &\leq \|\nabla M\|_\rho \|g*\|_{s*} , \\ \|(\nabla^2 M) w*\|_r &\leq \|\nabla^2 M\|_\tau \|w*\|_{q*} , \\ \|(\nabla M) g*\|_q &\leq \|\nabla M\|_\infty \|g*\|_q , \\ \|(\nabla^2 M) w*\|_q &\leq \|\nabla^2 M\|_\infty \|w*\|_q . \end{aligned}$$

Thus f^o and g^o satisfy the regularity assumptions asserted above and it remains to show that $Mw*$ and Mp may be identified with the solution of (3.10).

To do so it is sufficient to approximate the solution u, p in Proposition 3.6 by functions with compact support as follows: Using the same cut off technique as in the proof of Proposition 2.2 we consider (3.7) for $u_k = u\varphi_k$ and $p_k = p\varphi_k$ instead of u and p, obtaining in G the equations

$$-\Delta u_k + \nabla p_k = f_k, \quad \operatorname{div} u_k = g_k, \quad u_{k|\partial G} = 0,$$
$$f_k = f\varphi_k - 2\nabla u \nabla \varphi_k - u(\Delta\varphi_k) + p(\nabla\varphi_k), \qquad g_k = g\varphi_k + u\cdot(\nabla\varphi_k).$$

Now a calculation similar to that in the proof of 2.2 yields $f_k \longrightarrow f$ in $L^r(G)^3$ and in $L^q(G)^3$ as well as $g_k \longrightarrow g$ in $\hat{H}_0^{1,r}(G)$ and $\hat{H}_0^{1,q}(G)$, hence by 3.6 we have $u_k \longrightarrow u$ in $H^{2,q}(G)^3$, $p_k \longrightarrow p$ in $H^{1,q}(G)$, $\|\nabla^2(u_k - u)\|_r \longrightarrow 0$, $\|\nabla(u_k - u)\|_s \longrightarrow 0$, $\|\nabla(p_k - p)\|_r \longrightarrow 0$, and $\|p_k - p\|_s \longrightarrow 0$. This proves the theorem.

<u>Proof of corollary 1.5</u> : From the assumptions on f,g it follows that $f \in L^t(G)^3$ and $g \in H^{1,t}(G)$ for all $1 < t < \infty$. Hence the first part of theorem 1.2 yields that for any fixed $q > 3$ there is a unique solution u,p of (1.1) satisfying the asserted properties. Because this solution must coincide for different values of q, the first statement of 1.5 is proved.

To prove the second assertion, let α with $0 < \alpha < 1$ be fixed and consider a weight function M as in 1.5. A simple calculation yields $\|\nabla M\|_\rho < \infty$ for all $\rho > 3/(1-\alpha)$ and $\|\nabla^2 M\|_\tau < \infty$ for all $\tau > 3/(2-\alpha)$. Due to the assumptions made on f and g we may apply the second part of 1.2 for all indices $r*$, $s*$, $q*$ with $1 < r^* < 3/2 < s^* < 3 < q^* < \infty$, and obtain a unique solution u,p of (1.1) with $M(u - u_\infty) \in H^{2,q}(G)^3$ and $Mp \in H^{1,q}(G)$ for some $q = q(\rho, \tau, r*, s*, q*)$. Because for any suitable choice of these numbers the corresponding solution must coincide, and because

$$1/q = 1/r - 2/3 < 1/r - 1/s* = 1/\rho < (1-\alpha)/3,$$
$$1/s = 1/r - 1/3 < 1/r - 1/q* = 1/\tau < (2-\alpha)/3$$

yields $3/(1-\alpha) < q$, the corollary is proved.

Proof of corollary 1.6 : Setting $f = A_s u$ we obtain in G the identity

$$-\Delta u + \nabla p = f, \quad \operatorname{div} u = 0, \quad u_{|\partial G} = 0$$

for some $p \in L^s_{loc}(G)$ with $\|\nabla p\|_s < \infty$. Let us choose $M \in C^2(\overline{G})$ such that $M(x) = |x|^\alpha$ for sufficiently large $|x|$. Then it holds $f \in L^{r^*}(G)^3$ for all $r^* > 3/(2+\beta)$ and $Mf \in L^r(G)^3$ for all $r > 3/(2+\beta-\alpha)$, where $1 \le 3/(2+\beta) < 3/(2+\beta-\alpha) < 3/2$. Since $\|\nabla M\|_\rho < \infty$, $\|\nabla^2 M\|_\tau < \infty$ for all $\rho > 3/(1-\alpha)$ and $\tau > 3/(2-\alpha)$, respectively, and since $\|\nabla M\|_\infty < \infty$ and $\|\nabla^2 M\|_\infty < \infty$, we can apply 1.2 for a=0 and A=0 obtaining some suitable $q > 3$ with $Mu \in H^{2,q}(G)^3$, hence $|x|^\alpha |u(x)| \longrightarrow 0$ for $|x| \longrightarrow \infty$. This proves the corollary.

References

[1] Adams, R. A.: Sobolev Spaces. New York-San Francisco-London: Academic Press 1975

[2] Bogovski, M. E.: Solution of the first boundary value problem for the equation of continuity of an incompressible medium. Soviet Math. Dokl. 20 (1979) 1094-1098

[3] Borchers, W., Sohr, H.: On the semigroup of the Stokes operator for exterior domains in L^q-spaces. Math. Z. 196 (1987) 415-425

[4] Borchers, W., Sohr, H.: The equations divu=f and rotv=g with zero boundary conditions, to appear.

[5] Cattabriga, L.: Su un problema al contorno relativo al sistema di equazioni di Stokes. Sem. Mat. Univ. Padova 31 (1964) 308-340

[6] Choquet-Bruhat, Y.: Elliptic systems in $H_{s,\delta}$-spaces on manifolds which are Euklidic at infinity. Acta Math. 146 (1981) 129-150

[7] Finn, R., Chang, D.: On the solutions of a class of equations occurring in continuum mechanics with application to the Stokes paradox. Arch. Rat. Mech. 7 (19) 389-401

[8] Friedmann, A.: Partial differential equations. USA: Holt, Rinehart and Winston, Inc. 1969

[9] Fujiwara, D., Morimoto, H.: An L_r-theorem of the Helmholtz decomposition of vector fields. J. Fac. Sci. Univ. Tokyo, Sect. IA Math. 24 (1977) 685-700

[10] Galdi, G. P., Moremonti, P.: Monotonic decreasing and asymptotic behaviour of the kinetic energy for weak solutions of the Navier-Stokes equations in exterior domains. Arch. Rat. Mech. Anal. 94 (1986) 253-266

[11] Giga, Y., Sohr, H.: On the Stokes operator in exterior domains, to appear in J. Fac. Sci. Univ. Tokyo 1989

[12] Giroire, J.: Etude de quelques problemes aux limites exterieurs et resolution par equations integrales. Thèse de doctorat d'etat es sciences mathematiques. Paris 6: Universite Pierre et Marie Curie 1987

[13] Miyakawa, T.: On nonstationary solutions of the Navier-Stokes equations in an exterior domain. Hiroshima Math. J. 12 (1982) 115-140

[14] Miyakawa, T., Sohr, H.: On energy inequality, smoothness and large time behaviour in L_2 for weak solutions of Navier-Stokes equations. Math. Z. 199 (1988) 455-478

[15] Sequeira, A.: Couplage entre la méthode des éléments finis et la méthode des équations intégrales. Application au problème extè rieur de Stokes stationnaire dans le plan. Thèse de doctorat de 3ème cycle. Paris 6: Université Pierre et Marie Curie 1981

[16] Solonnikov, V. A.: Estimates for solutions of nonstationary Navier-Stokes equations. J. Soviet. Math. 8 (1977) 467-529

[17] Specovius-Neugebauer, M.: Exterior Stokes problem and decay at infinity. Math. Meth. Appl. Sci 8 (1986) 351-367

[18] Yosida, K.: Functional analysis. Berlin-Heidelberg-New York: Springer 1965

Paper is in final form. No similar paper has been or is being submitted elsewhere.
(received January 23, 1989.)

On necessary and sufficient conditions for the solvability of the equations rot $u = \gamma$ and div $u = \varepsilon$ with u vanishing on the boundary

Wolf von Wahl

Mathematisches Institut der Universität

Postfach 101251, 8580 Bayreuth, FRG

Let G be a bounded open set of $\mathbb{R}^3$ with smooth boundary. Let $\hat{G} = \mathbb{R}^3 - \bar{G}$ be the open kernel of its complement. In this paper we study the equation rot $u = \gamma$ on G or $\hat{G}$, $u|\partial G = 0$ or $u|\partial\hat{G} = 0$, and the equation div $u = \varepsilon$ on G or $\hat{G}$, $u|\partial G = 0$ or $u|\partial\hat{G} = 0$, first in the framework of classical solutions and then in L^p-spaces. Our conditions on γ, ε resp. are necessary and sufficient and they depend on the topological structure of the underlying domain. Our results grew our of our lecture notes on potential theory [9]; beside more common subjects we treat there the theory of the Dirichlet- and Neumann-Problem for inhomogeneous harmonic vector fields which the author had the pleasure to learn from a lecture given by R. Kress [5]. Kress has published his results in [6,7]. We make strong use of these results and also adopt the notation Kress has used in his concise lecture.

Precise knowledge on the solvability of the equations rot $u = \gamma$ or div $u = \varepsilon$ is important when studying the Navier-Stokes equations. The equation div $u = \varepsilon$ with u vanishing on $\partial\Omega$ was treated by Ladyzhenskaja and Solonnikov [8] within an L^2-framework. Then Bogovskij [1] gave a different access for the divergence equation working with an explicit formula on starlike domains. His method can be generalized to more general domains which was done by Erig [3]. Grenz [4] has treated the equation rot $u = \gamma$ in L^2 with u vanishing on ∂G if G is simply connected. Finally Borchers and Sohr [2] have treated the problems in question by a method being completely different from the present one. Their results are basically the same ones as ours. They provide also information on the higher regularity of u.

The open set $G \subset \mathbb{R}^3$ has the structure

$$G = \bigcup_{i=1}^{\hat{m}} G_i$$

with connected open bounded sets $G_i \subset \mathbb{R}^3, \bar{G}_i \cap \bar{G}_j = \emptyset$, $i \neq j$. ∂G_i consists of finitely many connected closed, smooth and oriented surfaces. $\hat{m}$ is the second Betti-number of the complement $\hat{G} = \mathbb{R}^3 - \bar{G}$. Each G_i may have finitely many handles; the total number of handles of G is called the first Betti-number of G and we denote it by n. By

Alexander's duality-theorem the total number of handles of $\hat{G}$ is also n. For $\hat{G}$ we have the representation

$$\hat{G} = \bigcup_{i=1}^{m} \hat{G}_i \cup \hat{G}_{m+1}$$

with the bounded components of connectedness $\hat{G}_i$, $i = 1, ..., m$, and the one unbounded component of connectedness $\hat{G}_{m+1}$. m is the second Betti-number of G. We introduce the vector-space $Z(G) = \{z | z \in C^\alpha(\bar{G}) \cap C^1(G), \text{ div } z = 0, \text{ rot } z = 0, (\nu, z) = 0 \text{ on } \partial G\}$ where ν is the outer normal on ∂G. This space has dimension n and is called the space of Neumann-fields. The corresponding space for $\hat{G}$, namely $Z(\hat{G}) = \{z | z \in C^\alpha(\bar{\hat{G}}) \cap C^1(\hat{G})$, div $z = 0$, rot $z = 0$, $(\hat{\nu}, z) = 0$ on ∂G, $|z(x)| \to 0$ as $|x| \to +\infty\}$, has also dimension n; $\hat{\nu}$ is the outer normal on $\partial\hat{G}$. That $Z(G)$ and $Z(\hat{G})$ have the same finite dimension is a consequence of Alexander's duality theorem in dimension 3. Bases of $Z(G)$ and $Z(\hat{G})$ can be constructed explicitly. It turns out that $z(x) = O(\frac{1}{|x|^2})$ as $|x| \to \infty$, $z \in Z(\hat{G})$.

Now let us consider the problem rot $u = \gamma$ in G, $u|\partial G = 0$ for $\gamma \in C^\alpha(\bar{G}) \cap C^1(G)$. It is easily proved that the following conditions are necessary for the existence of a classical solution:

$$(1) \quad \begin{cases} \operatorname{div} \gamma = 0 \quad \text{in } G\,, \\ (\nu, \gamma) = 0 \quad \text{on } \partial G\,, \\ \int\limits_G (\gamma, z)\, dx = 0\,, \quad z \in Z(G)\,. \end{cases}$$

We claim that these conditions are also sufficient. For this purpose we study a different problem, namely Dirichlet's problem for harmonic vector fields. We look for a vector field v with

$$(2) \quad \begin{cases} \operatorname{rot} v = \gamma \quad \text{in } G,, \\ \operatorname{div} v = \varepsilon \quad \text{in } G,, \\ -[\nu, v] = 0 \quad \text{on } \partial G,, \\ -\int\limits_{\partial\hat{G}} (\nu, v)\hat{h}^i d\Omega = 0, \quad 1 \le i \le m. \end{cases}$$

$[.,.]$ is the cross-product in $\mathbb{R}^3$. Here the $\hat{h}^i$ form a basis of the null-space of the Neumann-problem on $\hat{G}$. This means that $\hat{h}^i =$ const. $\neq 0$ on $\hat{G}_i$, $i = 1, ..., m$; on $\hat{G}_{m+1}$ the problem $\Delta u = 0$, $\frac{\partial u}{\partial n} = 0$ has only the solution $u \equiv 0$ since we require that $u(x) = O(\frac{1}{|x|})$ as $|x| \to +\infty$. Thus $\hat{h}^{m+1} \equiv 0$. ε is any sufficiently smooth given function which we use as a free parameter. As it follows from [5,6] the conditions (1) are necessary and sufficient to solve (2) uniquely, and moreover the integral equation ($r = |x - x'|$, $\varepsilon' = \varepsilon(x')$)

$$\varepsilon^* - \int\limits_{\partial G} (\frac{\partial}{\partial \nu}\frac{1}{r})\varepsilon^{*\prime} d\Omega' = \frac{1}{2\pi}(\nu, \text{ grad } \int\limits_G \frac{\varepsilon'}{r}\, dx' - \text{ rot} \int\limits_G \frac{\gamma'}{r}\, dx') \text{ on } \partial G$$

together with the conditions $\int\limits_{\partial\hat{G}} \varepsilon^* \hat{h}^i\, d\Omega = 0$, $i = 1, ..., m$, admits a unique solution ε^* which turns out to be $-(\nu, v)$. We intend to choose ε in such a way that $\varepsilon^* = -(\nu, v) = 0$. Then this particular v can be taken as a solution of our problem rot $u = \gamma$, $u|\partial G = 0$. To do so we solve Neumann's problem

$$(3) \quad \begin{cases} \Delta\tilde{u} &= 0 \text{ on } \hat{G}, \\ \frac{\partial\tilde{u}}{\partial\hat{\nu}} &= -\frac{1}{4\pi}(\hat{\nu}, \text{ rot } \int\limits_G \frac{\gamma'}{r}\, dx') \text{ on } \partial\hat{G} \end{cases}$$

by means of a single-layer potential. $\hat{\nu}$ is the outer normal on $\partial\hat{G}$. Since the pointwise identity $(\hat{\nu}, \text{ rot } \int\limits_G \frac{\gamma'}{r}\, dx') = -\text{ Div } [\hat{\nu}, \int\limits_G \frac{\gamma'}{r}\, dx']$ holds on $\partial\hat{G}_i$ ($\partial\hat{G}_i$ is made a Riemannian manifold in the usual way and *Div* is the divergence on this manifold) we arrive at

$$(4) \quad \int\limits_{\partial\hat{G}_i} (\hat{\nu}, \text{ rot } \int\limits_G \frac{\gamma'}{r}\, dx')\, d\Omega' = 0$$

if we use our assumption that the components of connectedness of $\partial\hat{G}_i$ are closed surfaces. (4) now shows that the necessary and sufficient condition for the solvability of (3) is fulfilled. According to the general elliptic theory the solution $\tilde{u}$ is in $C^{2+\alpha}$ in each boundary strip. Thus, by applying the usual extension operator, we can extend $\tilde{u}$ to the whole of $\mathbb{R}^3$ as a $C^{2+\alpha}$-function. Now we set

$$(5) \quad \varepsilon = -\Delta\tilde{u} \text{ on } \mathbb{R}^3.$$

Since $\tilde{u}$ decays in a proper way for $|x| \to +\infty$ which follows from the representation of $\tilde{u}$ as a single-layer potential, we obtain

$$(6) \quad \begin{cases} \tilde{u} &= \frac{1}{4\pi} \int\limits_G \frac{1}{r} \varepsilon' dx', \\ \frac{\partial\tilde{u}}{\partial\nu} &= \frac{1}{4\pi}(\nu, \text{ grad } \int\limits_G \frac{1}{r} \varepsilon'\, dx') = \frac{1}{4\pi}(\nu, \text{ rot } \int\limits_G \frac{1}{r} \gamma'\, dx') \end{cases}$$

according to the construction of $\tilde{u}$. If we take in (2) the function ε defined by (5) we arrive at $\varepsilon^* = 0$, $v|\partial G = 0$. Consequently we can set $u = v$ and we have constructed a right-hand-side inverse of rot (.). From the general theory of elliptic operators one can infer the estimate

$$(7) \quad ||\varepsilon||_{L^p(G)} \leq c(p)||\gamma||_{L^p(G)}$$

by making strong use of the representation of $\tilde{u}$ as a single-layer potential. Using the fundamental theorem of vector-analysis we obtain

$$(8) \quad u = -\text{ grad } \frac{1}{4\pi} \int\limits_G \frac{\varepsilon'}{r}\, dx' + \text{ rot } \frac{1}{4\pi} \int\limits_G \frac{\gamma'}{r}\, dx'.$$

Using Calderón-Zygmund's inequality and (7) it follows that

$$(9) \quad ||\nabla u||_{L^p(G)} \leq c||\gamma||_{L^p(G)}.$$

rot can be extended to the whole of $\overset{\circ}{H}{}^{1,p}(G) = \mathrm{D}(\mathrm{rot})$. We claim that the range $\Re(\mathrm{rot})$ is given by

$$\Re\,(\mathrm{rot}) \;=\; \mathrm{cl}\ \{\psi|\psi \in C_0^\infty(G),\ \ \mathrm{div}\ \psi = 0, \int_G (\psi, z)dx = 0,\ z \in Z(G),\}$$

where the closure cl of the set $\{...\}$ has to be taken with respect to the $L^p(G)$ -norm. Namely, let $f = \mathrm{rot}\ u$ for some $f \in L^p(G)$ and some $u \in \overset{\circ}{H}{}^{1,p}(G)$. We take a sequence (φ_μ) with $\varphi_\mu \in C_0^\infty(G), \mu \in \mathbb{N}$, $\varphi_\mu \to u$, $\mu \to \infty$, in $\overset{\circ}{H}{}^{1,p}(G)$. Then rot $\varphi_\mu \to f$, $\mu \to \infty$, in $L^p(\Omega)$,

$$f_\mu = \ \mathrm{rot}\ \varphi_\mu \in C_o^\infty(\Omega),\ \mathrm{div}\ f_\mu = 0, \int_G (f_\mu, z)dx = 0,\ z \in Z(G),\ \mu \in \mathbb{N}.$$

The situation is quite similar if we want to solve rot $u = \gamma$ in $\hat{G}$, $u|\partial\hat{G} = 0$. However we assume that $1 < p < 3$ since otherwise the expression $\int_G(\gamma, z)dx$, $z \in Z(\hat{G})$, may not be well defined (due to the asymptotic behaviour $z(x) = O(\frac{1}{|x|^2})$ as $|x| \to \infty$). In the preceeding proof one has to exchange the rôles of G and $\hat{G}$ as usual in potential theory (the general theory for inhomogeneous harmonic vector fields also works out on unbounded sets $\hat{G} \subset \mathbb{R}^3$). We end up with a solution u of rot u=γ, $u|\partial\hat{G} = 0$, having the property

(10) $\|\nabla u\|_{L^p(\hat{G})} \leq c\|\gamma\|_{L^p(\hat{G})}$.

If we take as domain of definition of rot the space $\mathrm{D}(\mathrm{rot}) = \overset{\circ}{\hat{H}}{}^{1,p}(\hat{G})$ = closure of the $C_0^\infty(\hat{G})$-vector fields with respect to the norm $\|\nabla \cdot\|_{L^p(\hat{G})}$ we obtain

$$\Re\,(\mathrm{rot}) \;=\; \mathrm{cl}\ \{\psi|\psi \in C_0^\infty(\hat{G}),\ \mathrm{div}\ \psi = 0, \int_{\hat{G}} (\psi, z)dx = 0,\ z \in Z(\hat{G})\}$$

as the range of the operator rot.

The method to solve div $u = \varepsilon$ in G, $u|\partial G = 0$ is in principle the same one. The following conditions are clearly necessary:

$$\int_{G_i} \varepsilon\, dx = 0,\ i = 1, ..., \hat{m}.$$

We claim that these conditions are also sufficient. Namely, we study Neumann's problem for inhomogeneous harmonic vector fields. We look for a vector field v with

$$(11)\quad \begin{cases} \mathrm{rot}\ v = \gamma \text{ in } G, \\ \mathrm{div}\ v = \varepsilon \text{ in } G, \\ -(\nu, v) = 0 \text{ on } \partial G, \\ -\int_{\partial G}([\nu, v], \hat{z}^i)d\Omega = 0,\ 1 \leq i \leq n, \end{cases}$$

where the $\hat{z}^i$ form a basis of $Z(\hat{G})$. The necessary and sufficient conditions to find a unique classical solution of (11) are as follows: γ is free of productivity in G, $\int_{G_i} \varepsilon\, dx = 0$, $i = 1,...,\hat{m}$. For the tangential component $\gamma^* = -[\nu, v]$ of v the following integral equation on ∂G holds:

$$(12)\quad \gamma^* + \Re\gamma^* = \frac{1}{2\pi}[\nu,\ \text{grad} \int_G \frac{1}{r}\varepsilon' dx' - \ \text{rot} \int_G \frac{1}{r}\gamma' dx']$$

where $\Re$ is a compact operator within the continuous tangential fields with Riesz-number 1. We can find a suitable γ such that

$$[\nu,\ \text{grad} \int_G \frac{1}{r}\varepsilon' dx'] = [\nu,\ \text{rot} \int_G \frac{1}{r}\gamma' dx'] \text{ on } \partial G$$

although this is somewhat more complicated than in the previous case. Then the unique solution γ^* of the integral equation (12) together with the side-conditions

$$-\int_{\partial G} (\gamma^*, \hat{z}^i)\, d\Omega = 0,\ 1 \le i \le n,$$

is $\gamma^* = 0$ and fulfills $\gamma^* = -[\nu, v]$, where v solves (11) with the vector field γ just constructed. Consequently $u = v$ is a solution of div $u = \varepsilon$, $u|\partial G = 0$. We can show that $||\gamma||_{L^p(G)} \le c||\varepsilon||_{L^p(G)}$ and by the fundamental theorem of vector analysis we obtain

$$||\nabla u||_{L^p(\Omega)} \le c||\varepsilon||_{L^p(\Omega)},\ 1 < p < +\infty.$$

If we take as domain of definition of div the space D(div) $= \mathring{H}^{1,p}(G)$ we get in the same way as for the operator rot that its range is

$$\Re\ (\text{div})\ =\ \text{cl}\ \{\varphi | \varphi \in C_0^\infty(G),\ \int_{G_i} \varphi\, dx = 0,\ i = 1,...,\hat{m}\},$$

where the closure is taken with respect to the $L^p(G)$-norm. This space is known to be the closed subspace $L^p_M(G) = \{u | u \in L^p(G), \int_{G_i} u\, dx = 0,\ i = 1,...,\hat{m}\}$ of $L^p(G)$.

If we want to solve div $u = \varepsilon$ in $\hat{G}$, $u|\partial\hat{G} = 0$ we have to exchange to the rôles of G and $\hat{G}$ in the previous setting. The result is as follows: If $1 < p < 3$, $\varepsilon \in L^p(\hat{G})$,

$$\int_{\hat{G}_i} \varepsilon\, dx = 0,\ i = 1,...,m,$$

then we can construct an $u \in \mathring{H}^{1,p}(\hat{G})$ with div $u = \varepsilon$. Observe that there is no condition on ε on the unbounded component of connectedness of $\hat{G}$ other than $\varepsilon \in L^p(\hat{G})$; this turns out to be important when treating the exterior problem for the instationary Navier-Stokes equations. We have again the estimate

$$||\nabla u||_{L^p(\hat{G})} \le c||\varepsilon||_{L^p(\hat{G})}$$

and by Sobolev we obtain $||u||_{L^q(\hat{G})} < +\infty$ for q with $\frac{1}{q} = \frac{1}{p} - \frac{1}{3}$. Thus u is integrable to the power q at infinity. It is also possible to solve div $u = \varepsilon$ in $\hat{G}$, $u|\partial\hat{G} = 0$, if $p \geq 3$ by an $u \in \hat{\mathring{H}}^{1,p}(\hat{G})$; this could be done by reducing this case to the already treated one over a bounded open set $G \subset \mathbb{R}^3$. Then, however, u may not decay at infinity. We have omitted this case since we do not need it for the Navier-Stokes equations.

Acknowledgement: The help of my colleague Prof. H. Sohr when constructing the ε in the proof of the solvability of rot $u = \gamma$ in G, $u|\partial G = 0$, is gratefully acknowledged. I owe to him the suggestion to choose ε as we did in (5).

References

[1] Bogovskij, M. E.: Solution of the first boundary value problem for the equation of continuity of an incompressible medium. Soviet Math. Doklady 20, 1094-1098(1979).

[2] Borchers, W. and Sohr, H.: The equations div u = f and rot v = g with homogeneous Dirichlet boundary conditions. Preprint (1988).

[3] Erig, W.: Die Gleichungen von Stokes und die Bogovskij-Formel. Diplomarbeit. Universität Paderborn(1982).

[4] Grenz, S.: Zerlegungssätze für Vektorfelder. Zerlegung von $H_0^{1,2}$ bezüglich der Rotation. Diplomarbeit. Universität Bayreuth (1987).

[5] Kress, R.: Vorlesungen über Potentialtheorie. Universität Göttingen (Wintersemester 1972/73).

[6] Kress, R.: Grundzüge einer Theorie der verallgemeinerten harmonischen Vektorfelder. Meth. Verf. math. Phys. 2, 49-83(1969).

[7] Kress, R.: Potentialtheoretische Randwertprobleme bei Tensorfeldern beliebiger Dimension und beliebigen Ranges. Arch. Rat. Mech. An. 47, 59-80(1972).

[8] Ladyzhenskaja, O. A. and Solonnikov, V. A.: Some problems of vector analysis and generalized formulations of boundary-value problems for the Navier-Stokes equations. Journal of Soviet Mathematics 10, 257-286(1978).

[9] Wahl, W. von: Vorlesung "Partielle Differentialgleichungen II (Potentialtheorie)". Vorlesungsausarbeitung. Universität Bayreuth (1987).

This paper is in final form and no similar paper has been or is being submitted elsewhere.

(received January 5, 1989).

ON OPTIMAL CONSTANTS IN SOME INEQUALITIES

Waldemar Velte
Institut für Angewandte Mathematik und Statistik
Universität Würzburg, Am Hubland, D-8700 Würzburg

0. INTRODUCTION

Our interest in the problem treated in this paper was originally stimulated by a paper of C.O.HORGAN and L.E.PAYNE [1] in which these authors considered three inequalities of different type, namely a version of KORN's second inequality, an inequality of FRIEDRICHS [3] involving pairs of conjugate harmonic functions, and finally an inequality which plays a role in the context of Navier-Stokes flows (BABUSKA and AZIZ [2]).

In [1],the authors proved simple relations between the optimal constants related to the three inequalities mentioned above, and they presented the exact values of the optimal constants for special domains. However, the results are restricted to inequalities for functions defined on (bounded, simply-connected) domains in $\mathbb{R}^2$.

Here, we will look at two inequalities for functions defined on three-dimensional domains, one of them being the inequality which plays a role in the context of Navier-Stokes flows. In particular, we will introduce related eigenvalue problems in variational form. In the special case of spherical domains, the optimal constants can then be given explicitly in terms of known eigenvalues.

Our considerations rest upon results which are known (or nearly known). It was our aim to bring together various results, partly old and partly more recent, in order to understand better the inequalities under consideration. In particular we want to draw attention to an orthogonal decomposition of the underlying space $[H_o^1(G)]^3$ into three subspaces, namely Ker(div), Ker(curl) and a third subspace H which plays a central role in the subsequent considerations.

I. INEQUALITIES UNDER CONSIDERATION, NOTATIONS

Throughout the paper, $G \subset \mathbb{R}^3$ will denote a bounded and simply-connected domain with Lipschitz boundary. We will consider only real valued functions belonging to the following standard function spaces:

The linear space $\mathcal{D}(G)$ of all C^∞-functions with compact support in G, the space $L^2(G)$, $(.,.)$, $\|.\|$ of all square integrable functions where

$$(u, v) = \int_G uv \, dx \quad , \quad \|u\| = (u, u)^{1/2} ,$$

and finally the Sobolev space $H_o^1(G)$ of all functions $u \in L^2(G)$ with (generalized) first derivatives in $L^2(G)$ and zero boundary values in the sense of traces.

However, instead of the standard norm $\|.\|_1$, we will introduce in $H_o^1(G)$ the equivalent norm $|.|_1$ defined by

$$(u,v)_1 = \int_G \operatorname{grad} u \operatorname{grad} v \, dx \quad , \quad |u|_1 = (u,u)_1^{1/2} .$$

Similarly, for vector functions $u = (u_1, u_2, u_3) \in [H_o^1(G)]^3$ we define

$$(u, v)_1 = \sum_{j=1}^{3} (u_j, v_j)_1 \quad , \quad |u|_1 = (u, u)_1^{1/2} .$$

Furtherly, we introduce the following closed linear subspaces:

$$L_o^2(G) = \{ u \in L^2(G) \mid (u, 1) = 0 \} ,$$

$$V = \operatorname{Ker}(\operatorname{div}) = \{ u \in [H_o^1(G)]^3 \mid \operatorname{div} u = 0 \} ,$$

$$W = \operatorname{Ker}(\operatorname{curl}) = \{ u \in [H_o^1(G)]^3 \mid \operatorname{curl} u = 0 \} .$$

By $V^\perp$ we will denote the orthogonal complement of V and by $W^\perp$ the orthogonal complement of W in $[H_o^1(G)]^3$, both with respect to the scalar produkt $(.,.)_1$.

We are mainly interested in the following two inequalities:

Inequality 1. There is a constant C, depending only on the domain G, such that

$$|v|_1^2 \leq C \,\|\operatorname{div} v\|^2 \quad \text{for all} \quad v \in V^{\perp} . \tag{1.1}$$

Inequality 2. There is a constant D, depending only on the domain G, such that

$$|w|_1^2 \leq D \,\|\operatorname{curl} w\|^2 \quad \text{for all} \quad w \in W^{\perp} . \tag{1.2}$$

In what follows, C and D will allways denote the optimal constants, namely

$$C = \sup \left\{ |v|_1^2 / \|\operatorname{div} v\|^2 \;\middle|\; v \in V^{\perp},\ v \neq 0 \right\},$$

$$D = \sup \left\{ |w|_1^2 / \|\operatorname{curl} w\|^2 \;\middle|\; w \in W^{\perp},\ w \neq 0 \right\}.$$

Sometimes the following version of (1.1) is used (see, for instance, [2], where the inequality is applied in the case of two-dimensional domains):

Inequality 1'. For any $p \in L_o^2(G)$ there exists $w \in [H_o^1(G)]^3$ satisfying $\operatorname{div} w = p$ together with

$$|w|_1^2 \leq C' \|p\|^2 . \tag{1.1'}$$

Proposition 1'. The optimal constants C and C' are equal.

Proof. The statement follows easily from the fact that the operator div is an isomorphism from $V^{\perp}$ onto $L_o^2(G)$ (see, for instance, [5] p. 33) together with the orthogonal decomposition $w = u + v$, where $u \in V$ and $v \in V^{\perp}$, so that $\operatorname{div} u = 0$, $\operatorname{div} v = p$ and $|w|_1 \geq |v|_1$.

Closely related to the inequality (1.1) is another inequality, which plays a role in the context of finite element approximations to Stokes or Navier-Stokes flows (see, for instance, [5] p. 35):

Inequality 1". There is a constant b , depending only on the domain G, such that for any $p \in L_o^2(G)$

$$b\|p\| \leq \sup \left\{ \frac{(\operatorname{div} w,\ p)}{|w|_1} \;\middle|\; w \in [H_o^1(G)]^3 ,\ w \neq 0 \right\}. \qquad (1.1)"$$

Proposition 1". The optimal constants in (1.1) and (1.1)" are related by

$$b^2 = 1 / C .$$

Proof. Using again the orthogonal decomposition $w = u + v$, where $u \in V$, $v \in V^\perp$, $\operatorname{div} w = \operatorname{div} v$ and $|w|_1 \geq |v|_1$, one gets for w with $\operatorname{div} w \neq 0$

$$(\operatorname{div} w,\ p) / |w|_1 \leq (\operatorname{div} v,\ p) / |v|_1 .$$

Now, for given $p \in L_o^2(G)$, the right hand side takes on its maximal value for $v \in V^\perp$ such that $\operatorname{div} v = p$. Thus the optimal constant b is given by

$$b = \inf \left\{ \|\operatorname{div} v\| / |v|_1 \;\middle|\; v \in V^\perp ,\ v \neq 0 \right\} ,$$

whence $b^2 = 1 / C$.

II. DECOMPOSITION OF THE SPACE $[H_o^1(G)]^3$

In the space $[H_o^1(G)]^3$, the scalar product $(u,\ v)_1$ can be represented as follows:

$$(u,\ v)_1 = (\operatorname{curl} u ,\ \operatorname{curl} v)_1 + (\operatorname{div} u ,\ \operatorname{div} v)_1 . \qquad (2.1)$$

For functions $u,\ v \in [\mathcal{D}(G)]^3$, formula (2.1) is readily established by partial integration. But, since $\mathcal{D}(G)$ is dense in $H_o^1(G)$, the identity (2.1) holds for all functions in $[H_o^1(G)]^3$, too.

In section I we have introduced the (closed) linear subspaces $V = \operatorname{Ker}(\operatorname{div})$ and $W = \operatorname{Ker}(\operatorname{curl})$. From (2.1) follows immediately that the two subspaces V and W are orthogonal with respect to the scalar product $(u,\ v)_1$. Thus the space $[H_o^1(G)]^3$ can be decomposed into three orthogonal subspaces as follows:

$$[H_o^1(G)]^3 = V \oplus W \oplus H \quad , \quad V^\perp = W \oplus H \tag{2.2}$$

In what follows, we will use a well known characterization of the subspace $V^\perp$: Consider pairs of functions $p \in L_o^2(G)$ and $v \in [H_o^1(G)]^3$ satisfying the relations

$$(v, w)_1 = (p, \operatorname{div} w) \quad \text{for all} \quad w \in [H_o^1(G)]^3 \quad . \tag{2.3}$$

For given function p , the right hand side is a bounded linear functional with respect to w . Thus, (2.3) has a unique solution v by Riesz's represented theorem. Now, for any $w \in V = \operatorname{Ker}(\operatorname{div})$ one gets from (2.3) the equation $(v, w)_1 = 0$. So the solution v belongs to $V^\perp$. More precisely can be shown that the mapping $p \to v$ is an isomorphism from $L_o^2(G)$ onto $V^\perp$ (see, for instance, [5] p. 36). This is the characterization we will use.

Now, we want to characterize the subspace H occuring in the decomposition (2.2).

<u>Theorem 1.</u> For given $p \in L_o^2(G)$, the uniquely determined solution $v \in V^\perp$ of (2.3) belongs to the subspace H if and only if the function p is harmonic.

<u>Proof.</u> Let us firstly suppose that $v \in H$. Then v is orthogonal to all functions $w \in W = \operatorname{Ker}(\operatorname{curl})$. In particular we get from (2.3) for all functions $w = \operatorname{grad} \phi$, $\phi \in \mathcal{D}(G)$, the relation $(p, \Delta \phi) = 0$. Hence, p is harmonic by Weyl's lemma.

Conversely, when p is harmonic, then follows $(v, \operatorname{grad} \phi)_1 = 0$. Since, however, the set $\{\operatorname{grad} \phi \mid \phi \in \mathcal{D}(G)\}$ is dense in W, one concludes that v is orthogonal to W and hence $v \in V^\perp \ominus W = H$.

<u>Remark.</u> The simple characterization of the subspace H given above will play a central role in the subsequent considerations. The author is not aware of a reference for this characterization, though it seems to be widely known. (So, H. Sohr pointed out to the author during the conference that he was aware of it.)

III. RELATED EIGENVALUE PROLEMS

According to (2.1), one has for all functions $v \in [H_o^1(G)]^3$ the identity

$$|v|_1^2 = \|\text{curl } v\|^2 + \|\text{div } v\|^2 \quad . \tag{3.1}$$

From inequality (1.1) we get

$$\|\text{div } v\|^2 \leq |v|_1^2 \leq C \, \|\text{div } v\|^2 \quad \text{for all} \quad v \in V^{\perp}$$

or equivalently

$$1 \leq \frac{|v|_1^2}{\|\text{div } v\|^2} \leq C \qquad \text{for all} \quad v \in V^{\perp} \, , \; v \neq 0 \, . \tag{3.2}$$

Here, the lower bound 1 is attained for all functions $v \in W$.

Now, let us introduce the following eigenvalue problem: Find $\lambda \in \mathbb{R}$ together with $v \in [H_o^1(G)]^3$, $v \neq 0$ such that

$$(v, w)_1 = \lambda \, (\text{div } v, \text{div } w) \qquad \text{for all} \quad w \in [H_o^1(G)]^3 \, . \tag{3.3}$$

Proposition 3. The eigenvalue problem (3.3) has the following properties:

(1) $\lambda = 1$ and $\lambda = \infty$ are eigenvalues of infinite multiplicity corresponding to the eigenspaces W and V respectively.

(2) Any other eigenvalue is contained in the intervall (0, C] , and the corresponding eigenfunctions belong to H.

Proof. (1) Let v be an eigenfunction to $\lambda = 1$. Putting $w = v$ in (3.3) and comparing with (3.1) one gets $\|\text{curl } v\| = 0$ and hence $w \in W$. Now, let v be an eigenfunction to $\lambda = \infty$ which means that

$$(\text{div } v, \text{div } w) = 0 \quad \text{for all} \quad w \in [H_o^1(G)]^3 \quad .$$

Then $\|\text{div } v\| = 0$ and hence $v \in V$.
(2) Consider now any other eigenvalue. Then, according to (3.2), the eigenvalue λ is contained in the intervall (1.C] .

In order to show that any corresponding eigenfunction belongs to H, let us write (3.3) in the equivalent form

$$(\operatorname{curl} v, \operatorname{curl} w) = (\lambda - 1)\,(\operatorname{div} v, \operatorname{div} w) \quad \text{for all} \quad w \in [H_o^1(G)]^3 ,$$

where $\lambda - 1 \neq 0$. Now, for any $w = \operatorname{grad} \phi$, $\phi \in \mathcal{D}(G)$ the left hand side becomes zero. Hence

$$(\operatorname{div} v, \Delta \phi) = 0 \quad \text{for all} \quad \phi \in \mathcal{D}(G).$$

Thus, $\operatorname{div} v$ is harmonic by Weyl's lemma, and v belongs to H by Theorem 1.

Similarly as above, one gets from (1.2) the inequalities

$$\|\operatorname{curl} v\|^2 \leq |v|_1^2 \leq D\, \|\operatorname{curl} v\|^2 \quad \text{for all} \quad v \in W^{\perp}$$

or equivalently

$$1 \leq \frac{|v|_1^2}{\|\operatorname{curl} v\|^2} \leq D \quad \text{for all} \quad v \in W^{\perp},\ w \neq 0 . \tag{3.4}$$

Here, the lower bound 1 is attained for all functions $w \in W$.

This time, we consider the following eigenvalue problem: Find $\mu \in \mathbb{R}$ together with $v \in [H_o^1(G)]^3$, $v \neq 0$ such that

$$(v, w)_1 = \mu\,(\operatorname{curl} v, \operatorname{curl} w) \quad \text{for all} \quad w \in [H_o^1(G)]^3 . \tag{3.5}$$

From the identity (2.1) follows easely, that the two eigenvalue problems (3.3) and (3.5) are equivalent: Any eigenfunction of (3.3) is also eigenfunction of (3.5) an vice versa, where the corresponding eigenvalues λ and μ are related by

$$\mu = 1 + \frac{1}{\lambda - 1} , \qquad \lambda = 1 + \frac{1}{\mu - 1} . \tag{3.6}$$

Thus one gets as corollary of Proposition 3:

<u>Proposition 4.</u> The eigenvalue problem (3.5) has the following properties:

(1) $\mu = 1$ and $\mu = \infty$ are eigenvalues of infinite multiplicity corresponding to the eigenspaces V and W respectively.

(2) Any other eigenvalue is contained in the interval $(1, D]$, and the corresponding eigenfunctions belong to H.

IV. Eigensolutions for spherical domains

The eigenvalue problem (3.3) in variational form is a weak form of the classical eigenvalueproblem

$$\Delta v = \lambda \operatorname{grad} \operatorname{div} v \quad \text{in } G, \quad v = 0 \quad \text{on } \partial G. \tag{4.1}$$

We are interested in eigensolutions to eigenvalues $\lambda \neq 1$, $\lambda \neq \infty$. For spherical regions of radius R the eigensolutions are known. They can be found, for instance, in the classical paper of E. and F. Cosserat [4] which appeared in 1898. The eigenfunctions are given in terms of harmonic polynomials $P^{(n)}$, where $P^{(n)}$ denotes the generic homogeneous harmonic polynomial of degree n. As is well known, there are $2n + 1$ linearly independent homogeneous harmonic polynomials of degree n.

The eigenfunctions $v^{(n)}$ and the eigenvalues λ_n are then given by

$$v^{(n)}(x) = (r^2 - R^2) \operatorname{grad} P^{(n)}(x), \quad r^2 = |x|^2,$$

$$\lambda_n = 2 + \frac{1}{n} \qquad (n \in \mathbb{N}),$$

where λ_n is an eigenvalue of multiplicity $2n + 1$.

Theorem 2. The eigenvalue problem (3.1) has the following properties: Besides of the two eigenvalues $\lambda = 1$ and $\lambda = \infty$ of infinite multiplicity there is a sequence of discrete eigenvalues of finite multiplicity,

$$\lambda_1 > \lambda_2 > \ldots > 2, \quad \lim_{n \to \infty} \lambda_n = 2.$$

The set of the corresponddting (orthogonal) eigenfunctions is complete in H. In the inequality (1.1)

$$|v|_1^2 \leq C \, \|\operatorname{div} v\|^2 \quad \text{for all } v \in V^{\perp},$$

the optimal constant is given by $\underline{C = 3}$.

Proof. The essential point is the completeness of the eigenfunctions $v^{(n)}$ in the subspace H. (Remember, that $v^{(n)}$ stands for the generic eigensolution in the eigenspace of dimension $2n+1$ which belongs to λ_n.) Suppose now, that $v \in H$ is arbitrarily given. Then, by Theorem 1, there is a harmonic function $p \in L_o^2(G)$ satisfying (2.3). Now, when $(v, v^{(n)})_1 = 0$ for all eigenfunctions $v^{(n)}$, then $(p, \operatorname{div} v^{(n)}) = 0$ holds, too. However, by Euler's formula for homogeneous functions of degree n one gets

$$\operatorname{div} v^{(n)} = 2\, x \operatorname{grad} P^{(n)}(x) = 2\, n\, P^{(n)}(x) .$$

Thus $(p, P^{(n)}) = 0$ for all homogeneous harmonic polynomials of degree $n \in \mathbb{N}$, whence $p = 0$ and $v = 0$.
Developing now $v \in H$ in a series of the (orthogonal) functions $v^{(n)}$, one gets for the Rayleigh quotient

$$\max_{v \in H} \left(|v|_1^2 / \|\operatorname{div} v\|^2 \right) = \lambda_1 = 3 , \quad \text{whence} \quad C = 3 .$$

Remembering the equivalence of (3.3) and (3.5) together with the relation (3.6) between the eigenvalues λ and μ, one gets as corollary from Theorem 2:

Theorem 3. The eigenvalue problem (3.5) has the following properties: Besides of the two eigenvalues $\mu = 1$ and $\mu = \infty$ of infinite multiplicity there is a sequence of discrete eigenvalues of finite multiplicity,

$$\mu_1 < \mu_2 < \dots < 2 , \qquad \lim_{n \to \infty} \mu_n = 2 .$$

In the inequality (1.2),

$$|v|_1^2 \le D \|\operatorname{curl} v\|^2 \qquad \text{for all} \quad w \in W^{\perp} ,$$

the optimal constant is given by $D = 2$.

Proof. Using again that the eigenfunctions $v^{(n)}$ yield a complete orthogonal system in H, one gets for the Rayleigh quotient

$$\sup_{v \in H} \left(|v|_1^2 / \|\operatorname{curl} v\|^2 \right) = 2 , \quad \text{whence} \quad D = 2 .$$

V. FINAL REMARKS

For more general bounded and simply-connected domains, the optimal constants C and D seem to be unknown. However, one has the following informations: The differential operator L given by

$$Lv = \Delta v - \lambda \operatorname{grad} \operatorname{div} v$$

fails to be elliptic for $\lambda = 1$ and $\lambda = \infty$. As we have seen, these values are eigenvalues of infinite multiplicity of (3.3), even for domains of general shape.

As to the theory of boundary value problems for systems of linear differential equations, one has to consider also the so called complementing condition ensuring that the boundary value problem under consideration is elliptic (or coercive). A version of the complementing condition which applies to domains with C^{∞} - boundaries can be found, for instance, in [7] or [8]. The latter condition is a local condition at the boundary. Now, since L has constant coefficients and is invariant with respect to translations and rigid rotations in $\mathbb{R}^3$, it turns out that the exceptional values where the complementing conditions fails are independent of the shape of the domain G .

In the case of zero boundary values it can be shown that, at least for domains with C^{∞} - boundaries, there are no exceptional values besides of $\lambda = 1$, $\lambda = 2$ and $\lambda = \infty$. Any other value of λ is either a regular value or an eigenvalue of finite multiplicity.

We will come back to these questions in another paper.

REFERENCES

[1] HORGAN, C.O. and L.E. PAYNE, On inequalities of Korn, Friedrichs and Babuska-Aziz. Arch. Rational Mech. Anal. 82, 165-179 (1983).

[2] BABUSKA, I. and A.K. AZIZ, Survey lectures on the mathematical foundation of the finite element method. In: AZIZ, A.K. (ed.), The Mathematical Foundation of the Finite Element Method with Applications to Partial Differential Equations. Academic Press, New York, 1972.

[3] FRIEDRICHS, K.O., On certain inequalities and characteristic value problems for analytic functions and for functions of two variables. Trans.Amer.Math.Soc. 41, 321 - 364 (1947).

[4) COSSERAT, E. and F. COSSERAT, Sur les equations de la theorie de l'élasticité. C.R. Acad. Sci. Paris 126, 1089 - 1091 (1898).

[5] GIRAULT, V. and P.-A. RAVIART, Finite element approximation of the Navier-Stokes equations. Lecture Notes in Mathematics, vol. 749 (Revised Reprint of the first edition), Springer-Verlag Berlin . Heidelberg . New York (1981).

[6] GIRAULT, V. and P.-A. RAVIART, Finite element methods for Navier-Stokes equations. Springer Series in Computational Mathematics vol. 5, Springer-Verlag Berlin . Heidelberg . New York (1986).

[7] AGMON, S., A. DOUGLIS and L. NIERENBERG, Estimates near the boundary for solutions of elliptic partial differential equations satisfying general boundary conditions II. Comm. Pure Appl. Math.. 17, 35 - 92 (1964).

[8] HÖRMANDER, L., Linear partial differential operators. Grundlehren der Math. Wiss. vol. 116. Springer-Verlag Berlin . Heidelberg . New York (1969), Chapter X.

This paper is in final form, and no similar paper has been or is submitted elsewhere.

(received August 28, 1989.)

BOUNDARY-VALUE PROBLEMS FOR NAVIER-STOKES EQUATIONS OF VISCOUS GAS

A.V.Kazhikhov
Lavrentyev Institute of Hydrodynamics, Novosibirsk, USSR

The present paper gives a survey concerned with the correctness of initial boundary-value problems for one-dimensional equations of viscous gas. First, we consider the problems with nonhomogeneous boundary conditions. Second, a progress is to be mentioned in solving the problems of justification of the approximate methods, such as difference and finite elements methods. In addition, we formulate some unsolved problems.

1. Nonhomogeneous boundary-value problems

Let us consider the Navier-Stokes equations for one-dimensional flow of viscous compressible heat-conductive gas. These equations can be written in Lagrangian coordinates as

$$\frac{\partial v}{\partial t} = \frac{\partial u}{\partial x},$$

$$\frac{\partial u}{\partial t} = \frac{\partial}{\partial x}\left(\frac{\mu}{v}\frac{\partial u}{\partial x}\right) - \frac{\partial p}{\partial x}, \tag{1}$$

$$\frac{\partial \theta}{\partial t} = \frac{\partial}{\partial x}\left(\frac{\varkappa}{v}\frac{\partial \theta}{\partial x}\right) + \frac{\mu}{v}\left(\frac{\partial u}{\partial x}\right)^2 - p\frac{\partial u}{\partial x}, \quad p = k\rho\,\theta, \quad v = \rho^{-1}.$$

Here ρ, u, θ are the density, velocity and absolute temperature, respectively; p the pressure, μ , $\varkappa$ and k are the physical coefficients (positive constants), $x \in \Omega = (0,1)$, $t \in (0,T)$, $0 < T < \infty$. Initially, when t=0, the unknown functions are assumed to be known:

$$v|_{t=o} = v_o(x), \quad u|_{t=o} = u_o(x), \quad \theta|_{t=o} = \theta_o(x), \quad x \in \Omega \tag{2}$$

where $(v_o, u_o, \theta_o) \in W_2^1(\Omega)$ and (v_o, θ_o) are the positive and bounded functions:

$$0 < m \le v_o(x) \le M < \infty, \quad m \le \Theta_o(x) \le M. \tag{3}$$

We consider the next conditions on the boundaries x=0 and x=1:

$$u|_{x=o} = q_o(t), \quad u|_{x=1} = q_1(t). \tag{4}$$

$$\Theta|_{x=o} = \chi_o(t), \quad \Theta|_{x=1} = \chi_1(t). \tag{5}$$

These functions have the properties

$$\begin{aligned} &(q_i(t), \chi_i(t)) \in W_2^1(0,T), \\ &l(t) \equiv \int_0^1 v_o(x)dx + \int_0^t [q_1(\tau) - q_o(\tau)]d\tau \ge l_o > 0 \\ &q_i(0) = u_o(i), \quad \chi_i(0) = \Theta_o(i), \quad i = 0,1. \end{aligned} \tag{6}$$

Theorem 1. The problem (1)-(6) has a unique strong solution.

II. The model of magnetic gas dynamics.

The system of equations has the form

$$\begin{aligned} &\frac{\partial v}{\partial t} = \frac{\partial u}{\partial x} \\ &\frac{\partial u}{\partial t} = \frac{\partial}{\partial x}\left(\frac{\mu}{v}\frac{\partial u}{\partial x}\right) - \frac{\partial p}{\partial x} - H\frac{\partial H}{\partial x} \\ &\frac{\partial \Theta}{\partial t} = \frac{\partial}{\partial x}\left(\frac{\varkappa}{v}\frac{\partial \Theta}{\partial x}\right) + \frac{\mu}{v}\left(\frac{\partial u}{\partial x}\right)^2 + \frac{\nu}{v}\left(\frac{\partial H}{\partial x}\right)^2 - p\frac{\partial u}{\partial x} \\ &\frac{\partial}{\partial t}(H v) = \frac{\partial}{\partial x}\left(\frac{\nu}{v}\frac{\partial H}{\partial x}\right), \quad p = k\rho\Theta, \quad v = \rho^{-1}. \end{aligned} \tag{7}$$

In addition to (1), magnetic field intensity H is the unknown function.

We prove the unique solvability of the initial boundary-value problems for the system (7) with conditions (2)-(6) and

$$H|_{x=o} = \psi_o(t), \quad H|_{x=1} = \psi_1(t) \tag{8}$$

or

$$\left.\frac{\partial H}{\partial x}\right|_{x=o} = \varphi_o(t), \quad \left.\frac{\partial H}{\partial x}\right|_{x=1} = \varphi_1(t) \tag{9}$$

and

$$H\big|_{t=0} = H_0(x). \tag{10}$$

for the function H .

III. Approximate methods.

In recent years a lot of interesting investigations have been made on the justification of the approximate methods for solving systems (1) and (7). The most significant results have been obtained by Sh.Smagulov, A.A.Amosov and A.A.Zlotnik (see [2]). They investigated a wide class of difference methods and proved the convergence of approximate solutions to the exact one.

IV. Unsolved problems.

Let us formulate some unsolved problems, which are of particular interest:

1) Viscosity as a function of temperature.

The existence of a global solution when the viscosity depends on the temperature is unknown even for simplest Burger's model:

$$\frac{\partial v}{\partial t} = \frac{\partial u}{\partial x}$$

$$\frac{\partial u}{\partial t} = \frac{\partial}{\partial x}\left(\frac{\mu(\Theta)}{v}\frac{\partial u}{\partial x}\right), \quad \frac{\partial \Theta}{\partial t} = \frac{\partial}{\partial x}\left(\frac{\varkappa(\Theta)}{v}\frac{\partial \Theta}{\partial x}\right) + \frac{\mu(\Theta)}{v}\left(\frac{\partial u}{\partial x}\right)^2$$

2) Van der Waals constitutive equation:

$$p = \frac{k\,\rho\,\Theta}{1 - b\,\rho} - a\,\rho^2 \qquad (a > 0,\ b > 0)$$

3) Cauchy problem with distinct limits at infinities:

$$u\big|_{t=0} = u_0(x),\ x \in (-\infty, +\infty)$$

$$u_- = \lim_{x \to -\infty} u_o(x) \neq u_+ = \lim_{x \to +\infty} u_o(x)$$

4) Unilateral problem

$$\frac{\partial v}{\partial t} = \frac{\partial u}{\partial x}$$

$$\frac{\partial u}{\partial t} = \frac{\partial}{\partial x}\left(\frac{\mu}{v}\frac{\partial u}{\partial x}\right) - \frac{\partial p(v)}{\partial x}$$

with the initial conditions

$$v|_{t=o} = v_o(x), \quad u|_{t=o} = u_o(x), \; x \in (0,1)$$

and the boundary conditions

$$u|_{x=o} = 0$$

$$\left(\frac{\mu}{v}\frac{\partial u}{\partial x} - p\right)\Big|_{x=1} \leq 0 \qquad \int_o^t u(x,\tau)d\tau \;\Big|_{x=1} \leq 0$$

$$\left(\frac{\mu}{v}\frac{\partial u}{\partial p} - p\right) \int_o^t u(x,\tau)d\tau\Big|_{x=1} = 0$$

The following problem is suggested to be a model example:

$$\frac{\partial u}{\partial t} = \frac{\partial^2 u}{\partial x^2}, \; x \in \Omega, \; t \in (0,T)$$

$$u|_{t=0} = u_o(x), \; u|x=0 = 0$$

$$\frac{\partial u}{\partial x} \leq 0, \quad \int_o^t u\,d\tau \leq 0, \; u_x \cdot \int_o^t u\,d\tau = 0 \text{ at } x = 1.$$

5) In addition, it is appropriate to mention the problems with axial and central symmetry.

References

1. Antonsev S.N., Kazhikhov A.V., Monakhov V.N. Boundary-value problems in mechanics of nonhomogeneous fluids(Russian),Novosibirsk,1983,320pp.
2. Smagulov Sh. Mathematical aspects in approximate methods for Navier-Stokes equations.Doct.dissertation(Russian),Novosibirsk,1988,380pp.

This paper is in its final form and no similar paper is being or has been submitted elsewhere

(received February 21, 1989.)

ON THE ONE-DIMENSIONAL NAVIER-STOKES EQUATIONS FOR COMPRESSIBLE FLUIDS

Alberto Valli
Dipartimento di Matematica, Università di Trento
I- 38050 Povo (Trento)

1. Introduction.

We want to present some results concerning compressible viscous fluids which have been essentially obtained in a joint paper with I. Straškraba ([10]).

Consider the one-dimensional Navier-Stokes equations for barotropic compressible viscous fluids in Lagrangian mass coordinates

$$
(1.1)\qquad
\begin{aligned}
&u_t + p(v)_x - \mu\,(v^{-1}u_x)_x = f(t,\int_0^x v(t,z)dz) && \text{in }]0,+\infty[\times]0,1[\ \equiv Q_\infty\\
&v_t = u_x && \text{in } Q_\infty\\
&u(t,0) = u(t,1) = 0 && \text{in } [0,+\infty[\\
&u(0,x) = u_0(x)\ ,\quad v(0,x) = v_0(x) && \text{in } [0,1]\\
&\int_0^1 v(t,z)dz = 1 && \text{in } [0,+\infty[\ ,
\end{aligned}
$$

where the velocity of the fluid u and the specific volume $v > 0$ are the unknowns, and p is the pressure (a known function satisfying $p'(\xi) < 0$ for $\xi > 0$), μ the viscosity (a given positive constant), f the assigned external force field for unit mass.

We are looking for solutions satisfying the additional conditions

$$
(1.2)\qquad 0 < \inf_{Q_\infty} v \le \sup_{Q_\infty} v < +\infty\ ,
$$

i.e. we are requiring that vacuum or infinite density cannot occur up to infinity.

Let us recall some previous results concerning this problem.

(i) Extending a procedure introduced by Kanel' [5] for the Cauchy problem, Kazhikhov [6], [7] proved that there exists a unique solution to (1.1), (1.2) for $f = 0$; moreover, the stationary solution $u = 0$, $v = 1$ is asymptotically stable in an exponential way. The existence result can be easily generalized to the case $f \in L^1(0,+\infty;L^\infty(0,1)) \cap L^2(0,+\infty;L^\infty(0,1))$; hence, if $f = f(y) \in L^\infty(0,1)$, one can prove the existence of a solution to (1.1), (1.2) on $]0,T[\times]0,1[$ for an arbitrary $T > 0$, i.e. the existence of a solution to (1.1). On the contrary, it is not possible to infer that in this case (1.2) is satisfied.

(ii) Shelukhin [9] showed the existence of a unique solution to (1.1), (1.2) if $f \in C^1(\overline{Q_\infty})$ and the pressure p satisfies the condition

---

This paper is in final form and no similar paper has been or is being submitted elsewhere.

$$\exists\, c \geq 1:\ c^{-1}\xi^{-1} \leq -p'(\xi) \leq c\,\xi^{-1} \quad \text{for each } \xi > 0\,,$$

for instance $p(\xi) = -A \log \xi$, A a given positive constant.

(iii) Assuming that $f \in L^1(0,1)$, Beirão da Veiga [2] proved that a stationary solution to $(1.1)_1$-$(1.1)_3$, $(1.1)_5$ and (1.2) (let us notice that such a solution must satisfy $u = 0$) exists if and only if a suitable compatibility condition between the external force f and the pressure p is satisfied. Let us make precise this condition. Define

$$\tilde{p}(\xi) \equiv p(\xi^{-1})\,, \qquad \pi(\xi) \equiv \int_1^{\xi} \eta^{-1}\tilde{p}'(\eta)d\eta\,, \qquad \Phi \equiv \pi^{-1}\,, \qquad F(y) \equiv \int_0^y f(z)dz\,, \tag{1.3}$$

$$\pi(0^+) \equiv a\,, \qquad \pi(+\infty) \equiv b\,, \qquad m_0 \equiv \min_{[0,1]} F\,, \qquad M_0 \equiv \max_{[0,1]} F\,. \tag{1.4}$$

Then the necessary and sufficient condition for the existence of a (unique) stationary solution is given by

$$a - m_0 < b - M_0\,, \tag{1.5}$$

$$\int_0^1 \Phi(a-m_0+F(z))dz < 1 < \int_0^1 \Phi(b-M_0+F(z))dz\,. \tag{1.6}$$

Some remarks about condition (1.5)-(1.6) are needed:

(i) it is satisfied if f is small enough in $L^1(0,1)$, or if $p(\xi) = -A \log \xi$, or if $p(\xi) = A\,\xi^{-1}$.

(ii) if $p(\xi) = A\,\xi^{-\gamma}$, $\gamma > 1$, and $f(y) = \lambda$, (1.5)-(1.6) becomes $|\lambda| < A\,[\gamma/(\gamma-1)]^{\gamma}$. One can easily see that if $|\lambda| = A\,[\gamma/(\gamma-1)]^{\gamma}$, then vacuum appears in one of the extreme points of $]0,1[$.

Some natural questions now arise: it is possible to find a solution to (1.1), (1.2) for each initial data, under general assumptions on $f = f(y)$ and p? If not, can we give a necessary and sufficient condition to get such a solution? From stability considerations, the result of Beirão da Veiga [2] seems to suggest that (1.5)-(1.6) has to be a necessary condition at least.

Indeed, in Straškraba-Valli [10] the following theorem is proved:

Theorem 1.1. Assume that $f = f(y) \in H^{1,\infty}(0,1)$ and $p \in C^2(0,+\infty)$. If a (regular enough) solution to (1.1), (1.2) exists, then a stationary solution to $(1.1)_1$-$(1.1)_3$, $(1.1)_5$ and (1.2) exists too. Hence, from the result of Beirão da Veiga [2], condition (1.5)-(1.6) must be satisfied. In other words, if (1.5)-(1.6) is not satisfied, any (regular enough) solution to (1.1) must asymptotically develop either vacuum or infinite density. Recall moreover that, from (the extension of) the result of Kazhikhov [7] mentioned above, vacuum or infinite density cannot develop in finite time.

Some further results concerning this argument can be found in Lovicar-Straškraba-Valli [8]. In particular, assuming $f = f(y)$, it is shown that the only solution to $(1.1)_1$-$(1.1)_3$, $(1.1)_5$ and (1.2) on the whole real line satisfying the additional condition $u \in L^\infty(\mathbf{R};L^2(0,1))$ is the stationary one.

Let us remark moreover that it is not known if (1.5)-(1.6) is a sufficient condition to find a solution to (1.1), (1.2). Some recent results obtained by Beirão da Veiga [3], [4] (see also these Proceedings) give a partial positive answer to this question. In fact, in [3] it is proved that, if $\sup_{]0,+\infty[\times]0,1[} |f| \leq r \leq r_0$, r_0 small enough, then the solution to (1.1), (1.2) exists for every initial data satisfying

$$\|u_0\|_0 < A(r)\,, \quad \|(\log v_0)_x\|_0 < A(r)\,,$$

where A is a real positive decreasing function depending only on μ and p, and such that $A(0^+) = +\infty$. (Here and in the sequel we will denote by $\|\cdot\|_k$ the usual norm in the Sobolev space $H^k(0,1)$).
In [4], assuming that (1.5)-(1.6) is satisfied, the stationary solution is proved to be asymptotically stable in an exponential way for small perturbations; in particular, the solution to (1.1), (1.2) exists for each initial data near the stationary solution.

Finally, as a consequence of a few numerical results (performed in the case $p(\xi) = A\,\xi^{-\gamma}$ and $f(y) = \lambda$ by employing a finite difference scheme proposed by Amosov-Zlotnik [1]), it seems possible to conjecture that the solution to (1.1), (1.2) does exist if (1.5)-(1.6) holds, and that the corresponding stationary solution is globally stable. Moreover, if (1.5)-(1.6) is violated, the numerical computations suggest that a (globally stable) stationary solution yet exists, but now the density is zero in a subset of $]0,1[$. However, no mathematical proof of these assertions is yet available.

2. Idea of the proof.

Let us go back to Eulerian coordinates, i.e.

$$y = \int_0^x v(t,z)dz \,, \qquad \rho(t,y) = 1/v(t,x) \,.$$

A stationary solution $\rho* = \rho*(y)$ has to satisfy

$$\tilde{p}(\rho*)_y = \rho* f \,, \qquad \int_0^1 \rho*(z)dz = 1 \,.$$

or, equivalently, a stationary solution $\rho*(y)$ exists if we find a real constant $c*$ such that

$$\pi(\rho*(y)) = F(y) + c* \,. \tag{2.1}$$

(See (1.3) for the definition of π and F).

Consider now the operator Ψ defined in the following way:

$$D(\Psi) \equiv \{ w \in H^1(0,1) \mid 0 < c_1 \le w(y) \le c_2 < +\infty \,,\ \int_0^1 w(z)dz = 1 \} \,, \tag{2.2}$$

$$\Psi(w) \equiv \pi o w \,. \tag{2.3}$$

It is easily seen that Ψ has a closed range in $H^1(0,1)$, since we can control the norm of w in $H^1(0,1)$ in term of the norm of $\Psi(w)$ in the same space, and the immersion of $H^1(0,1)$ in $C^0([0,1])$ is compact.

We claim: if one finds a function $\rho(t,y)$ and a sequence $t_n \to +\infty$ such that

$$y \to \rho(t_n,y) \text{ belongs to } D(\Psi) \text{ for each } n \ge 1 \,, \tag{2.4}$$

$$\| \pi(\rho(t_n,\cdot))_y - f \|_0 \to 0 \text{ as } n \to +\infty \,, \tag{2.5}$$

then there exists a solution to (2.1) belonging to $D(\Psi)$, hence, going back to the Lagrangian mass coordinates, there exists a stationary solution to $(1.1)_1$-$(1.1)_3$, $(1.1)_5$ and (1.2).

In fact, recalling that $f = F_y$, by using Poincaré's inequality we find

$$\| \pi(\rho(t_n,\cdot)) - \int_0^1 \pi(\rho(t_n,z))dz - F + \int_0^1 F(z)dz \|_1 \to 0 \text{ as } n \to +\infty \,.$$

Moreover, we can select a subsequence t_{n_k} such that

$$\int_0^1 \pi(\rho(t_{n_k},z))dz \to k_0 \text{ as } k \to +\infty$$

for a suitable real constant k_0.
Hence

$$\| \pi(\rho(t_{n_k},\cdot)) - k_0 - F + \int_0^1 F(z)dz \|_1 \to 0 \text{ as } k \to +\infty ,$$

and, since Ψ has a closed range, there exists $\rho^* \in D(\Psi)$ such that

$$\pi(\rho^*(y)) = F(y) + c^* \;,\; c^* \equiv k_0 - \int_0^1 F(z)dz \;.$$

In Lagrangian mass coordinates, (2.5) means

(2.6) $$\| v(t_n,\cdot)^{1/2}[\, p(v(t_n,\cdot))_x - f(\int_0^{\cdot} v(t_n,z)dz)\,] \|_0 \to 0 \text{ as } n \to +\infty .$$

Hence it is sufficient to show that a (regular enough) solution to (1.1), (1.2) satisfies

$$\| p(v(t,\cdot))_x - f(\int_0^{\cdot} v(t,z)dz) \|_0 \to 0 \text{ as } t \to +\infty ,$$

or, using $(1.1)_1$

(2.7) $$\| u_t - \mu(v^{-1}u_x)_x \|_0 \to 0 \text{ as } t \to +\infty .$$

Introducing, as in Shelukhin [9],

$$\psi \equiv u - \mu v^{-1} v_x = u - \mu(\log v)_x ,$$

from $(1.1)_2$ we can rewrite (2.7) as

(2.8) $$\| \psi_t \|_0 \to 0 \text{ as } t \to +\infty .$$

Indeed, we will prove that $\psi_t \in H^1(0,+\infty;L^2(0,1))$.

Let us remark finally that ψ satisfies the following equation

(2.9) $$\psi_t - \mu^{-1}p'(v)v\psi = f(\int_0^x v(t,z)dz) - \mu^{-1}p'(v)vu ,$$

since $p_x = p'(v)v_x = -\mu^{-1}p'(v)v\psi + \mu^{-1}p'(v)vu$. From the assumptions on v, we can infer that

$$-\mu^{-1}p'(v)v \geq c_3 > 0$$

for a suitable constant c_3.

3. Proof of the Theorem.

The proof requires several steps.

(i) $u \in L^\infty(0,+\infty;L^2(0,1)) \cap L^2(0,+\infty;H^1(0,1))$, $v_t \in L^2(0,+\infty;L^2(0,1))$.

Multiply $(1.1)_1$ by u and integrate in $]0,1[$. We get, integrating by parts

$$(3.1)\qquad \frac{d}{dt}\int_0^1[2^{-1}u^2 + P(v) + \lambda - F(\int_0^x v)] + \mu\int_0^1 v^{-1}u_x^2 = 0 \ ,$$

where

$$(3.2)\qquad P(\xi) \equiv \int_1^{\xi}[p(1) - p(\eta)]d\eta \ .$$

(Here and in the sequel we will omit the differential dx for integrals in $]0,1[$ or subintervals of it).
In fact

$$\int_0^1 p(v)_x u = -\int_0^1 p(v)u_x = -\int_0^1 p(v)v_t = \frac{d}{dt}[\int_0^1 P(v)] - p(1)\int_0^1 v_t = \frac{d}{dt}\int_0^1 P(v) \ ,$$

$$\int_0^1 f(\int_0^x v)u = \int_0^1 F'(\int_0^x v)\int_0^x v_t = \frac{d}{dt}\int_0^1 F(\int_0^x v) \ .$$

Let us remark that $P(\xi) \geq 0$ (since $p'(\xi) < 0$) and $\lambda - F(\int_0^x v) \geq 0$ if we choose λ large enough, say $\lambda \geq \| f \|_{L^1(0,1)}$.

(ii) $\psi \in L^\infty(0,+\infty;L^2(0,1))$, $v_x \in L^\infty(0,+\infty;L^2(0,1))$.

Multiply (2.9) by ψ and integrate in $]0,1[$. We have

$$\frac{d}{dt}\int_0^1 2^{-1}\psi^2 - \mu^{-1}\int_0^1 p'(v)v\psi^2 = \int_0^1 f(\int_0^x v)\psi - \mu^{-1}\int_0^1 p'(v)vu\psi \ .$$

Hence, by using Young's inequality,

$$\frac{d}{dt}\int_0^1 \psi^2 + c\int_0^1 \psi^2 \leq c\,(\sup_{]0,1[} f^2 + \int_0^1 u^2) \leq c \ ,$$

since $u \in L^\infty(0,+\infty;L^2(0,1))$, and the result concerning ψ follows from comparison theorems for ordinary differential equations. (Here and in the sequel c will denote a positive constant, which in general can be different from time to time).
Recalling that $v_x = \mu^{-1}v(u - \psi)$, the thesis follows.

(iii) $u \in L^\infty(0,+\infty;H^1(0,1))$.

Multiply $(1.1)_1$ by u_{xx} and integrate in $]0,1[$. Integrating by parts, and recalling that u and v_x belong to $L^\infty(0,+\infty;L^2(0,1))$, from Young's inequality one obtains

$$\frac{d}{dt}\int_0^1 u_x^2 + c_4\int_0^1 u_{xx}^2 \leq c\,[\sup_{]0,1[} f^2 + (1 + \sup_{]0,1[} u_x^2)\int_0^1 v_x^2] \leq c + 2^{-1}c_4\int_0^1 u_{xx}^2 \ ,$$

since by interpolation we have

$$(3.3)\qquad \sup_{]0,1[} u_x^2 \leq 2\, \|u\|_0^{1/2}\, \|u_{xx}\|_0^{3/2} \ .$$

On the other hand

$$\int_0^1 u_{xx}^2 \geq 4^{-1} \int_0^1 u_x^2 ,$$

hence the thesis follows from comparison theorems for ordinary differential equations.

(iv) If $u \in L^2(0,+\infty;H^2(0,1))$, then $u_t \in L^\infty(0,+\infty;L^2(0,1)) \cap L^2(0,+\infty;H^1(0,1))$.

Differentiate $(1.1)_1$ by t, multiply by u_t and integrate in $]0,1[$. From $(1.1)_2$, $(1.1)_3$ one has

$$[f(\int_0^x v)]_t = f'(\int_0^x v) \int_0^x v_t = f'(\int_0^x v)\, u \;;$$

hence integrating by parts one gets

$$\frac{d}{dt} \int_0^1 u_t^2 + c \int_0^1 u_{tx}^2 \leq c \int_0^1 (u_x^2 + u_x^4) .$$

Since $u \in L^2(0,+\infty;H^1(0,1))$, integrating with respect to time on $]0,t[$ it is easily obtained

$$\int_0^1 u_t^2 + \int_0^t \int_0^1 u_{tx}^2 \leq c_5 \, (1 + \int_0^t \int_0^1 u_x^4) .$$

Finally, using $u \in L^\infty(0,+\infty;H^1(0,1))$ and (3.3), one gets from Young's inequality

$$c_5 \int_0^1 u_x^4 \leq c_5 \, (\sup_{]0,1[} u_x^2) \int_0^1 u_x^2 \leq c(\varepsilon) \int_0^1 u^2 + \varepsilon \int_0^1 u_{xx}^2 ,$$

which gives

$$\int_0^1 u_t^2 + \int_0^t \int_0^1 u_{tx}^2 \leq c(\varepsilon) + \varepsilon \int_0^t \int_0^1 u_{xx}^2 . \tag{3.4}$$

(Here ε is an arbitrary positive constant).

(v) If $u \in L^2(0,+\infty;H^2(0,1))$, then $\psi_t \in L^\infty(0,+\infty;L^2(0,1)) \cap L^2(0,+\infty;L^2(0,1))$.

Differentiate (2.9) by t, multiply by ψ_t and integrate over $]0,t[\times]0,1[$. After simple calculations, recalling that $v_t = u_x$, one obtains

$$\int_0^1 \psi_t^2 + \int_0^t \int_0^1 \psi_t^2 \leq c\, [1 + \int_0^t \int_0^1 (u^2 + u_t^2 + u^2 u_x^2 + \psi^2 u_x^2)] .$$

Since u and ψ belong to $L^\infty(0,+\infty;L^2(0,1))$, by (i), (3.3) and (3.4) one has

$$\int_0^1 \psi_t^2 + \int_0^t \int_0^1 \psi_t^2 \leq c(\varepsilon) + \varepsilon \int_0^t \int_0^1 u_{xx}^2 . \tag{3.5}$$

(vi) $u \in L^2(0,+\infty;H^2(0,1))$.

Consider equation $(1.1)_1$ in the form

$$u_t - \mu\,(v^{-1}u_x)_x = \psi_t \ .$$

Proceeding as in (iii) one gets at once

$$\int_0^1 u_x^2 + \int_0^t\int_0^1 u_{xx}^2 \le c\,[1 + \int_0^t\int_0^1 \psi_t^2 + \int_0^t\int_0^1 u_x^2 v_x^2] \ ;$$

moreover, using $v_x \in L^\infty(0,+\infty;L^2(0,1))$ and (3.3),

$$\int_0^t\int_0^1 u_x^2 v_x^2 \le \int_0^t [(\sup_{]0,1[}\, u_x^2)\int_0^1 v_x^2\,] \le c(\varepsilon)\int_0^t\int_0^1 u^2 + \varepsilon \int_0^t\int_0^1 u_{xx}^2 \ .$$

The thesis follows now by (i) and (3.5), choosing ε small enough.

(vii) $\psi_{tt} \in L^2(0,+\infty;L^2(0,1))$.

This is at once obtained by direct computation. In fact

$$\psi_{tt} = f\,'(\int_0^x v)\ u - p''(v)u_x v_x - p'(v)u_{xx} \ .$$

We have thus obtained $\psi_t \in H^1(0,+\infty;L^2(0,1))$, and consequently (2.8) holds.

References.

[1] A. A. Amosov, A. A. Zlotnik, *A difference scheme for the equations of one-dimensional motion of a viscous barotropic gas, its properties and "global" error estimates*, Dokl. Akad. Nauk SSSR, **288** (1986), 270-275 [Russian] = Soviet Math. Dokl., **33** (1986), 633-638.

[2] H. Beirão da Veiga, *An L^p-theory for the n-dimensional, stationary, compressible Navier-Stokes equations, and the incompressible limit for compressible fluids. The equilibrium solutions*, Commun. Math. Phys., **109** (1987), 229-248.

[3] H. Beirão da Veiga, *Long time behavior for one dimensional motion of a general barotropic viscous fluid*, submitted to Arch. Rational Mech. Anal..

[4] H. Beirão da Veiga, *The stability of one dimensional stationary flows of compressible viscous fluids*, Ann. Inst. H. Poincaré. Anal. Non Linéaire, to appear.

[5] Ya. I. Kanel', *On a model system of equations of one-dimensional gas motion*, Differ. Uravn.,**4** (1968), 721-734 [Russian] = Differ. Equations, **4** (1968), 374-380.

[6] A. V. Kazhikhov, *Correctness "in the large" of mixed boundary value problems for a model system of equations of a viscous gas*, Din. Sploshnoj Sredy, **21** (1975), 18-47 [Russian].

[7] A. V. Kazhikhov, *Stabilization of solutions of an initial-boundary-value problem for the equations of motion of a barotropic viscous fluid*, Differ. Uravn. , **15** (1979), 662-667 [Russian] = Differ. Equations, **15** (1979), 463-467.

[8] V. Lovicar, I. Straškraba, A. Valli, *On bounded solutions of one-dimensional Navier-Stokes equations*, to appear.

[9] V. V. Shelukhin, *Bounded, almost-periodic solutions of a viscous gas equation*, Din. Sploshnoj Sredy, **44** (1980), 147-163 [Russian].

[10] I. Straškraba, A. Valli, *Asymptotic behaviour of the density for one-dimensional Navier-Stokes equations*, Manuscripta Math., **62** (1988), 401-416.

ON THE NUMERICAL ANALYSIS OF THE NONSTATIONARY NAVIER-STOKES EQUATIONS

(The role of regularity and stability theory)

Rolf Rannacher
Institut für Angewandte Mathematik
Universität Heidelberg
Im Neuenheimer Feld 294, D-6900 Heidelberg

1. Introduction and Summary

In a series of joint papers with J.G.Heywood, [8; Parts I-IV], a rigorous mathematical analysis was given of some discretization methods for solving the Navier-Stokes problem

$$\partial_t u + u \cdot \nabla u = f + \nu \Delta u - \nabla p , \quad \nabla \cdot u = 0 , \quad \text{in } \Omega \times (0,\infty) , \tag{1.1}$$

$$u_{|t=0} = a , \quad u_{|\partial\Omega} = b . \tag{1.2}$$

Here, u and p describe the velocity and pressure, respectively, of a viscous incompressible fluid, with viscosity ν, in a bounded two- or three-dimensional domain Ω. For simplicity, the boundary values are assumed to be zero, $b \equiv 0$. In this article we review the basic principles underlying the analysis of [8] and discuss its implications particularly on finite element Galerkin time stepping methods for computing laminar flows. The main conclusions can be summarized as follows: First, in analysing discretization schemes for solving problem (1.1), (1.2), it is not appropriate to simply assume as much regularity for the solution as may appear convenient for proving optimal order error estimates. One rather has to take into account a break-down of the solution's regularity, at the initial time $t = 0$. This requires to establish "smoothing" error estimates involving constants which may blow up as $t \to 0$. Second, in general, an exponential growth of these error constants, as $t \to \infty$, is unavoidable even for arbitrarily smooth data because of the possible instability of the solution. In order to obtain meaningful error estimates, i.e., those which provide realistic bounds in terms of the maximum local truncation errors on long intervals of time, one has to consider further assumptions about the stability of the solution to be approximated. These assumptions are to be sharp enough for technical purposes, while being expected to fit the descriptions of stability seen in various physical phenomena. Further, they should, at least in principle, be amenable to numerical verification. These goals seem to be reached through the introduction of the concept of "contractive stability to a tolerance".

2. Discretization of the Navier-Stokes Problem

Suppose that the finite element Galerkin method is used for the discretization of problem (1.1), (1.2) w.r.t. the spatial variables. This results in a highly *stiff* system of ordinary differential equations which may be written in the following abstract form: Find $u^h(t)\in \mathbf{J}_h$, s.t.

$$\partial_t u^h + A^h u^h + N^h(u^h)u^h = f^h , \quad t>0, \quad u^h(0) = a^h . \tag{2.1}$$

Here, the condition "$\nabla\cdot u^h = 0$" is thought to be built into the trial space $\mathbf{J}_h$ for the velocity u^h, such that the pressure is formally eliminated from the problem. The operator A^h is a discrete analogue of the "Stokes operator" $-\nu\tilde{\Delta} = -\nu P\Delta$, where P denotes the usual projection onto the space of solenoidal vector function in $\mathbf{L}^2(\Omega)$. Further, the nonlinearity is assumed to be approximated in such a way that the conservative character of the problem is preserved,

$$\int_\Omega N^h(u^h)u^h\cdot u^h \, dx = 0 ; \tag{2.2}$$

see, e.g., [23], [2], and [8; Part I]. The discretization of (2.1) w.r.t. time is usually by finite differences. The simplest example is the first-order "*backward Euler scheme*":
Find $\{U_n\}_{n>0}\subset \mathbf{J}_h$, s.t.

$$d_t U_n + A^h U_n + N^h(U_n)U_n = F_n , \quad t_n>0 , \quad U_0 = a^h , \tag{2.3}$$

where $t_n = nk$, k the (constant) time step size, $d_t U_n = (U_n - U_{n-1})/k$, and $F_n = f^h(t_n)$. In practice, one prefers the second-order "*Crank-Nicolson scheme*":
Find $\{U_n\}_{n>0}\subset \mathbf{J}_h$, s.t.

$$d_t U_n + A^h U_{n-1/2} + N^h(U_{n-1/2})U_{n-1/2} = F_{n-1/2} , \quad t_n > 0 , \quad U_0 = a^h , \tag{2.4}$$

where $U_{n-1/2} = (U_n + U_{n-1})/2$. Here, the nonlinerity is evaluated by the implicit midpoint rule, in order to preserve the conservative character of the problem. Possible alternatives are the "*backward differencing two-step scheme*" and the "*diagonally implicit Runge-Kutta scheme*" which are also of second order (see [14], and also [19]). Even more attractive are the so-called "*splitting schemes*" (e.g., the "*fractional step θ-scheme*" of [3]) which, in the course of the time stepping, separate the nonlinearity from the incompressibility constraint (see also [29]). This requires the fast solution of generalized Stokes problems which can be achieved very efficiently by using the pressure correction method together with a preconditioned cg-iteration, by direct multi-grid iteration, or by a boundary element approach (for the latter see [26]). All these schemes are designed to work in the case of moderate Reynolds numbers to which this article is exclusively devoted. In the case of higher Reynolds numbers, the time stepping should be combined with numerical damping procedures, e.g., ordinary upwinding, "streamline" upwinding (see [12]), or "characteristic" upwinding (see [16], and also [26]). Criteria for the choice of a time stepping scheme, besides its accuracy (order of truncation error) and its complexity (computational costs), are particularly its numerical stability properties. The global stability for "smooth" data is a property shared by nearly all reasonable discretization schemes for the Navier-Stokes equations. The local stability for "rough" data is a somewhat delicate matter and requires a scheme to possess the "smoothing property"; see, e.g., [25] and [17]. The global stability for "rough" forcing terms is closely related to the "discrete regularity property" of a scheme. These properties will be discussed in more detail, below. The results of [8; Part II] on the role of regularity and stability theory in the numerical analysis of problem (1.1), (1.2) are of more conceptional character and apply to a wide range of discretization methods, e.g., like those described in [28], [30], [32], and [33].

3. Local Error Estimates and the Smoothing Property

For the discretization scheme (2.4) one expects asymptotic error estimates of the form

$$\|U_n - u(t_n)\| \leq E(t_n)\{h^m + k^2\}, \quad t_n > 0, \tag{3.1}$$

where $\|\cdot\|$ denotes the $\mathbf{L}^2$-norm over Ω, and m refers to the order of the spatial discretization. This has been proven for most of the afore-mentioned methods (with the exception of the splitting schemes) even in the context of general evolution equations provided that the solution to be approximated is *sufficiently smooth*; see [25], and the literature cited therein. For the solution u of the Navier-Stokes problem, one obtains quantitative regularity bounds from the following sequence of differential identities

$$\begin{aligned}
\frac{1}{2}\frac{d}{dt}\|u\|^2 + \nu\|\nabla u\|^2 &= (f,u)\,,\\
\frac{1}{2}\frac{d}{dt}\|\nabla u\|^2 + \nu\|\tilde{\Delta} u\|^2 &= (u\cdot\nabla u,\tilde{\Delta} u) - (f,\tilde{\Delta} u)\,,\\
\frac{1}{2}\frac{d}{dt}\|u_t\|^2 + \nu\|\nabla u_t\|^2 &= -(u_t\cdot\nabla u,u_t) + (f_t,u_t)\,,\\
\frac{1}{2}\frac{d}{dt}\|\nabla u_t\|^2 + \nu\|\tilde{\Delta} u_t\|^2 &= (u_t\cdot\nabla u,\tilde{\Delta} u_t) + (u\cdot\nabla u_t,\tilde{\Delta} u_t) - (f_t,\tilde{\Delta} u_t)\,,\\
\frac{1}{2}\frac{d}{dt}\|u_{tt}\|^2 + \nu\|\nabla u_{tt}\|^2 &= (u_{tt}\cdot\nabla u,u_{tt}) + \cdots\,,\\
&\text{e.t.c.}\,,
\end{aligned} \tag{3.2}$$

which are easily derived by standard energy arguments (see [5]). Assuming a bound

$$\sup_{t\geq 0}\|\nabla u(t)\| < M\,, \tag{3.3}$$

on the Dirichlet norm of u, the estimates (3.2) together with the usual elliptic regularity results imply that $u \in C^\infty(\overline{\Omega}\times(0,\infty))$, if all the data are smooth. However, for the purposes of numerical analysis one needs regularity estimates which hold uniformly for $t \to 0$. To get such information from the above equations requires starting values for all the quantities $\|u\|$, $\|\nabla u\|$, $\|u_t\|$, $\|\nabla u_t\|$, $\|u_{tt}\|$, etc., at $t=0$. However, there is a problem already with $\|\nabla u_t(0)\|$. In [8;Part I] it is shown that the assumption that, e.g., $\|\nabla u_t(t)\|$ remains bounded, as $t \to 0$, implies that the over-determined Neumann problem

$$\Delta p_0 = \nabla\cdot(f_0 - a\cdot\nabla a), \quad \text{in } \Omega, \qquad \nabla p_0 = f_0 - (\partial_t b)_0 + \Delta a - a\cdot\nabla a, \quad \text{on } \partial\Omega, \tag{3.4}$$

for the initial pressure p_0, has a solution in the Sobolev space $H^1(\Omega)$. This amounts to a compatibility condition on the data of problem (1.1), (1.2) which, in view of its non-local character, is virtually uncheckable, and is also rarely satisfied in practice. Such nonstandard compatibility conditions for the Navier-Stokes problem are well known in the literature; see [21], [13], [6], [24], [20], and also [27]. In order to quantitatively describe the "natural" regularity of the solution u, as $t \to 0$, and as $t \to \infty$, in [8;Part I] a sequence of time weighted a priori estimates is proven using the weight functions $\tau(t) = \min(t,1)$ and $e^{\alpha t}$, with fixed $\alpha>0$

$$\tau(t)^{2n+m-2}\left\{\|\partial_t^n u(t)\|_{H^m}^2 + \|\partial_t^n p(t)\|_{H^{m-1}/\mathbf{R}}^2\right\} \leq K, \tag{3.5}$$

$$e^{-\alpha t}\int_0^t e^{\alpha s}\tau(s)^{2n+m-2}\left\{\|\partial_t^n u(t)\|_{H^{m+1}}^2 + \|\partial_t^n p(t)\|_{H^m/\mathbf{R}}^2\right\} ds \leq K. \tag{3.6}$$

In [8] it is taken whatever is needed of this, as assumptions in the numerical analysis of problem (1.1), (1.2). It is shown there that the discretization scheme (2.4) admits a "local" error estimate of the form

$$\|U_n - u(t_n)\| \leq K e^{Kt_n}\left(\frac{h^m}{\tau_n^{m/2-1}} + \frac{k^2}{\tau_n}\right), \quad t_n > 0, \tag{3.7}$$

for each of the orders $m \in \{2, 3, 4, 5, (6)\}$ of spatial approximation. This estimate holds under the assumption that $a \in \mathbf{J}_1(\Omega)\cap H^2(\Omega)$, and does not require the non-local compatibility condition (3.4) to be satisfied. The maximum order $m = 6$ seems to be a natural upper bound for a smoothing estimate of the type (3.7). It is shown in [11], by counter-examples, that for the finite element Galerkin method applied to nonlinear problems of the type

$$\partial_t u - \Delta u = f(u), \quad t > 0, \quad u_{|t=0} = a, \tag{3.8}$$

the second order error estimate

$$\|(u^h - u)(t)\| \leq E(t;\|a\|)\, h^2, \quad t > 0, \tag{3.9}$$

is generally best possible, for initial data $a \in L^2(\Omega)$. The simplest of these examples is the following one which, although concerning a very particular spectral Galerkin approximation, is representative also for the finite element Galerkin method.

Example (from [11]): $x \in (-\pi,\pi)$, $t > 0$,

$$\partial_t u - \partial_x^2 u = 4\min(v^2,1), \quad u_0(x) = 0,$$

$$\partial_t v - \partial_x^2 v = 0, \qquad v_0(x) = m^{-r}\cos(mx).$$

For any fixed $m \in \mathbf{N}$, and $r \in \mathbf{N} \cup \{0\}$, the exact solution is

$$u(x,t) = m^{-2r-2}(1-e^{-2m^2t})(1+e^{-2m^2t}\cos(2mx)), \qquad v(x,t) = m^{-r}e^{-m^2t}\cos(mx).$$

For discretizing this problem by the Galerkin method, we choose the trial space

$$S^m = \{1, \cos(x), \sin(x), \dots, \cos([m-1]x), \sin([m-1]x)\},$$

and let P^m denote the L^2-projection onto S^m. Since $P^m v_0 \equiv 0$, taking as usual $P^m u_0$, $P^m v_0$ as the initial values for the Galerkin approximation results in $u^m(t) \equiv v^m(t) \equiv 0$. Consequently, for fixed $t > 0$, there holds

$$\|(u^m-u)(t)\| = \|u(t)\| \sim \sqrt{2\pi}\, m^{-2r-2} = \sqrt{2}\, \|v_0\|_r\, h^{2r+2}, \tag{3.10}$$

where $h = m^{-1}$. This shows that, indeed, for $v_0 \in H^2(-\pi,\pi)$, i.e., for $r = 2$, the best one can expect is smoothing up to order $O(h^6)$.

There is another remarkable aspect of the estimate (3.7) which particularly concerns the Crank-Nicolson scheme. This scheme, due to its weak damping properties (not *strongly* A-stable), possesses only a reduced smoothing property. For initial data $a \in \mathbf{L}^2(\Omega)$, one has $\|U_n-u(t_n)\| \to 0$ (as $h, k \to 0$) for fixed $t = t_n > 0$, but without any specific order of convergence. If $a \notin \mathbf{L}^2(\Omega)$ (e.g., $a = \delta_x$ Dirac functional), then even $\|U_n\| \to \infty$ (as $h, k \to 0$). However, the optimal smoothing behavior is recovered if one keeps the relation $k \sim h^2$. This undesirable step size restriction can be avoided simply through starting the computation with a few (two or three) backward Euler steps (2.3); see [15] and [17]. Surprisingly, such a modification is not necessary for slightly more regular initial data, $a \in \mathbf{H}^2(\Omega)$. In this case, the Crank-Nicolson scheme admits a smoothing error estimate of the form

$$\|U_n-u^h(t_n)\| \le K(T;\|\Delta a\|)\frac{k^2}{t_n}, \quad t_n \in (0,T]. \tag{3.11}$$

For the heat equation this is easily seen by a standard spectral argument (see [1] and [17]). The extension of the estimate (3.11) to the nonlinear Navier-Stokes problem is one of the major results of [8; Part IV].

4. Global Error Estimates and the Discrete Regularity Property

In order to remove the exponential growth factor in the error estimate (3.7), one has to make some assumptions about the stability of the solution to be approximated. A discussion of several concepts of hydrodynamic stability which are appropriate for the puposes of numerical analysis is given in [8; Part II] and [9]. First, there is the classical (conditional) exponential stability which may be described as follows: The (bounded) solution u of problem (1.1), (1.2) is called "*exponentially stable*" if there exist positive constants δ and T, such that any secondary solution v of the equations (1.1), starting at an initial time $t_* \geq 0$ with an initial value $v_* \in \mathbf{J}_1(\Omega)$, $\|\nabla v_*\| < \delta$, exists on $[t_*, t_*+T]$ as a *strong* solution and satisfies

$$\|(v-u)(t_*+T)\| \leq \frac{1}{2}\|v_*-u(t_*)\|. \tag{4.1}$$

This "half life" formulation is equivalent to the traditional "exponential decay" formulation of exponential stability. It has the adventage of being *topology independent*, as it allows equivalent definitions w.r.t. any H^m-norm, for $m=0,1,2,\ldots$. This is crucial for use in the numerical analysis of problem (1.1), (1.2), since there one needs to have control of the perturbations in the H^1-norm. Further, arguing along the continuous dependence properties of problem (1.1), (1.2), one obtains the "principle of linearized stability", the $\mathbf{H}^1$-openness of the set of initial values giving rise to exponentially stable solutions, and the following characterization of the possible limits of such solutions: If the force f is time periodic, of period T, and if u is exponentially stable, then there exists a function u_∞ of period nT (n some integer), such that

$$\|\nabla(u-u_\infty)(t)\| \leq ce^{-\alpha t}. \tag{4.2}$$

The limit function u_∞ is steady if f is steady.

It is shown in [8; Part II] (see also [7]) for the spatial semi-discretization of problem (1.1), (1.2) by a second-order finite element Galerkin method that, if the solution u is exponentially stable, then the error constant remains bounded as $t \to 0$,

$$\sup_{t>0}\|(u^h-u)(t)\| \leq K h^2. \tag{4.3}$$

This is extended to higher order spatial approximation, $m \in \{3, 4, 5, (6)\}$, in [8; Part III],

$$\|(u^h-u)(t)\| \leq K\frac{h^m}{\tau^{m/2-1}}, \quad t > 0, \tag{4.4}$$

where $\tau = \min(t,1)$. As a remarkable by-product one obtains that, for an exponentially stable solution u, the corresponding spatially semi-discrete approximation u^h is also exponentially stable, with stability parameters which can be chosen independent of h. The argument of [8; Part II] also applies to the first-order time discretization, e.g., by the backward Euler scheme, yielding an estimate of the form

$$\sup_{t_n>0} \|U_n - u(t_n)\| \le K\{h^2 + k\} . \tag{4.5}$$

However, the corresponding analysis of the second-order Crank-Nicolson scheme which is given in [8; Part IV] turns out to be much more involved. The proof of the global error estimate

$$\|U_n - u(t_n)\| \le K_* \left(\frac{h^m}{\tau_n^{m/2-1}} + \frac{k^2}{\tau_n} \right) , \tag{4.6}$$

is based on an induction argument w.r.t. the sequence of time intervals $I_r = [0,rT]$, $r \in \mathbf{N}$. In the course of this induction argument, the "local" error estimate (3.7), taken over I_2 , provides the starting step for the induction

$$\|U_n - u(t_n)\| \le K \left(\frac{h^m}{\tau_n^{m/2-1}} + \frac{k^2}{\tau_n} \right), \quad 0 < t_n \le 2T , \tag{4.7}$$

where $K_1 = Ke^{2cT} \le K_*$. Assuming that the estimate (4.6) is already known to hold over the interval I_r , for some $r \ge 2$, it is then shown to hold also over I_{r+1} . To this end, one considers the perturbed solution $v(t)$ of equations (1.1) which starts at the time $t_* = rT$, with the initial value $v_* \sim u(rT)$, where $v_* \in \mathbf{J}_1(\Omega) \cap \mathbf{H}^2(\Omega)$ is a *continuous* function associated to the *discrete* function U_{rN} . Then, the stability assumption guarantees that, for sufficiently small step size k , the solution $v(t)$ exists on $[rT,(r+1)T]$, and satisfies

$$\|(v-u)((r+1)T)\| \le \frac{1}{2} K_* \{h^m + k^2\} . \tag{4.8}$$

Next, viewing U_n as an approximation to $v(t_n)$, for $t_n > rT$, the "local" estimate (3.7) yields

$$\|U_{(r+1)N} - v((r+1)T)\| \le K_2\{h^m + k^2\} . \tag{4.9}$$

The constant K_2 depends on the norm $\|\tilde{\Delta} v(rT)\| \sim \|\tilde{\Delta}_h U_{rN}\|$, where $\tilde{\Delta}_h : S^h \to S^h$ is a discrete analogue of the Stokes operator $\tilde{\Delta}$. Therefore, to make the induction argument work, one has to get control on $\|\tilde{\Delta}_h U_{rN}\|$, through the induction assumption (4.6). However, since the latter provides a bound only for $\sup_{0 \le n \le rN} \|U_n\|$, a "discrete regularity estimate" of the form

$$\sup_{0 \le n \le rN} \|\tilde{\Delta}_h U_n\| \le K_3 \left\{ \|\tilde{\Delta}_h U_0\| + \sup_{0 \le n \le rN} \|U_n\|^{\gamma} + M(f) \right\} , \tag{4.10}$$

is needed, with a number $\gamma > 0$, and a constant K_3 independent of h , k , and $\{U_n\}$. The estimate (4.10) resembles a well-known a priori estimate for the solutions of a weakly nonlinear parabolic equation of the form $\partial_t u - \Delta u = f(u)$, with a bounded function $f(\cdot)$. Due to its stronger type of nonlinearity, for the Navier-Stokes equations (particularly in three dimensions) one only has the weaker estimate

$$\sup_{0 \le t \le rT} \|\tilde{\Delta} u(t)\| \le K_3 \left\{ \|\tilde{\Delta} a\| + \sup_{0 \le t \le rT} \|\nabla u(t)\|^{\gamma} + M(f) \right\} , \tag{4.11}$$

with $\gamma = 6$. A corresponding discrete analogue can be proven for the solution $u_h(t)$ of the semi-discrete problem (2.1), by using an adaption of the energy technique which leads to

(4.11). A similar argument gives an a priori estimate also for the fully discrete solution $\{U_n\}$ obtained by the backward Euler scheme. However, for the Crank-Nicolson scheme, an estimate of the type (4.11) requires that $k \le ch$ (as $k, h \to 0$). This then also leads to the global error estimate (4.6), but under the step size condition

$$k \le c_*(1+\Lambda)^{-2}, \quad \Lambda = k/h \tag{4.12}$$

The necessity of this restriction for the Crank-Nicolson scheme can be shown by looking at the linear model equation $\partial_t u - \Delta u = f$. In this case, the usual spectral argument applies and yields a discrete a priori estimate ([18])

$$\|U_N\| \le e^{-\min\{\lambda,\Lambda\}t_N} \|U_0\| + \max\left(\frac{1}{\sqrt{\lambda}}, \frac{1}{\sqrt{\Lambda}}\right) \max_{0\le n\le N} \|F_n\|_{-1}, \tag{4.13}$$

where λ is the smallest eigenvalue of the operator $-\Delta_h$, and $\|\cdot\|_{-1}$ denotes the usual H^{-1}-norm. This cannot be improved and shows that the Crank-Nicolson scheme has only a conditional stability w.r.t. "rough" forcing terms. Again, it is possible to avoid the undesirable restriction on the relation of k and h, by enhancing the stability properties of the scheme through introducing additional damping. This can be achieved either by shifting the Crank-Nicolson steps slightly to the "implicit side",

$$d_t U_n + \frac{1+k}{2} A^h U_n + \frac{1-k}{2} A^h U_{n-1} + N^h(U_{n-1/2})U_{n-1/2} = F_n, \tag{4.14}$$

or by adding a sequence of backward Euler steps (2.3) to the scheme, one per unit of time (see [18] and [8; Part IV]). However, it should be noted that numerical tests on the one-dimensional Burgers equation (commonly viewed as a model for the Navier-Stokes equations) did not show any evidence of a dependency of the error behavior of the Crank-Nicolson scheme on the relation of k and h. This is related to the fact, that also in the estimate (4.13), the quantity Λ does not occur, if the right hand side F_n is smoother, say, $\max_{0\le n\le N} \|F_n\| \le c$. For the nonlinear Navier-Stokes problem, this question has not been settled yet. Finally, it is suspected that the other time stepping methods (particularly the fractional step θ-scheme), mentioned above, which are all *strongly* A-stable, allow for global error estimates of the form (4.6), without any restriction on the relation of k and h. This is the subject of continuing research.

5. The Role of Stability Assumptions and A-posteriori Error Estimates

Finally, we address the question to what extent the stability properties used in [8; Part II] are crucial for the global approximation behavior of discretization schemes. The concept of exponential stability is very restrictive and does not fit most of the phenomena occuring in practically "stable" flows. In particular, it generally does not apply to flows which are time periodic and/or involve a rotational symmetry. Such flows may be stable, but only modulo time shifts and/or spatial rotations. This leads one to the notion of "*quasi-exponential stability (modulo time shifts and/or spatial rotations)*" which generalizes that of "*exponential stability*" in that it implies the decay property (4.1) for the perturbations only in the weaker form

$$\|(v-\tilde{u})(t_*+T)\| \leq \tfrac{1}{2}\|v_*-u(t_*)\| . \tag{5.1}$$

Here, $\tilde{u}(x,t) = \omega^{-1}[u(\omega[x],t+s)]$ is the rotated image of the unperturbed flow taken at a rotated and shifted space-time point. The spatial rotations $\omega(\cdot)$ and time shifts s are to satisfy

$$\|\omega\| + |s| \leq B\|v_*-u(t_*)\| . \tag{5.2}$$

This definition applies even during transient periods while the pattern of a flow slowly changes shape. Examples of flows possessing stability properties of this type can be seen in the experiments described in [31], and possibly also in von Karman vortex shedding and in doubly periodic Taylor cells. Again, the above stability concept is topology independent and admits a "principle of linearized stability". The notion of "quasi-exponential stability (modulo time shifts)" sharpens the "uniform orbital stability" of the classical theory of dynamical systems (see, e.g., [4]) as it provides proportional bounds (in terms of the initial perturbation) also for the time shifts. It is known (see [22]) that in approximating (non-stiff) ordinary differential equations uniform error bounds can be obtained if the solution to be approximated is merely "uniformly asymptotically stable". However, these error bounds are of only qualitative nature and do not guarantee the proportionality of the "global" discretization error w.r.t. the maximum "local" truncation error. This latter property is crucial for the use of higher order schemes and also for the automatic step size control based on *a posteriori* estimates of the truncation error. In [8; Part II] it is shown that the global uniform error estimate (4.5) carries over to a quasi-exponentially stable (e.g., modulo time shifts) solution, in the form

$$\sup_{t_n>0} \|U_n-u(t_n+s_n)\| \leq K\{h^2+k\} , \tag{5.3}$$

where the time shifts s_n satisfy

$$\sup_{t_n>0} |d_t s_n| \leq K\{h^2+k\} . \tag{5.4}$$

The uniform estimate (5.4), on the rate of precession of the discrete solution, essentially depends on the presumed proportional bounds for the time shifts in the concept of quasi-exponential stability. This has been confirmed by numerical experiments at simple dynamical systems.

Example (from [10]): Consider the dynamical system (from [4]),

$$x' = px - qy , \quad y' = qx + py , \tag{5.5}$$

$$p = r^{-2}(1-r^2)^{\mu} , \quad q = 1 + (1-r^2)^{\beta} , \quad r = (x^2+y^2)^{1/2} , \tag{5.6}$$

with appropriate constants $\mu, \beta > 0$. This system describes a motion on the unit circle with constant angular velocity, $x(t) = \cos(t)$, $y(t) = \sin(t)$. Writing (5.5) in polar coordinates (r,ϕ) and introducing the new variable $z = 1-r^2$ leads to the system

$$z' = -2z^{\mu}, \quad \phi' = 1 + z^{\beta}. \tag{5.7}$$

Setting $z_0 = z(0)$ and $\phi_0 = \phi(0)$, the exact solution of (5.7) is:

$$\mu = 1, \beta > 0: \qquad z(t) = z_0 e^{-2t}, \qquad \phi(t) = \phi_0 + t + \frac{1}{2\beta}(1-e^{-2\beta t})z_0^{\beta}, \tag{5.8}$$

$$\mu > 1, \beta \neq 2: \qquad z(t) = z_0(1+2(\mu-1)z_0^{\mu-1}t)^{-1/(\mu-1)},$$

$$\phi(t) = \phi_0 + t + \frac{1}{2(\mu-1-\beta)} z_0^{\beta+1-\mu}\left(\left(1+2(\mu-1)z_0^{\mu-1}t\right)^{1-\beta/(\mu-1)}-1\right). \tag{5.9}$$

From these representations, we conclude the following decay properties of perturbations of the equilibrium solution $r_*(t) = 1$, $\phi_*(t) = t$ $(0<\varepsilon<1)$:

$$\mu = 1, \beta = 2: \qquad |r(t)-1| \le e^{-2t}|r_0-1|, \qquad |\phi(t)-t| \le |\phi_0| + B|r_0-1|, \tag{5.10}$$

$$\mu = 3, \beta = 3: \qquad |r(t)-1| \le |r_0-1|^{1-\varepsilon}t^{-\varepsilon/2}, \qquad |\phi(t)-t| \le |\phi_0| + B|r_0-1|, \tag{5.11}$$

$$\mu = 3, \beta = 1, \qquad |r(t)-1| \le |r_0-1|^{1-\varepsilon}t^{-\varepsilon/2}, \qquad |\phi(t)-t| \le |\phi_0| + B|r_0-1|^{1/3}t^{2/3}. \tag{5.12}$$

Hence, the equilibrium solution of (5.5) is

- quasi-exponentially stable (modulo time shifts), for $\mu = 1$, $\beta = 2$,
- uniformly orbitally stable (<u>with</u> prop. bounds for the time shifts), for $\mu = 3$, $\beta = 3$,
- uniformly orbitally stable (<u>without</u> prop. bounds for the time shifts), for $\mu = 3$, $\beta = 1$.

The discretization of the (non-stiff) system (5.5), is by an explicit two stage Runge-Kutta method (Heun's method) of order two:

$$d_t U_n = \frac{\Delta t}{2}(k_1[U_{n-1}]+k_2[U_{n-1}]), \quad k_1[x] = F(t_{n-1},x), \quad k_2[x] = F(t_n,x+\Delta t\, k_1[x]). \tag{5.13}$$

For describing the results obtained with this scheme, we introduce the notation

$$R_n = \sqrt{X_n^2+Y_n^2}, \qquad \Phi_n = \arctan(Y_n/X_n), \qquad X_n \sim x(t_n), \; Y_n \sim y(t_n),$$

where it is assumed that the points (X_n,Y_n), (X_{n-1},Y_{n-1}) are in the same quadrant. Further,

$$\Delta R_n = |R_n-1|, \qquad \Delta R_\infty = \sup_{n\ge1} \Delta R_n,$$

$$\Delta d_t\Phi_n = |d_t\Phi_n-1|, \qquad \Delta d_t\Phi_\infty = \sup_{n\ge1} \Delta d_t\Phi_n,$$

are the radial and phase errors, respectively, and

$$K_r = \sup_{n\geq 1} \frac{\Delta R_n}{\sigma_n} , \qquad K_{d_t\phi} = \sup_{n\geq 1} \frac{\Delta d_t\Phi_n}{\sigma_n}$$

are the corresponding error constants, where the maximum truncation error σ_n is defined by

$$\sigma_n = \max_{1\leq\mu\leq n} \|\sigma(t_\mu)\| , \qquad \sigma(t_\mu) = d_t u(t_\mu) - \frac{\Delta t}{2}(k_1[u(t_{\mu-1})]+k_2[u(t_{\mu-1})]) .$$

Table 1. Case $\mu = 1$, $\beta = 2$ (quasi-exponentially stable)

$k = 2^{-i}$	ΔR_∞	K_r	$\Delta d_t\Phi_\infty$	$K_{d_t\phi}$
i = 1	.0734444	.4229	.0844557	.4863
i = 2	.0181514	.4147	.0124759	.3061
i = 3	.0042562	.3768	.0027225	.2661
i = 4	.0010216	.3561	.0006581	.2554
i = 5	.0002498	.3457	.0001631	.2520
i = 6	.0000617	.3405	.0000407	.2508

Table 2. Case $\mu = 3$, $\beta = 3$ (unif. orbitally stable with prop. bounds for the time shifts)

$k = 2^{-i}$	ΔR_∞	K_r	$\Delta d_t\Phi_\infty$	$K_{d_t\phi}$
i = 1	.0297036	0.675	.0299923	.7891
i = 2	.0423374	2.917	.0100922	.8468
i = 3	.0263402	7.159	.0025899	.9145
i = 4	.0146029	15.862	.0006502	.9552
i = 5	.0075676	32.878	.0001627	.9770
i = 6	.0038464	66.842	.0000406	.9884

Table 3. Case $\mu = 3$, $\beta = 1$ (unif. orbitally stable without prop. bounds for the time shifts)

$k = 2^{-i}$	ΔR_∞	K_r	$\Delta d_t\Phi_\infty$	$K_{d_t\phi}$
i = 1	.0266434	0.186	.1510406	1.059
i = 2	.0383402	1.229	.0993711	3.186
i = 3	.0250625	3.352	.0559672	7.485
i = 4	.0140562	7.604	.0296119	16.020
i = 5	.0074178	16.098	.0152621	33.021
i = 6	.0038073	33.073	.0077104	66.979

Finally, we come to the question of whether it is possible to detect the stability of a flow through a numerical computation. For steady flows, thanks to the principle of linearized stability, the well-known eigenvalue criterion applies. This may lead to an *a-posteriori* criterion for the discrete solution of the nonstationary Navier-Stokes problem which guarantees the existence of a closely neighboring steady-state continuous solution; see [34] for such a construction in the context of the spectral Stokes-Galerkin discretization. However, in the transient case exponential stability is not amenable to numerical checking since this would require to look at arbitrarily small perturbations. In experiments one often observes flows which appear to be fairly stable in their large scale features, though with some fluttering and small scale irregularities; see, e.g., the experiments reported in [31]. In [8; Part II] it is proposed to describe this "partial" stability by the notion of "contractive stability to a tolerance" (or more generally by that of "quasi-contractive stability to a tolerance"). By this we mean that there must exist positive constants δ, $\rho<\delta$, A, and T, such that any secondary solution v of the equations (1.1), starting at an initial time $t_* \geq 0$ with an initial value $v_* \in \mathbf{J}_1(\Omega)$, $\|\nabla v_*\| < \delta$, satisfies

$$\|\nabla(v-u)(t_*+T)\| \leq \rho, \quad \sup_{[t_*,t_*+T]}\|\nabla v\| \leq A. \tag{5.14}$$

If the solution of problem (1.1), (1.2) is merely assumed to have bounded rotation and to be contractively stable to the tolerance ρ, then one obtains corresponding global error bounds to the tolerance 2ρ, for example,

$$\sup_{[0,\infty)}\|\nabla(u^h-u)\| \leq Kh + 2\rho. \tag{5.15}$$

It will be the subject of future research to extend this kind of analysis to the case of "turbulent" flow where only certain time-averaged quantities can be assumed to be stable (see, e.g., the results on turbulent pipe flow in [32]). From the point of view of numerical analysis the concept of "contractive stability", or, more precisely, its discrete analogue is most important as it leads to an *a posteriori* error analysis reversing, e.g., the role of u and u^h in the estimate (5.15); see [8; Part II]. This can be summarized as follows: If, for some h, the discrete solution u^h is found to be contractively stable, where with a certain function $G(\cdot, \dots, \cdot)$,

$$h < G(\rho, \delta, A, T, \sup_{[0,\infty)}\|\nabla u^h\|, \text{data}), \tag{5.16}$$

then, there exists a smooth solution u of problem (1.1), (1.2) satisfying (5.15). In turn, if the solution of problem (1.1), (1.2) has bounded rotation and is contractively stable, then there exists h_0, such that for $h < h_0$, the discrete solutions u^h are contractively stable, such that (5.16) holds true. Furthermore, using the continuous dependence properties of discrete solutions and the finite dimensionality of $\mathbf{J}_h$, it follows that, at least in principle, (5.16) can be verified for fixed h by a finite amount of computation per unit of time. Briefly speaking, these results tell us the following: For each particular set of data and any interval of time $[0,T]$, the existence of a stable strong solution over $[0,T]$ of the Navier-Stokes problem can be verified by a *finite* amount of numerical computation.

References.

[1] Chen,C.M., Thomee,V.: The lumped mass finite element method for a parabolic problem, J.Austral.Math.Soc.Ser. B 26, 329-354 (1985).

[2] Girault,V., Raviart,P.A.: Finite Element Methods for Navier-Stokes Equations, Springer 1986.

[3] Glowinski,R., Periaux,J.: Numerical methods for nonlinear problems in fluid dynamics, Proc.Intern.Seminar on Scientific Supercomputers, Paris, North-Holland 1987.

[4] Hahn,W.: Stability of Motion, Springer 1967.

[5] Heywood,J.G.: The Navier-Stokes equations: On the existence regularity and decay of solutions, Indiana Univ.Math.J. 29, 636-682(1980).

[6] Heywood,J.G.: Classical solutions of the Navier-Stokes equations, Proc. IUTAM Symp., Paderborn 1979,Springer, LNM 771, 1980, pp. 235-248.

[7] Heywood,J.G.: Stability, regularity and numerical analysis of the nonstationary Navier-Stokes problem, Nonlinear PDE in Applied Science, U.S.-Japan Seminar, Tokyo, July 1982, Lecture Notes in Num.Appl.Anal. 5, 377-397(1982).

[8] Heywood,J.G., Rannacher,R.: Finite element approximation of the nonstationary Navier-Stokes problem, **I.** Regularity of solutions and second order error estimates for spatial discretization, SIAM J.Numer.Anal. 19, 275-311 (1982), **II.** Stability of solutions and error estimates uniform in time, SIAM J.Numer.Anal. 23, 750-777 (1986), **III.** Smoothing property and higher order error estimates for spatial discretization, SIAM J.Numer.Anal. 25, 489-512 (1988), **IV.** Error analysis for second order time discretization, SIAM J.Numer. Anal., to appear.

[9] Heywood,J.G., Rannacher,R.: An analysis of stability concepts for the Navier-Stokes equations, J.Reine Angew.Math. 372, 1-33 (1986).

[10] Heywood,J.G., Rannacher,R.: On the global numerical approximability of dynamical systems, Research Report, University of Saarbrücken, August 1986.

[11] Johnson,C., Larsson,S., Thomee,V., Wahlbin,L.B.: Error estimates for spatially discrete approximations of semilinear parabolic equations with nonsmooth initial data, Math. Comp. 49, 331-357(1987).

[12] Johnson,C., Saranen,J.:Streamline diffusion methods for the incompressible Euler and Navier-Stokes equations, Math. Comp.47, 1-18 (1986).

[13] Ladyshenskaya,O.A.: The Mathematical Theory of Viscous Incompressible Flow, Gordon and Breach 1969

[14] Lambert,J.D.: Computational Methods in Ordinary Differential Equations, John Wiley 1973.

[15] Luskin,M., Rannacher,R.: On the smoothing property of the Crank-Nicolson scheme, Applicable Anal. 14, 117-135(1982).

[16] Pironneau,O.: On the transport-diffusion algorithm and its application to the Navier-Stokes equations, Numer.Math. 38, 309-332 (1982).

[17] Rannacher,R.: Finite element solution of diffusion problems with irregular data, Numer. Math. 43, 309-327(1984).

[18] Rannacher,R.: On the stabilization of the Crank-Nicolson scheme for long time calculations, Technical Report, University of Saarbrücken, September 1986.

[19] Rannacher,R.: Numerical analysis of nonstationary fluid flow (a survey), Proc. III German-Italian Symp. "The Applications of Mathematics in Industry and Technology", Siena, June 18-22, 1988, Teubner-Kluver, to appear.

[20] Rautmann,R.: On optimum regularity of Navier-Stokes solutions at time $t = 0$, Proc. Conf. Oberwolfach, November 21-27, 1982, Math.Z. 184, 141-149 (1983)

[21] Solonnikov,V.A.: Estimates of solutions of a nonstationary linearized system of Navier-Stokes equations, A.M.S. Transl. 75, 1-116 (1968).

[22] Stetter,H.J.: Analysis of Discretization Methods for Ordinary Differential Equations, Springer 1973.

[23] Temam,R.: Theory and Numerical Analysis of the Navier-Stokes Equations, North-Holland 1977.

[24] Temam,R.: Behaviour at time $t = 0$ of the solutions of semi-linear evolution equations, J.Differential Equations 43, 73-92 (1982).

[25] Thomee,V.: Galerkin-Finite Element Methods for Parabolic Problems, Springer, LNM 1054, 1984.

<u>Contributions to this Oberwolfach-Conference</u>

[26] Hebeker,F.K.: On Lagrangean methods and Volterra integral equations of the first kind for incompressible Navier-Stokes problems, these Proceedings.

[27] Masuda,K.: Remarks on compatibility conditions for solutions of Navier-Stokes equations, J.Fac.Sci.Univ. Tokyo, Sect. IA, Math. 34, 159-164 (1987).

[28] Polezhaev,V.I.: The problems of mathematical modelling on the basis of the unsteady Navier-Stokes equations.

[29] Rautmann,R.: Eine konvergente Produktformel für linearisierte Navier-Stokes-Probleme, Proceedings GAMM-Conf. 1988, ZAMM 69, T181-T183 (1989).

[30] Rivkind,V.J.: Numerical methods for the Navier-Stokes equations with an unknown boundary between two viscous incompressible fluids, these Proceeedings.

[31] Roesner,K.G., Bar-Joseph, P., Solan, A.: Numerical simulation and experimental verification of cavity flows, these Proceedings.

[32] Rozhdestvensky,B.L.: Numerical simulation of turbulent incompressible fluid flows in channels and pipes by nonstationary solutions of Navier-Stokes equations.

[33] Tarunin,E.L.: Investigation of implicit schemes in terms of the variables Ψ, ω with the help of the Babenko-Gelfand principle.

[34] Titi,E.S.: Numerical criteria for detecting stable stationary and time periodic solutions to the Navier-Stokes equations.

This paper is in final form and no similar paper is being or has been submitted elsewhere
(received February 21, 1989).

NUMERICAL METHODS FOR THE NAVIER-STOKES EQUATIONS WITH AN UNKNOWN BOUNDARY BETWEEN TWO VISCOUS INCOMPRESSIBLE FLUIDS

Valeriy Ya. Rivkind
Research Institute of Mathematics and Mechanics
Petrodvorets, Bibliotechnaya Square 2
Leningrad, USSR

The question of existence and of uniqueness of solutions for problems of stationary flow of several viscous incompressible fluids with unknown separation boundaries was considered in [1],[2],[3], etc. Under the assumption that characteristic sizes and velocities of the flow are small while viscosity and surface tension are large (i.e. Reynolds and Weber numbers are small) the existence of a classical solution was proved. In this case the solution belongs together with all its derivatives to Hölder classes in every domain of the flow, and the unknown separation boundary belongs to Hölder classes together with its derivatives up to the third order. In [3] and in [5] results of this kind were obtained for the problem of a viscous liquid drop motion in the flow of another viscous incompressible fluid. Moreover, in [6] the case when an unknown separation boundary between two fluids intersects the solid surface was studied, and in [7] the case of the multi-layer film flow was considered. As usual the convergence of a numerical method can be proved in suitable Sobolev spaces, for which purpose theorems of existence and uniqueness are also deduced in these spaces. In [8] on the basis of the results of [4] the convergence of approximate methods for problems of this type has been proved, which in particular applies to finite element, Galerkin methods and so on. In [8] it is supposed that the velocities are square integrable together with their derivatives up to the second order in domains of the flow of each fluid ($u \in W_2^2(\Omega_i)$), pressure gradient is square integrable ($\nabla p \in L_2(\Omega_i)$), and the unknown boundary Γ has square integrable derivatives up to the order 5/2 ($\Gamma \in W_2^{5/2}$). As coordinate functions for velocities we use some divergence free piecewise polynomial local functions which approximate the velocities in the energy space with order O(h) (h is the grid step) and are , bounded in the space $W_2^2(\Omega_i)$. We use similar approximations for the pressure and piecewise polynomial approximations for the boundary (with order O(h) in the space W_2^1 and bounded in the space $W_2^{5/2}$). Such functions can be obtained from known local piecewise polynomial functions ϕ_h approximating the original ones in the classes $W_2^2(\Omega_i)$ with order $O(h^k)$ (see (8)). Then $\vec{u}_h = (\partial\phi_h/\partial y, -\partial\phi_h/\partial x)$ are constructed. The validity of our assumptions about $\vec{u}_h$ is derived from the fact that the system for $\vec{u}_h$ satisfies the Markov inequality [8]. The convergence in the $W_2^1(\Omega)$ norm with order $O(h^k)$ is implied by the convergence in $W_2^2(\Omega)$with order $O(h^{k-1})$. Similar results are valid for the spatial case as well. Moreover, from the condition that the operator which corresponds to the exact problem is the contracting operator for $(\vec{u},\nabla p,\Gamma)$ in a ball K_ρ of the space $(W_2^2(\Omega_1) \cap W_2^2(\Omega_2)) \times (L_2(\Omega_1) \cap L_2(\Omega_2)) \times W_2^{5/2}$, it follows that the approximate operator is also a contracting operator in the space of approximations, and similar estimates are obtained for the rate of convergence. The fact that both operators (for exact and approximate problems) are contractions allows us to make use of the method of successive approximations, which can be realized in two different ways. Let us consider them in more details.

The flow in each fluid $(x \in \Omega_1)$, $(x \in \Omega_2)$ is described by the full system of Navier-Stokes equations

$$- \tilde{\nu}\, \Delta \vec{u} + (\vec{u},\nabla)\vec{u} = - \frac{1}{\rho} \nabla \rho + f \qquad\qquad (1)$$
$$\operatorname{div} \vec{u} = 0 ,$$

where $\tilde{\nu} = \nu_1$ if $x \in \Omega_1$, and $\tilde{\nu} = \nu_2$ if $x \in \Omega_2$, $\vec{u}$ is the velocity, ρ is the density, ν is the kinematic viscosity, f is the mass force. On the unknown boundary between the fluids the conditions of impermeability and continuity of the velocities are satisfied:

$$(\vec{u},\vec{n})|_\Gamma = 0 \qquad\qquad (2)$$
$$[\vec{u}]|_\Gamma = 0,$$

as well as the conditions of the continuity of tangential stresses $[\vec{r}\cdot\sigma\cdot\vec{n}]|_\Gamma = 0$, (3) and jump condition for the normal stress which equals the capillary forces

$$- \gamma\left(\frac{1}{R_1} + \frac{1}{R_2}\right)\Big|_\Gamma = [\vec{n}\cdot\sigma\cdot\vec{n}]|_\Gamma , \qquad\qquad (4)$$

where [ ] denotes the jump of the function on Γ, $\vec{n}$ and $\vec{r}$ are unit vectors, normal and tangential to Γ, correspondingly, and σ is the stress tensor with the components $\sigma_{ij} = - p\delta_{ij} + 2\mu\left(\frac{\partial u_i}{\partial x_j} + \frac{\partial u_j}{\partial x_i}\right)$, $\frac{1}{R_1} + \frac{1}{R_2}$ is the mean curvature of Γ, γ is the surface tension coefficient between the fluids. In every individual problem there are some extra restrictions imposed on the solution. E.g., for the problem of the axis-symmetrical displacement motion of a rotating drop along the axis of symmetry under the force of gravity, the following condition in needed:

$$u_\parallel |_{|x| \to \infty} = u_\infty ; \quad u_\phi |_{|x| \to \infty} = u_{\phi_0} , \qquad\qquad (5)$$

here u_∞ is such that

$$c_d \frac{\rho u_\infty^2 \, d^2}{4} = (\rho_1 - \rho_2) V g, \qquad\qquad (6)$$

V is a volume of the drop, c_d is the resistance coefficient, $u_{\parallel} = (u_r, u_\Theta)$; r,Θ,ϕ are the spherical coordinates, $\vec{u} = (u_r, u_\Theta, u_\phi)$ is the velocity vector with components in these coordinates.

In the first type of the successive approximation method, in a way similar to [9],[10] the problem is being splitted at each step into the auxiliary problem (1)-(3), (5), (6) with a fixed boundary $\Gamma^{(k)}$, from which we find the fields of velocities and pressures $(u^{(k)}, p^{(k)})$, as well as the jump of the normal stresses on $\Gamma^{(k)}$(the right-hand side in Equation (4)). Next, a new approximation $\Gamma^{(k+1)}$ is found from (4), and the process is repeated. In the auxiliary problem the boundary $\Gamma^{(k)}$ is transformed to a canonic surface $\tilde{\Gamma}^{(k)}$ by a non-orthogonal transformation of the coordinates (for the drop $\tilde{\Gamma}^{(k)}$is a sphere). During the computations $\tilde{\Gamma}^{(k)}$ and $\Omega_2^{(k)}$ remain untransformed, while Eq. (1) and boundary conditions (2)-(4) are changed. Finite elements (or coordinate functions in the Galerkin method, grid schemes etc.) are constructed in such a way that at each step of the successive approximations conditions(2) are satisfied in the new coordinates. Such a method provides convergence with a rate of a geometrical progression with the ratio $q < 1$ ([2]). In such a way problems of the axis-symmetrical motion of a deformed drop in a fluid under the force of gravity ([2],[4],[5]), the problem of film flow etc. were solved.

In another type of the iteration method (see [12],[13],[14]etc.),the unknown boundary is found simultaneously with the velocity vector field and the pressure. Here finite elements for the Navier-Stokes equations are determined via the unknwon coordinates of the boundary. During a computatinal cyclus the finite elements can deform significantly, and the ratio between their sides becomes too large, so the condition numbers of corresponding algebraic systems increases significantly.

In the above methods it is necessary to solve some equations of the Navier-Stokes type in an arbitrary domain with the boundary conditions transformed to the new coordinates at every step of iteration. In the first modification of the method the alternating Schwartz method method of the separation of domains)is used without an intersection of the domains. It is convenient to construct it in such a way that the Navier-Stokes equations would be computed independently in every Ω_i $((\vec{u},\vec{n})|_\Gamma = 0)$ and then combined at the boundary. This "combination" is performed in such a way that the continuity condition $[\vec{r}\cdot\sigma\cdot\vec{n}]|_\Gamma = 0$ always holds while the condition of continuity of the tangential velocities is achieved in the process of iterations:

$$\vec{r}\cdot\sigma_1^{(k)}\cdot\vec{n} = \vec{r}\cdot\sigma_1^{(k-1)}\cdot\vec{n}\ |_\Gamma + \alpha\ [\vec{u}^{(k-1)}]|_\Gamma;\ [\vec{r}\cdot\sigma^{(k)}\cdot\vec{n}]|_\Gamma = 0\ , \tag{7}$$

$\vec{u}^{(k)}$ and $\sigma^{(k)}$ being the values of the velocities and stresses determined in the k-th iteration. Such method provides the convergence of the order c/k^N, where $N > 1$ is a number determined by a smoothness of the solution in search. In a finite-dimensional space the Schwartz method converges with a rate of a geometric progression; but the constant which appears in the estimate depends on the dimension of the space. (The proof of the convergence is similar to the one given in 4 .) For the solution of algebraic systems which arise here,usual iterative methods were used (Uzava algorithm, multigrid methods, etc.); the rate of their convergence was investigated by several authors. With this approach a number of problems was solved: such as the problem of the axis-symmetrical stationary motion of deformed drops of viscous incompressible liquid in a flow of another incompressible fluid, of the stationary motion of drops in non-isothermal fluid when thermocapilary phenomena are taken into accont, the problem of the stationary heat and mass exchange of drops and bubbles ([2],[4] etc.). There were discovered main features of drops' and bubbles' deformation, of the structure of hydrodynamical flows inside and outside them depending on physical parameters. Some peculiarities of the flows which were not observed for stationary flows past solid bodies (characteristic separation of the boundary layer, etc.) were also discovered.

The system (1) is non-dimensionalised, the characteristic radius of the drop, the asymptotic flow velocity u_∞ and the speed of rotation of the drop $\omega_0 d$ being taken for characteristic parametres. Because of the axial symmetry the problem of a rotating drop's motion is reformulated using stream functions, a stream function ψ^*, connected with the original ψ_2 by a relation $\psi_2 = 1/2r^2 \sin^2\Theta + \psi^*$ being introduced in an external domain.

Numerical computation of the stationary problem is performed by the successive iterations method described above, which is re-formulated for the stream functions. The computations have been done for severed mathematical and physical test problems checking the interior convergence, as well as comparising different numerical schemes.

The parametres are non-dimensional numbers

$$Ta = \frac{\omega_0 d^2}{\nu} ; \quad Re = \frac{u_\infty d}{\nu} ; \quad Ro = \frac{Re}{Ta} ; \quad We = \frac{\rho_1 d_1 u_\infty^2}{2\gamma} ; \quad \mu = \frac{\nu_1 \rho_1}{\nu_2 \rho_2} ,$$

and it was assumed that $u_\phi|_{x \to \infty} = 0$, $u_\phi|_\Gamma = u_{\phi_0}$. Computations showed that even at small Re (e.g., $Re_i = 1$, $i = 1,2$; $Ta = 3$; $\mu = 10$) the rotation causes a recirculating eddy behind the drop, which is the larger the greater is the role of rotation. At Ro = 1 the eddy becomes essentially smaller, and at Ro = 5 disappears completely. Such a phenomenon is also known for a displacement rotating motion of a rotating solid ball.

Together with the recirculating flows near the backward edge of the drop at Ro = 2/3, and intermediate μ ($\mu = 10 \div 100$) a stagnation zone is formed near its upward edge which has a tendency to grow into a recirculating eddy before the drop with the descreasing Ro and the eddy accumulation. Such tendencies were observed also under strong deformation of drops without rotation, but here they are caused by the rotation forces. The peculiarity of a structure inside the drop is also due to these forces. Here under comparatively small μ (μ=10) inside the drop a significant secondary eddy is formed at the backward zone and a small stagnation zone is formed in the upward zone of the drop. In the absence of rotation such phenomena were observed only at significant deviations of μ from 1 and at large values of Re or in inhomogenous temperature field. Together with the change of the flow structure the deformation of drops is also changing, increasing parallel to the increase of Ta. By the computations the increase of the resistance coefficient with the increase of Ta was also observed.

Investigation of the stability of the drop's shape under directed rotating movement can be reduced to the problem of perturbations' developement for the solutions of stationary equations which can be observed in solving the full non-stationary system of the Navier-Stokes equations. Unlike stationary problems, here a numerical scheme is used in which the free boundary Γ at a time step is found from the kinematic condition

$$\frac{\partial\Gamma}{\partial t} + (\vec{u},\nabla\Gamma) = 0. \qquad (8)$$

This equation is solved by a grid method of the predictor-corrector type or by a finite element method with the subsequent filtration of parasitic high-frequency components (the number of harmonics being determined by a physical meaning of the problem). Next, on the level t the non-stationary system of the Navier-Stokes equations (4) with the boundary conditions (3)-(5) is computed, and then we move to the next time level. Here equations (3) and (4) are differentiated with respect to tangent variables, and the derivatives of the pressure arising in this way are determined from the Navier-Stokes equations.

Although there still is no justification of this method for non-linear non-stationary problems, computations of a number of test and physical problems (a vertical film, a horizontal non-isothermal film, and so on - see [12]) showed that it is numerically possible to use such algorithms.

Because of the computational complication in the non-linear non-stationary problem of the displacement motion of a rotating drop first of all numbers Ta*, We*, Re*,μ*, are determined at which the solution of a linearized problem is unstable. For this purpose linear equations are constructed for Γ_1, u_1 from the non-stationary Navier-Stokes equations and the conditions (4)-(6), (*): $\Gamma = \Gamma_0 + \varepsilon\Gamma_1 e^{\lambda t}$, $u = u_0 + \varepsilon u_1 \cdot e^{\lambda t}$, where u_0, Γ_0 represent the determined numerical solution of the non-stationary problem (the terms ε^2 are neglected, ε being a small parametar). In such a way we arrive to a problem of finding an eigenvalue λ, with a minimal positive real part. The QR-algorithm for this problem gave the following results:

$$Re_1^* = Re_2^* = 50,\ Ta^* = 2,\ \mu^* = 10,\ We^* = 4.2\ ;$$
$$Re_1^* = Re_2^* = 50,\ Ta^* = 2,\ \mu^* = 0.1, We^* = 3.8\ ;$$
$$Re_1^* = Re_2^* = 50,\ Ta^* = 1,\ \mu^* = 10,\ We^* = 4.53.$$

The following tendencies in the behaviour of We^* are seen: It is decreasing with the increase of Ta and the decrease of μ. In a rather large range they are grouped near $We^* = 4$, which corresponds to the data of the physical experiment ([15],[16]).

At such We^*, Ta^*, μ^*, Re^* computations of the full non-stationary problem were performed, in which a larger growth of initial perturbations than at other values of these parametres was observed.

References

[1] V.Ya. Rivkind, N.B. Fridman. On the Navier-Stokes equations with discountinuous coefficients. Zapiski nauchn. seminarov LOMI. 1978, Leningrad, v.38, pp. 137-152. (in Russian).

[2] V.Ya. Rivkind. Investigation of a problem of the stationary movement of a drop in the flow of a viscous incompressible fluid. Doklady AN SSSR, 1976, 227, N5, Moscow (in Russian).

[3] V.Ya. Rivkind. Investigation of some problems of the flow of multi-layer viscous incompressible fluids, Trudy Vseoyuznoi konferencli po uravneniyam v chastnyh proizvodnyh (Proceedings of the All-Union conference on partial differential equations, dedicated to the 75-th anniversary of I.G. Petrovskii). Nauka, Moscow, 1978 (in Russian).

[4] V.Ya. Rivkind. Computational methods for fluid-flows of viscous incompressible fluids with free boundaries. Chislenn. Meth.Meh. Sploshnoi sredy, 12, N4, pp. 106-115 (1981), Novosibirsk (in Russian).

[5] Bemelmans J. Liquid drops in a viscous fluid under the influence of gravity and surface tension. Manuscripta math., 36, 1981, 105-123.

[6] Yerunova I.B. Investigation of a problem of the stationary movement of two fluids in a vessl. Doklady AN SSSR, 1984, v.279, N1.

[7] J. Sokolovskiy. Eine verallgemeinerte Leitlinienmethode zur Berechnung wehrschichtiger Strömungen nichtlinearviskoser Fluide. J.Appl. Maths and Phys. (ZAMP) v.39, März 1988, p.221.

[8] V.Ya. Rivkind. Approximate Methods for solving problems of viscous fluid with a free boundary. Numerical Methods and Applications. Sofia, 1985.

[9] V.V. Puhnachev. A non-stationarymproblem for the Navier-Stokes equations with a free boundary in plane. Prikl.mat. i tehn. fizika, 1973, N3, Novosibirsk (in Russian).

[10] V.A. Solonnikov, V.E. Scadilov. On one boundary value problem for the stationary Navier-Stokes equations system. Trudy mat. inst. AN SSSR, 1973, 125, Leningrad (in Russian).

[11] W.J. William, L.E. Scriven. Separating flow near a static contact line: Slip at a wall and shape of a free surface. J. Comp. Phys. 34, 287-313 (1980).

[12] A.V. Ilyin. Numerical investigation of problems of thin fluid layers movements. Ph.D. Thesis, Leningrad, 1986 (in Russian).

[13] K.J. Ruschak, A method for incorporating free boundaries with surface tension in finite-element fluid flow simulators. Inst. J.num.meth. engng.15, N5, 639-648 (1980).

[14] S.F. Kistler and L.E. Scriven. Coating flow theory by finite element and asymptotic analysis of the Navier-Stokes system. Int. J. num. meth. fluids 4, 207-229 (1984).

[15] Wang T.G., Saffren M.M., Elleman D.D. Drop dynamics in space. In: Mater.Sci. Space appl. Space Process. New York, 1977, 151-172.

[16] Bushlanov V.P., Vasenin I.M. Stability of a rotating viscous drop. Teplofiz. i fiz. gidrodinamika. Novosibirsk, 1978, pp. 9-14 (in Russian).

(received January 23,1989.)

CURL-CONFORMING FINITE ELEMENT METHODS for NAVIER-STOKES EQUATIONS with NON-STANDARD BOUNDARY CONDITIONS in $\Re^3$

V. GIRAULT
Laboratoire d'Analyse Numérique
Université PIERRE et MARIE CURIE
PARIS , FRANCE

Abstract.

This paper is devoted to the steady-state, incompressible Navier-Stokes equations with non standard boundary conditions of the form:

$$\mathbf{u} \times \mathbf{n} = \mathbf{0} \quad , \quad p + (1/2)\,\mathbf{u} \cdot \mathbf{u} = 0 \ ,$$

or

$$\mathbf{u} \cdot \mathbf{n} = 0 \ , \quad \mathbf{curl}\ \mathbf{u} \cdot \mathbf{n} = 0 \ , \quad \mathbf{curl}\ \mathbf{curl}\ \mathbf{u} \cdot \mathbf{n} = 0 \ ,$$

or

$$\mathbf{u} \cdot \mathbf{n} = 0 \ , \quad \mathbf{curl}\ \mathbf{u} \times \mathbf{n} = \mathbf{0} \ .$$

The problem is formulated in the primitive variables : velocity and pressure, and the divergence-free condition is imposed weakly by the equation

$$(\nabla q \, , \mathbf{v}) = 0 \ .$$

Thus, while more regularity is required for the pressure, owing to the boundary conditions, the velocity needs only have a smooth **curl**. Hence, the velocity is approximated with **curl** conforming finite elements and the pressure with standard continuous finite elements. The error analysis gives optimal results .

This paper is in final form and no similar paper has been or is being submitted elsewhere.

I. Introduction.

This paper is devoted to the numerical solution of the steady-state Navier-Stokes equations :

(1.1) $$-\nu \Delta \mathbf{u} + \sum_{j\leq 3} u_j \partial \mathbf{u}/\partial x_j + \nabla p = \mathbf{f} \quad , \quad \operatorname{div} \mathbf{u} = 0 \,, \ \text{in } \Omega \,,$$

with boundary conditions :

(1.2) $$\mathbf{u} \times \mathbf{n} = \mathbf{0} \quad , \quad p + (1/2)\mathbf{u}\cdot \mathbf{u} = 0 \ \text{ on } \Gamma \,,$$

or

(1.3) $$\mathbf{u}\cdot \mathbf{n} = 0 \ , \ \ \mathbf{curl}\ \mathbf{u}\cdot \mathbf{n} = 0 \ , \ \ (\mathbf{curl}\ \mathbf{curl}\ \mathbf{u})\cdot \mathbf{n} = 0 \ \text{ on } \Gamma \,,$$

or

(1.4) $$\mathbf{u}\cdot \mathbf{n} = 0 \quad , \quad \mathbf{curl}\ \mathbf{u} \times \mathbf{n} = \mathbf{0} \ \text{ on } \Gamma \,,$$

where Ω is a bounded, convex domain of $\Re^3$ with a polyhedral boundary Γ and $\mathbf{n}$ is the exterior unit normal to Γ. The quantity :

$$\tilde{p} = p + (1/2)\mathbf{u}\cdot \mathbf{u}$$

denotes the dynamical pressure; it reduces to the pressure if we replace (1.1) by the linear Stokes equations.

These sets of boundary conditions lend themselves readily to a variational formulation where the Laplacian operator is expressed by a $(\mathbf{curl}\cdot , \mathbf{curl}\cdot)$ term and the incompressibility condition by an equation of the form : $(\nabla q , \mathbf{v}) = 0$. This suggests to use a partially non-conforming finite element method, where just the **curl** of the velocity is continuous at interface boundaries whereas the pressure is globally continuous. Thus, we shall approximate the velocity with **curl**-conforming elements of Nedelec and the pressure with standard conforming finite elements. The resulting schemes have already been suggested by Bossavit [6, 7] in a paper devoted to Whitney elements.

None of the above boundary conditions corresponds to the classical Dirichlet condition : $\mathbf{u} = \mathbf{0}$. It is possible to prescribe it indirectly by imposing :

$$\mathbf{u} \times \mathbf{n} = \mathbf{0} \ \text{ on } \Gamma \ \text{ and } \ (\nabla q , \mathbf{u}) = 0 \ \text{ for all } q \text{ in } H^1(\Omega) \,,$$

but unfortunately, the resulting finite-dimensional problem appears to be ill-posed.

The convexity assumption on Ω is a well-known theoretical consequence of the fact that Γ is not smooth. There is no practical evidence that it is necessary and this assumption is disregarded in practice: instead we can assume that Ω is simply-connected and Γ is connected. We have completely left out the cases where Γ is smooth or not connected or Ω not simply-connected. A domain with "holes" or a multiply-connected domain can be handled with the techniques of Bendali, Dominguez & Gallic [3]. We refer to Dubois [12] for a good treatment of the potential problem (i.e. a linear problem with the boundary conditions (1.2)) on a domain with a curved and multiply-connected boundary. In particular, this reference includes an interesting application of curved finite elements near the boundary. We also refer to Verfürth [25] for a different approximation of the same potential problem on a curved domain. As far as the theory is concerned, the reader will find in Bègue, Conca, Murat & Pironneau [2] a very comprehensive study of Navier-Stokes equations with non-standard (and non-homogeneous) boundary conditions on a variety of domains. These authors include a conforming approximation of the Hood-Taylor type for the velocity; the corresponding theoretical analysis is done by Franca & Hugues [14]. Also, we refer to a previous work (Girault [15]) for a "vector potential-vorticity" approximation of similar Navier-Stokes type problems, including the standard one.

2. Curl-conforming variational formulations.

From now on, the assumptions on the domain are :

Ω is bounded, simply connected, with a polyhedral, connected boundary Γ.

In the sequel, we shall use the standard Sobolev spaces $W^{m,p}(\Omega)$ (or $H^m(\Omega)$ if $p = 2$) :

$$W^{m,p}(\Omega) = \{v \in L^p(\Omega) \, ; \, \partial^\alpha v \in L^p(\Omega) \ \forall \, |\alpha| \leq m\} ,$$

equipped with the following semi-norm and norm :

$$|v|_{m,p,\Omega} = \{\sum_{|\alpha|=m} \int_\Omega |\partial^\alpha v(x)|^p \, dx\}^{1/p} \quad , \quad \|v\|_{m,p,\Omega} = \{\sum_{k \leq m} |v|^p_{k,p,\Omega}\}^{1/p} .$$

As usual, we shall omit p when $p = 2$ and denote by $(\cdot \, , \cdot)$ the scalar product of $L^2(\Omega)$. Also, recall the familiar notation :

$$H_0^1(\Omega) = \{v \in H^1(\Omega) \, ; \, v = 0 \ \text{on} \ \Gamma\} .$$

We shall work mainly with the following Hilbert spaces relative to the divergence and **curl** operators :

$$H(\text{div};\Omega) = \{\mathbf{v} \in L^2(\Omega)^3 \, ; \, \text{div}\, \mathbf{v} \in L^2(\Omega)\} \ , \ H_0(\text{div};\Omega) = \{\mathbf{v} \in H(\text{div};\Omega) \, ; \, \mathbf{v} \cdot \mathbf{n} = 0 \ \text{on} \ \Gamma\} ,$$

$$H(\mathbf{curl};\Omega) = \{\mathbf{v} \in L^2(\Omega)^3 \, ; \, \mathbf{curl}\, \mathbf{v} \in L^2(\Omega)^3\} \ , \ H_0(\mathbf{curl};\Omega) = \{\mathbf{v} \in H(\mathbf{curl};\Omega) \, ; \, \mathbf{v} \times \mathbf{n} = \mathbf{0} \ \text{on} \ \Gamma\} ,$$

normed respectively by :

$$\|\mathbf{v}\|_{H(\text{div};\Omega)} = \{\|\mathbf{v}\|^2_{0,\Omega} + \|\text{div}\, \mathbf{v}\|^2_{0,\Omega}\}^{1/2} ,$$

$$\|\mathbf{v}\|_{H(\mathbf{curl};\Omega)} = \{\|\mathbf{v}\|^2_{0,\Omega} + \|\mathbf{curl}\ \mathbf{v}\|^2_{0,\Omega}\}^{1/2} .$$

The reader will find in Duvaut & Lions [13] (for the case of a smooth boundary) and Girault & Raviart [16] (for the case of a polygonal or polyhedral boundary) the theoretical foundations of these spaces. One of the major difficulties encountered with them is due to the roughness of Γ. The following regularity theorems, proved by Bernardi [4], Dauge [11], Girault & Raviart [16], Grisvard [17] and Nedelec [19] will play here a crucial part. The first lemma is due to Dauge.

Lemma 2.1.

Let $\mathbf{f}$ *be given in* $L^p(\Omega)^3$ *and let* Ω *be* ***convex****. If*

$$6/(3 + \sqrt{5}) < p < 6/(3 - \sqrt{5}) ,$$

then there exists a unique function w *in* $W^{1,p}(\Omega)/\mathfrak{R}$ *(resp.* $W_0^{1,p}(\Omega)$ *) such that :*

$$(\nabla w \, , \nabla v) = (\mathbf{f} \, , \nabla v) \quad \forall \, v \in W^{1,q}(\Omega)/\mathfrak{R} \ (\textit{resp.}\ W_0^{1,q}(\Omega)) ,$$

with
$$1/p + 1/q = 1$$

and
$$|w|_{1,p,\Omega} \leq C \|\mathbf{f}\|_{0,p,\Omega} .$$

The above range for p includes in particular $p = 4$ and $p = 4/3$, which are the two exponents involved with Navier-Stokes equations.

Theorem 2.1.

Each function $\mathbf{u}$ *in* $L^2(\Omega)^3$ *that satisfies :* $\text{div}\, \mathbf{u} = 0$ *in* Ω *(resp. and* $\mathbf{u} \cdot \mathbf{n} = 0$ *on* Γ*) has a unique vector potential* $\mathbf{v}$ *in* $L^2(\Omega)^3$ *characterized by :*

$$\mathbf{curl}\ \mathbf{v} = \mathbf{u} \ , \ \text{div}\, \mathbf{v} = 0 \ \textit{in} \ \Omega \ , \ \mathbf{v} \cdot \mathbf{n} = 0 \ \textit{on} \ \Gamma$$

$$(\textit{resp.}\ \mathbf{curl}\ \mathbf{v} = \mathbf{u} \ , \ \text{div}\, \mathbf{v} = 0 \ \textit{in} \ \Omega \ , \ \mathbf{v} \times \mathbf{n} = \mathbf{0} \ \textit{on} \ \Gamma) .$$

If in addition Ω *is* ***convex****, then* $\mathbf{v}$ *belongs to* $H^1(\Omega)^3$ *. Moreover, there exists a real* $s > 2$ *depending on the angles of* Γ *such that :*

(2.1) $$\mathbf{v} \in W^{1,t}(\Omega)^3 \ \textit{whenever}\ \mathbf{u}\ \textit{belongs to}\ L^t(\Omega)^3 \ \forall \ t \ \textit{in} \ [2 \, , s] .$$

Theorem 2.2.

Let Ω *be* **convex**. *All functions* $\mathbf{v}$ *in* $L^2(\Omega)^3$ *that satisfy :*

$$\operatorname{div} \mathbf{v} = 0 \ , \quad \mathbf{curl}\ \mathbf{v} \in L^2(\Omega)^3 \ , \quad \mathbf{v}\cdot\mathbf{n} = 0 \text{ or } \mathbf{v}\times\mathbf{n} = \mathbf{0} \ \text{ on } \Gamma \ ,$$

belong to $H^1(\Omega)^3$ *and*

$$\| \mathbf{v} \|_{1,\Omega} \leq C \| \mathbf{curl}\ \mathbf{v} \|_{0,\Omega} \ . \tag{2.2}$$

In addition, if **curl v** *belongs to* $L^s(\Omega)^3$ *with the real* $s > 2$ *of Theorem 2.2, then for each* t *in* $[2, s]$ *we have* $\mathbf{v}$ *in* $W^{1,t}(\Omega)^3$ *and*

$$\| \mathbf{v} \|_{1,t,\Omega} \leq C(t) \| \mathbf{curl}\ \mathbf{v} \|_{0,t,\Omega} \ . \tag{2.3}$$

Now, assume that the right-hand side $\mathbf{f}$ of (1.1) belongs to $L^{4/3}(\Omega)^3$. Using Lemma 2.1 and the techniques of Bègue *et al* [2], we can show that system (1.1) with either boundary conditions (1.2), (1.3) or (1.4) has at least one solution :

$$(\mathbf{u}, p) \ \text{in} \ H^1(\Omega)^3 \times W^{1,4/3}(\Omega) \ \text{with} \ \mathbf{curl}\ \mathbf{curl}\ \mathbf{u} \ \text{in} \ L^{4/3}(\Omega)^3 \ .$$

Then in view of the identities :

$$-\Delta\mathbf{u} = \mathbf{curl}\ \mathbf{curl}\ \mathbf{u} \ , \quad \sum_{j\leq 3} u_j\, \partial\mathbf{u}/\partial x_j = \mathbf{curl}\ \mathbf{u}\times\mathbf{u} + (1/2)\nabla(\mathbf{u}\cdot\mathbf{u}) \ ,$$

we can easily prove the following theorems.

Theorem 2.3.

Let Ω *be* **convex** *and suppose that the right-hand side* $\mathbf{f}$ *of the Navier-Stokes equations (1.1) satisfies :*

$$\mathbf{f} \in L^{4/3}(\Omega)^3 \ . \tag{2.4}$$

Then problem (1.1), (1.2) has the weak variational formulation :

Find $\mathbf{u}$ *in* $H(\mathbf{curl};\Omega) \cap L^4(\Omega)^3$ *and* $\tilde{p}$ *in* $W^{1,4/3}(\Omega)$ *such that :*

$$\nu\,(\mathbf{curl}\ \mathbf{u}, \mathbf{curl}\ \mathbf{v}) + (\mathbf{curl}\ \mathbf{u}\times\mathbf{u}, \mathbf{v}) + (\nabla\tilde{p}, \mathbf{v}) = (\mathbf{f}, \mathbf{v}) \quad \forall\ \mathbf{v} \in H_0(\mathbf{curl};\Omega) \cap L^4(\Omega)^3 \ , \tag{2.5}$$

$$(\nabla q, \mathbf{u}) = 0 \quad \forall q \in H_0^1(\Omega) \ , \tag{2.6}$$

$$\mathbf{u}\times\mathbf{n} = \mathbf{0} \ , \quad \tilde{p} = 0 \ \text{on} \ \Gamma \ , \tag{1.2}$$

where

$$\tilde{p} = p + (1/2)\,\mathbf{u}\cdot\mathbf{u} \ .$$

Theorem 2.4.

Let $\mathbb{X}$ *be the space :*

$$\mathbb{X} = \{\mathbf{v} \in H(\mathbf{curl};\Omega) \, ; \, \mathbf{curl}\ \mathbf{v}\cdot\mathbf{n} = 0 \ \text{on} \ \Gamma\} \ . \tag{2.7}$$

Under hypothesis (2.4), problem (1.1), (1.3) has the weak variational formulation :

Find $\mathbf{u}$ *in* $\mathbb{X} \cap L^4(\Omega)^3$ *and* $\tilde{p}$ *in* $W^{1,4/3}(\Omega)/\mathfrak{R}$ *such that :*

$$\nu\,(\mathbf{curl}\ \mathbf{u}, \mathbf{curl}\ \mathbf{v}) + (\mathbf{curl}\ \mathbf{u}\times\mathbf{u}, \mathbf{v}) + (\nabla\tilde{p}, \mathbf{v}) = (\mathbf{f}, \mathbf{v}) \quad \forall\ \mathbf{v} \in \mathbb{X} \cap L^4(\Omega)^3 \ , \tag{2.8}$$

$$(\nabla q, \mathbf{u}) = 0 \quad \forall q \in H^1(\Omega) \ , \tag{2.9}$$

where

$$\tilde{p} = p + (1/2)\,\mathbf{u}\cdot\mathbf{u} \ .$$

Remark 2.1.

Since **curl v** belongs to $H(\mathrm{div};\Omega)$, the boundary condition $\mathbf{curl}\ \mathbf{v}\cdot\mathbf{n} = 0$ on Γ is well defined. In addition, owing to Theorem 2.1, all functions $\mathbf{v}$ of $\mathbb{X}$ have the decomposition :

$$\mathbf{v} = \mathbf{w} + \nabla q \ , \tag{2.10}$$

with q in $H^1(\Omega)$, $\mathbf{w}$ in $H^1(\Omega)^3$, $\mathrm{div}\ \mathbf{w} = 0$ in Ω and $\mathbf{w}\times\mathbf{n} = \mathbf{0}$ on Γ. Conversely, all functions of the form (2.10) belong to $\mathbb{X}$. ♦

Theorem 2.5.

Under hypothesis (2.4), problem (1.1), (1.4) has the weak variational formulation :

Find $\mathbf{u}$ *in* $H(\mathbf{curl};\Omega)\cap L^4(\Omega)^3$ *and* $\tilde{p}$ *in* $W^{1,4/3}(\Omega)/\mathfrak{R}$ *such that:*

(2.11) $\nu\,(\mathbf{curl}\ \mathbf{u}\ ,\ \mathbf{curl}\ \mathbf{v}) + (\mathbf{curl}\ \mathbf{u}\times\mathbf{u}\ ,\ \mathbf{v}) + (\nabla\tilde{p}\ ,\mathbf{v}) = (\mathbf{f}\ ,\ \mathbf{v}) \quad \forall\ \mathbf{v}\in H(\mathbf{curl};\Omega)\cap L^4(\Omega)^3$,

$$(\nabla q\ ,\mathbf{u}) = 0 \quad \forall q\in H^1(\Omega)\ , \tag{2.12}$$

where

$$\tilde{p} = p + (1/2)\,\mathbf{u}\cdot\mathbf{u}\ .$$

The variational formulation (2.5), (2.6) imposes no extra boundary condition on $\mathbf{u}$ or $\tilde{p}$. But on the other hand, formulation (2.8), (2.9) implies :

$$(\mathbf{curl}\ \mathbf{u}\ ,\ \mathbf{curl}\ \mathbf{v}) = (\mathbf{curl}\ \mathbf{curl}\ \mathbf{u}\ ,\ \mathbf{v}) \quad \forall\ \mathbf{v}\in\mathbb{X}\ .$$

Now, in view of (2.10), this yields :

$$(\mathbf{curl}\ \mathbf{curl}\ \mathbf{u}\ ,\ \nabla q) = 0 \quad \forall q\in W^{1,4}(\Omega)\ ,$$

which is a weak way to prescribe : $\mathbf{curl}\ \mathbf{curl}\ \mathbf{u}\cdot\mathbf{n} = 0$ on Γ. Similarly, by choosing $\mathbf{v} = \nabla q$ with q in $W^{1,4}(\Omega)$, (2.8) becomes :

$$(\nabla\tilde{p}\ ,\nabla q) = (\mathbf{f}\ ,\nabla q) - (\mathbf{curl}\ \mathbf{u}\times\mathbf{u}\ ,\nabla q) \quad \forall q\in W^{1,4}(\Omega)\ ,$$

which is a weak form of :

$$\partial\tilde{p}/\partial\mathbf{n} = (\mathbf{f} - \mathbf{curl}\ \mathbf{u}\times\mathbf{u})\cdot\mathbf{n} \quad \text{on } \Gamma\ .$$

Likewise, formulation (2.11), (2.12) implies :

$$\mathbf{curl}\ \mathbf{u}\times\mathbf{n} = \mathbf{0}\ , \quad \partial\tilde{p}/\partial\mathbf{n} = (\mathbf{f} - \mathbf{curl}\ \mathbf{u}\times\mathbf{u})\cdot\mathbf{n} \quad \text{on } \Gamma\ .$$

Each variational formulation decouples into a system for the velocity and a Poisson equation for the pressure. Indeed, for (2.5), (2.6), we introduce the space of divergence free functions :

$$\mathbb{V}_0 = \{\mathbf{v}\in H_0(\mathbf{curl};\Omega)\ ;\ (\nabla q\ ,\mathbf{v}) = 0 \quad \forall q\in H_0^1(\Omega)\}\ .$$

Then problem (2.5), (2.6), (1.2) is equivalent to the two problems :

Find $\mathbf{u}$ *in* $\mathbb{V}_0\cap L^4(\Omega)^3$ *such that:*

$$\nu\,(\mathbf{curl}\ \mathbf{u}\ ,\ \mathbf{curl}\ \mathbf{v}) + (\mathbf{curl}\ \mathbf{u}\times\mathbf{u}\ ,\ \mathbf{v}) = (\mathbf{f}\ ,\ \mathbf{v}) \quad \forall\ \mathbf{v}\in\mathbb{V}_0\cap L^4(\Omega)^3\ . \tag{2.13}$$

Find $\tilde{p}$ *in* $W_0^{1,4/3}(\Omega)$ *such that:*

$$(\nabla\tilde{p}\ ,\nabla q) = (\mathbf{f}\ ,\nabla q) - (\mathbf{curl}\ \mathbf{u}\times\mathbf{u}\ ,\nabla q) \quad \forall q\in W_0^{1,4}(\Omega)\ . \tag{2.14}$$

Likewise, defining the space :

$$\mathbb{U}_0 = \{\mathbf{v}\in\mathbb{X}\ ;\ (\nabla q\ ,\mathbf{v}) = 0 \quad \forall q\in H^1(\Omega)\}\ ,$$

$$(\text{resp. } \mathbb{U} = \{\mathbf{v}\in H(\mathbf{curl};\Omega)\ ;\ (\nabla q\ ,\mathbf{v}) = 0 \quad \forall q\in H^1(\Omega)\}\ )\ ,$$

we split problem (2.8), (2.9) (resp. (2.11), (2.12)) into :

Find $\mathbf{u}$ *in* $\mathbb{U}_0 \cap L^4(\Omega)^3$ *(resp.* $\mathbb{U} \cap L^4(\Omega)^3$*) such that :*

(2.15) $$\nu\,(\mathbf{curl\ u}\,,\,\mathbf{curl\ v}) + (\mathbf{curl\ u} \times \mathbf{u}\,,\,\mathbf{v}) = (\mathbf{f}\,,\,\mathbf{v}) \ \ \forall\ \mathbf{v} \in \mathbb{U}_0 \cap L^4(\Omega)^3 \quad (\text{resp. } \forall\ \mathbf{v} \in \mathbb{U} \cap L^4(\Omega)^3)\ .$$

Find $\tilde{p}$ *in* $W^{1,4/3}(\Omega)/\mathfrak{R}$ *such that :*

(2.16) $$(\nabla \tilde{p}\,,\,\nabla q) = (\mathbf{f}\,,\,\nabla q) - (\mathbf{curl\ u} \times \mathbf{u}\,,\,\nabla q) \ \ \forall\ q \in W^{1,4}(\Omega)\ .$$

It is important to stress that, although all the above formulations imply that $\mathbf{u}$ belongs to H^1, this regularity is not explicit : all they require is that **curl u** belong to L^2. This accounts for the idea of using **curl**-conforming finite elements.

Remark 2.2.

When $\mathbf{f}$ belongs to $L^2(\Omega)^3$, the Stokes problem :

(2.17) $$-\nu\,\Delta\mathbf{u} + \nabla p = \mathbf{f} \quad , \quad \operatorname{div}\mathbf{u} = 0 \ \text{ in } \Omega\ ,$$

(2.18) $$\mathbf{u} \times \mathbf{n} = \mathbf{0} \quad , \quad p = 0 \ \text{ on } \Gamma\ ,$$

has the equivalent formulation :

Find $\mathbf{u}$ *in* $\mathbb{V}_0$ *such that :*

(2.19) $$\nu\,(\mathbf{curl\ u}\,,\,\mathbf{curl\ v}) = (\mathbf{f}\,,\,\mathbf{v}) \ \ \forall\ \mathbf{v} \in \mathbb{V}_0\ .$$

Find p *in* $H_0^1(\Omega)$ *such that :*

(2.20) $$(\nabla p\,,\,\nabla q) = (\mathbf{f}\,,\,\nabla q) \ \ \forall\ q \in H_0^1(\Omega)\ .$$

It is easy to prove that this problem has a unique solution that satisfies the following bounds :

$$|p|_{1,\Omega} \le \|\mathbf{f}\|_{0,\Omega} \ , \quad \|\mathbf{curl\ u}\|_{0,\Omega} \le (C_1/\nu)\,\|\mathbf{f}\|_{0,\Omega}\ ,$$

$$\|\mathbf{u}\|_{1,\Omega} \le (C_2/\nu)\,\|\mathbf{f}\|_{0,\Omega} \ , \quad \|\mathbf{curl\ u}\|_{1,\Omega} \le (C_3/\nu)\,\|\mathbf{f}\|_{0,\Omega}\ .$$ ◆

Remark 2.3.

Similarly, when $\mathbf{f}$ belongs to $L^2(\Omega)^3$, the Stokes problems :

(2.21) $$-\nu\,\Delta\mathbf{u} + \nabla p = \mathbf{f} \quad , \quad \operatorname{div}\mathbf{u} = 0 \ \text{ in } \Omega\ ,$$

(2.22) $$\mathbf{u}\cdot\mathbf{n} = 0 \quad , \quad \mathbf{curl\ u}\cdot\mathbf{n} = 0 \quad , \quad (\mathbf{curl\ curl\ u})\cdot\mathbf{n} = 0 \ \text{ on } \Gamma\ ,$$

(2.23) $$(\text{resp. } \mathbf{u}\cdot\mathbf{n} = 0 \quad , \quad \mathbf{curl\ u} \times \mathbf{n} = \mathbf{0} \ \text{ on } \Gamma)\ ,$$

have the equivalent formulations :

Find $\mathbf{u}$ *in* $\mathbb{U}_0$ *(resp.* $\mathbf{u}$ *in* $\mathbb{U}$*) such that :*

(2.24) $$\nu\,(\mathbf{curl\ u}\,,\,\mathbf{curl\ v}) = (\mathbf{f}\,,\,\mathbf{v}) \ \ \forall\ \mathbf{v} \in \mathbb{U}_0 \ \ (\text{resp. } \mathbf{v} \in \mathbb{U})\ .$$

Find p *in* $H^1(\Omega)/\mathfrak{R}$ *such that :*

(2.25) $$(\nabla p\,,\,\nabla q) = (\mathbf{f}\,,\,\nabla q) \ \ \forall\ q \in H^1(\Omega)\ ;$$

and consequently, p satisfies also the Neumann boundary condition $\partial p/\partial \mathbf{n} = \mathbf{f}\cdot\mathbf{n}$ on Γ. The solution $(\mathbf{u}\,,p)$ of each problem is unique and bounded like above. ◆

3. Three families of finite element spaces.

The above equations involve three operators : **curl**, **grad** and div, even if the latter is implicit. Each one will be discretized with a different finite element space. The **curl**-conforming finite elements we shall use were developed by Nedelec in [18] and [19] and lately related to Whitney elements by Bossavit in [6, 7]. They are thoroughly studied in these references and in Girault & Raviart [16]. Let us define them and describe briefly their main properties (nearly all proved by Nedelec).

Let P_k be the space of polynomials of three variables of degree at most k, and $\tilde{P}_k$ the subspace of homogeneous polynomials of degree exactly k. For a *fixed* integer $k \geq 1$, we define the following subspace of $P_k{}^3$:

(3.1) $$\mathbb{R}_k = (P_{k-1})^3 \oplus \{\mathbf{p} \in \tilde{P}_k{}^3 \, ; \mathbf{p}(x)\cdot \mathbf{x} = 0\} \; .$$

Then we associate with $\mathbb{R}_k$ the following degrees of freedom.

Definition 3.1.

Let κ be a tetrahedron with faces denoted by $\mathscr{f}$ and edges denoted by $\mathscr{e}$, $\mathbf{t}$ being the direction vector of $\mathscr{e}$ and let $\mathbf{u}$ belong to $W^{1,t}(\kappa)^3$ for some $t > 2$. We define the three sets of moments of $\mathbf{u}$ on κ :

$$\{\textstyle\int_{\mathscr{e}} (\mathbf{u}\cdot \mathbf{t})\, q \, d\mathscr{e} \; ; \forall\, q \in P_{k-1}(\mathscr{e}) \, , \text{ for the six edges } \mathscr{e} \text{ of } \kappa\} \; ,$$

$$\{\textstyle\int_{\mathscr{f}} (\mathbf{u} \times \mathbf{n})\cdot \mathbf{q} \, ds \, ; \forall\, \mathbf{q} \in (P_{k-2})^2(\mathscr{f}) \, , \text{ for the four faces } \mathscr{f} \text{ of } \kappa\} \; ,$$

$$\{\textstyle\int_{\kappa} \mathbf{u}\cdot \mathbf{q} \, dx \, ; \forall\, \mathbf{q} \in (P_{k-3})^3(\kappa)\} \; . \; ■$$

This set of moments is *unisolvent and* ***curl****-conforming* on $\mathbb{R}_k$. Hence it determines the following interpolation operator :

(3.2) $r_\kappa(\mathbf{u})$ is the unique polynomial of $\mathbb{R}_k$ that has the same moments on κ as $\mathbf{u}$.

As an example, the polynomials of $\mathbb{R}_1$ are of the form :

$$\mathbf{c} + \mathbf{a} \times \mathbf{x} \quad \forall\, \mathbf{a}, \mathbf{c} \in \Re$$

and their degrees of freedom reduce to the average flux along the edges :

$$\textstyle\int_{\mathscr{e}} (\mathbf{u}\cdot \mathbf{t})\, d\mathscr{e} \; , \text{ for the six edges } \mathscr{e} \text{ of } \kappa \, .$$

The above elements are related to the following div-conforming finite elements, introduced by Nedelec in [18], generalizing to $\Re^3$ the elements of Raviart & Thomas [22] (*cf.* also Brezzi, Douglas & Marini [8]). For the above integer k, we define the subspace of $P_k{}^3$:

(3.3) $$\mathbb{D}_k = (P_{k-1})^3 \oplus \{p(x)\mathbf{x} \, ; p \in \tilde{P}_{k-1}\}$$

and we associate with $\mathbb{D}_k$ the following degrees of freedom :

Definition 3.2.

Let κ be a tetrahedron with faces denoted by $\mathscr{f}$ and let $\mathbf{u}$ belong to $H^1(\kappa)^3$. We define the two sets of moments of $\mathbf{u}$ on κ :

$$\{\textstyle\int_{\mathscr{f}} (\mathbf{u}\cdot \mathbf{n})\, q \, ds \, ; \forall\, q \in P_{k-1}(\mathscr{f}) \, , \text{ for the four faces } \mathscr{f} \text{ of } \kappa\} \; ,$$

$$\{\textstyle\int_{\kappa} \mathbf{u}\cdot \mathbf{q} \, dx \, ; \forall\, \mathbf{q} \in (P_{k-2}(\kappa))^3\} \; . \; ■$$

This set of moments is *unisolvent and div-conforming on* $\mathbb{D}_k$; the corresponding interpolation operator is

(3.4) $\omega_\kappa(\mathbf{u})$ is the unique polynomial of $\mathbb{D}_k$ that has the same moments on κ as $\mathbf{u}$.

As far as the gradient is concerned, we shall take the space of polynomials P_k with the following degrees of freedom.

Definition 3.3.

Let κ be like above and let p belong to $H^2(\kappa)$. We define the four sets of moments of p on κ:

$$p(a_i) \text{ for the four vertices } a_i \text{ of } \kappa\ ,$$

$$\{\textstyle\int_e p\, q\, de\ ;\ \forall\, q \in P_{k-2}(e),\ \text{for the six edges } e \text{ of } \kappa\}\ ,$$

$$\{\textstyle\int_f p\, q\, ds\ ;\ \forall\, q \in P_{k-3}(f),\ \text{for the four faces } f \text{ of } \kappa\}\ ,$$

$$\{\textstyle\int_\kappa p\, q\, dx\ ;\ \forall\, q \in P_{k-4}(\kappa)\}\ .$$

This set of moments is *unisolvent and gradient-conforming on* P_k and its corresponding interpolation operator is:

(3.5) $I_\kappa(p)$ is the unique polynomial of P_k that has the same moments on κ as p.

The above choice is suggested by the following essential relations between the two pairs $(\omega_\kappa, r_\kappa)$ and (I_κ, r_κ):

$$\omega_\kappa(\mathbf{curl\ u}) = \mathbf{curl}\ r_\kappa(\mathbf{u}) \quad \forall\ \mathbf{u} \in H^2(\kappa)^3\ , \tag{3.6}$$

$$r_\kappa(\nabla p) = \nabla I_\kappa(p) \quad \forall\ p \in W^{2,t}(\kappa)^3 \text{ for some } t > 2\ . \tag{3.7}$$

Remark 3.1.

Definition 3.1 requires a $W^{1,t}$-regularity because the edge moments are not defined if $\mathbf{u}$ has no more than an H^1-regularity. Of course, there are other interpolation operators (like those defined by Clément [10]) that require less regularity than r_κ, but r_κ seems to be the only one that preserves the vanishing of curls.

The finite element spaces themselves are constructed by triangulating (entirely) $\bar{\Omega}$ with tetrahedra. Let $\mathfrak{I}_h$ be a triangulation of $\bar{\Omega}$ made of tetrahedra κ with diameters bounded by h. For each integer $k \geq 1$, we define the following finite element spaces:

$$\mathbb{M}_h = \{\mathbf{u}_h \in H(\mathbf{curl};\Omega)\ ;\ \mathbf{u}_{h\,|\kappa} \in \mathbb{R}_k \quad \forall\, \kappa \in \mathfrak{I}_h\}\ , \tag{3.8a}$$

$$\mathbb{M}_{0h} = \mathbb{M}_h \cap H_0(\mathbf{curl};\Omega)\ , \tag{3.8b}$$

$$\mathbb{X}_h = \mathbb{M}_h \cap \mathbb{X} = \{\mathbf{u}_h \in \mathbb{M}_h\ ;\ \mathbf{curl\ u}_h \cdot \mathbf{n}_{|\Gamma} = 0\}\ , \tag{3.8c}$$

$$\mathbb{D}_h = \{\mathbf{v}_h \in H(\mathrm{div};\Omega)\ ;\ \mathbf{v}_{h\,|\kappa} \in \mathbb{D}_k \quad \forall\, \kappa \in \mathfrak{I}_h\}\ , \tag{3.9a}$$

$$\mathbb{D}_{0h} = \mathbb{D}_h \cap H_0(\mathrm{div};\Omega)\ , \tag{3.9b}$$

$$\mathbb{Q}_h = \{q_h \in C^0(\bar{\Omega})\ ;\ q_{h\,|\kappa} \in P_k \quad \forall\, \kappa \in \mathfrak{I}_h\}\ , \tag{3.10a}$$

$$\mathbb{Q}_{0h} = \mathbb{Q}_h \cap H_0^1(\Omega)\ . \tag{3.10b}$$

They are related together by:

$$\{\mathbf{u}_h \in \mathbb{M}_h\ ;\ \mathbf{curl\ u}_h = \mathbf{0}\} = \{\mathbf{u}_h \in \mathbb{X}_h\ ;\ \mathbf{curl\ u}_h = \mathbf{0}\} = \{\nabla q_h\ ;\ q_h \in \mathbb{Q}_h\}\ ,$$

$$\{\mathbf{u}_h \in \mathbb{M}_{0h} \, ; \, \mathbf{curl}\, \mathbf{u}_h = \mathbf{0}\} = \{\nabla q_h \, ; \, q_h \in \mathbb{Q}_{0h}\} \, ,$$

$$\{\mathbf{f}_h \in \mathbb{D}_h \text{ (resp. } \mathbb{D}_{0h}) \, ; \, \mathrm{div}\, \mathbf{f}_h = 0\} = \{\mathbf{curl}\, \mathbf{u}_h \, ; \, \mathbf{u}_h \in \mathbb{M}_h \text{ (resp. } \mathbb{M}_{0h})\} \, .$$

Remark 3.2.

This last characterization implies that all functions $\mathbf{v}_h$ of $\mathbb{X}_h$ are of the form :

$$\mathbf{v}_h = \mathbf{w}_h + \nabla q_h \text{ with } \mathbf{w}_h \in \mathbb{M}_{0h} \text{ and } q_h \in \mathbb{Q}_h \, .$$

Thus it is easy to construct a basis for the space $\mathbb{X}_h$: it suffices to take all the basis functions of $\mathbb{M}_{0h}$ and retain only the basis functions of $\mathbb{Q}_h$ that do not vanish on Γ . ♦

With these spaces, the finite dimensional analogues of $\mathbb{V}_0$, $\mathbb{U}_0$ and $\mathbb{U}$ are :

$$\mathbb{V}_{0h} = \{\mathbf{v}_h \in \mathbb{M}_{0h} \, ; \, (\nabla q_h, \mathbf{v}_h) = 0 \quad \forall\, q_h \in \mathbb{Q}_{0h}\} \, ,$$

$$\mathbb{U}_{0h} \text{ (resp. } \mathbb{U}_h) = \{\mathbf{v}_h \in \mathbb{X}_h \text{ (resp. } \mathbb{M}_h) \, ; \, (\nabla q_h, \mathbf{v}_h) = 0 \quad \forall\, q_h \in \mathbb{Q}_h\} \, .$$

Next we define the interpolation operators : r_h from $W^{1,t}(\Omega)^3$ for some $t > 2$ onto $\mathbb{M}_h$, ω_h from $H^1(\Omega)^3$ onto $\mathbb{D}_h$ and I_h from $H^2(\Omega)$ onto $\mathbb{Q}_h$ by :

(3.11) $$r_h\mathbf{u} = r_\kappa(\mathbf{u}) \text{ on } \kappa \quad \forall\, \kappa \in \mathfrak{T}_h \text{ (similarly for } \omega_h \text{ and } I_h) \, .$$

It follows from the above considerations that they have the following crucial properties :

$$\mathbf{u} \times \mathbf{n} = \mathbf{0} \Rightarrow r_h\mathbf{u} \times \mathbf{n} = \mathbf{0} \, , \quad \mathbf{curl}\, \mathbf{u} = \mathbf{0} \Rightarrow \mathbf{curl}\, r_h\mathbf{u} = \mathbf{0} \, ,$$

$$\mathbf{u} = \nabla p \Rightarrow r_h\mathbf{u} = \nabla(I_h p) \, ;$$

$$\mathbf{u}\cdot\mathbf{n} = 0 \Rightarrow \omega_h\mathbf{u}\cdot\mathbf{n} = 0 \, , \quad \mathrm{div}\, \mathbf{u} = 0 \Rightarrow \mathrm{div}\, \omega_h\mathbf{u} = 0 \, .$$

The major approximation properties of ω_h and r_h are listed in the following theorem. (The operator I_h plays only an auxiliary part here and is a straightforward generalization of the plane interpolation operator introduced by Thomas [24]). Previously, let us recall the notion of a *regular* (resp. *uniformly regular*) *triangulation* (*cf* Ciarlet [9]) :

there exists a constant $\sigma > 0$ *(resp. and a constant* $\tau > 0$ *) independent of* h *and* κ *such that*

$$h_\kappa / \rho_\kappa \le \sigma \quad (\textit{resp. } \tau h \le h_\kappa \le \sigma \rho_\kappa) \quad \forall\, \kappa \in \mathfrak{T}_h \, ,$$

where h_κ denotes the diameter of the tetrahedron κ and ρ_κ the maximum diameter of the balls inscribed in κ.

Theorem 3.1.

Assume that the triangulation $\mathfrak{T}_h$ *is regular and let* k *be the integer of (3.1). Then the interpolation operators* r_h *and* ω_h *satisfy the following stability estimates for all* $k \ge 1$:

(3.12) $$\| \mathbf{u} - r_h\mathbf{u} \|_{0,\Omega} + h \| \mathbf{curl}\, (\mathbf{u} - r_h\mathbf{u}) \|_{0,\Omega} \le C(t)\, h\, | \mathbf{u} |_{1,t,\Omega} \quad \forall\, \mathbf{u} \in W^{1,t}(\Omega)^3 \text{ for some } t > 2 \, ,$$

(3.13) $$\| \mathbf{u} - \omega_h\mathbf{u} \|_{0,\Omega} + h \| \mathrm{div}\, (\mathbf{u} - \omega_h\mathbf{u}) \|_{0,\Omega} \le C\, h\, | \mathbf{u} |_{1,\Omega} \quad \forall\, \mathbf{u} \in H^1(\Omega)^3 \, .$$

Moreover, when $\mathbf{u}$ *belongs to* $H^k(\Omega)^3$ *with* $k \ge 2$ *then*

(3.14) $$\| \mathbf{u} - \omega_h\mathbf{u} \|_{0,\Omega} + \| \mathbf{u} - r_h\mathbf{u} \|_{0,\Omega} \le C\, h^k\, | \mathbf{u} |_{k,\Omega} \, ,$$

and when $\mathbf{u}$ *belongs to* $H^{k+1}(\Omega)^3$ *with* $k \geq 1$, *we have*

(3.15) $$\| \operatorname{div}(\mathbf{u} - \omega_h \mathbf{u}) \|_{0,\Omega} + \| \mathbf{curl}\,(\mathbf{u} - r_h \mathbf{u}) \|_{0,\Omega} \leq C\, h^k |\mathbf{u}|_{k+1,\Omega} .$$

All the above constants are independent of h .

Remark 3.3.

The approximation of the space $\mathbb{X}$ by $\mathbb{X}_h$ follows readily from Remarks 2.1 and 3.2. As each $\mathbf{v}$ in $\mathbb{X}$ is of the form $\mathbf{w} + \nabla q$, we can define $s_h\mathbf{v}$ in $\mathbb{X}_h$ by :

$$s_h \mathbf{v} = r_h \mathbf{w} + \nabla I_h q$$

when $\mathbf{w}$ and q are sufficiently smooth; otherwise, we replace r_h and I_h by local regularization operators of the type defined by Clément [10] or Bernardi [5] that preserve (in the case of r_h) the boundary constraint $\mathbf{w} \times \mathbf{n} = \mathbf{0}$. Then, clearly, $s_h\mathbf{v}$ belongs to $\mathbb{X}_h$ and the approximation properties of s_h stem from those of r_h and I_h . ◆

With these spaces, we discretize the Navier-Stokes system (1.1) , (1.2) by :

Find $\mathbf{u}_h$ *in* $\mathbb{V}_{0h}$ *and* $\tilde{p}_h$ *in* $\mathbb{Q}_{0h}$ *such that :*

(3.16) $$\nu\,(\mathbf{curl}\ \mathbf{u}_h , \mathbf{curl}\ \mathbf{v}_h) + (\mathbf{curl}\ \mathbf{u}_h \times \mathbf{u}_h , \mathbf{v}_h) + (\nabla \tilde{p}_h , \mathbf{v}_h) = (\mathbf{f} , \mathbf{v}_h) \quad \forall\ \mathbf{v}_h \in \mathbb{M}_{0h} .$$

Similarly, the Navier-Stokes problem (1.1) , (1.3) (resp. (1.4)) is discretized by :

Find $\mathbf{u}_h$ *in* $\mathbb{U}_{0h}$ *(resp.* $\mathbb{U}_h$ *) and* $\tilde{p}_h$ *in* $\mathbb{Q}_h/\mathfrak{R}$ *such that :*

(3.17) $$\nu\,(\mathbf{curl}\ \mathbf{u}_h , \mathbf{curl}\ \mathbf{v}_h) + (\mathbf{curl}\ \mathbf{u}_h \times \mathbf{u}_h , \mathbf{v}_h) + (\nabla \tilde{p}_h , \mathbf{v}_h) = (\mathbf{f} , \mathbf{v}_h) \quad \forall\ \mathbf{v}_h \in \mathbb{X}_h \ \textit{(resp. } \mathbb{M}_h \textit{)} .$$

Obviously each of these problems can be split into a discrete equation for the velocity and a discrete Poisson equation for the pressure like in (2.13) - (2.16). But this is useful only if a convenient basis of "divergence free" velocities (i.e. a basis for $\mathbb{V}_{0h}$, $\mathbb{U}_{0h}$ or $\mathbb{U}_h$) is available. Otherwise, the "divergence free" constraint can be approximated by a penalty or Uzawa algorithm in the usual way.

Remark 3.4.

The Stokes problems (2.19) , (2.20) and (2.24) , (2.25) can be discretized like above by deleting the non-linear convection term from (3.16) and (3.17). In the resulting linear schemes, the pressure is entirely dissociated from the velocity, i.e. it can be computed without knowing the velocity. ◆

4. Nonlinear analysis.

The first difficulty in the numerical analysis of the above schemes is to establish an analogue of Sobolev's inequality and compact imbeddings between the spaces $\mathbb{V}_{0h}$, $\mathbb{U}_{0h}$ or $\mathbb{U}_h$ and $L^4(\Omega)^3$. The proofs we give here generalize (and are inspired by) a Poincaré inequality established by Nedelec in [19] for the space $\mathbb{V}_{0h}$. They are straightforward but not elementary because they rely heavily on the regularity theorems stated at the beginning of Section 2.

Theorem 4.1.

Let Ω *be a convex polyhedron and* $\mathfrak{I}_h$ *a uniformly regular triangulation of* Ω. *For each space* $\mathbb{V}_{0h}$, $\mathbb{U}_{0h}$ *or* $\mathbb{U}_h$ *there exists a constant* C , *independent of* h , *such that :*

(4.1) $$\| \mathbf{u}_h \|_{0,4,\Omega} \leq C\, \| \mathbf{curl}\ \mathbf{u}_h \|_{0,\Omega} \quad \forall\ \mathbf{u}_h \in \mathbb{V}_{0h}, \mathbb{U}_{0h} \ \textit{or}\ \mathbb{U}_h .$$

Proof

Let us write the proof for the space $\mathbb{V}_{0h}$. The inequality (4.1) is valid in $\mathbb{V}_0$ but this does not carry over directly to $\mathbb{V}_{0h}$ because its functions $\mathbf{u}_h$ are not divergence free. However, $\mathbf{u}_h$ can be related to a divergence free function by the solution z in $H_0^1(\Omega)$ of the Dirichlet problem :

$$(\nabla z , \nabla \mu) = (\mathbf{u}_h , \nabla \mu) \quad \forall \mu \in H_0^1(\Omega) .$$

The difference $\mathbf{w} = \mathbf{u}_h - \nabla z$ belongs to $\mathbb{V}_0$ and $\mathbf{curl\ w} = \mathbf{curl\ u}_h$. As $\mathbf{curl\ u}_h$ is in $L^\infty(\Omega)^3$, it follows from Theorem 2.2 that there exists a real $s > 2$ such that $\mathbf{w}$ belongs to $W^{1,s}(\Omega)^3$ and

$$\| \mathbf{w} \|_{1,s,\Omega} \leq C_1(s) \| \mathbf{curl\ u}_h \|_{0,s,\Omega} .$$

Therefore, we can apply the interpolation operator r_h to $\mathbf{w}$, and $\mathbf{u}_h$ splits into :

$$\mathbf{u}_h = r_h\mathbf{w} + \nabla z_h \quad \text{with } z_h \in \mathbb{Q}_{0h} .$$

Hence $$\| \mathbf{u}_h \|_{0,4,\Omega} \leq \| r_h\mathbf{w} - \mathbf{w} \|_{0,4,\Omega} + \| \mathbf{w} \|_{0,4,\Omega} + \| \nabla z_h \|_{0,4,\Omega} .$$

Since on the one hand

(4.2) $$\| \mathbf{w} - r_h\mathbf{w} \|_{0,4,\Omega} \leq C_2(s)\, h^{1/4} \| \mathbf{curl\ u}_h \|_{0,\Omega} ,$$

and on the other hand

$$\| \mathbf{w} \|_{0,4,\Omega} \leq C_3 \| \mathbf{w} \|_{1,\Omega} ,$$

we see that it suffices to estimate $\| \nabla z_h \|_{0,4,\Omega}$.

We propose to evaluate first $\| \nabla z_h \|_{0,\Omega}$ and derive a bound for $\| \nabla z_h \|_{0,4,\Omega}$ by an inverse inequality. As $\mathbf{u}_h$ belongs to $\mathbb{V}_{0h}$ and $\mathbf{w}$ to $\mathbb{V}_0$, we have :

$$(\nabla z_h , \nabla \mu_h) = (\mathbf{w} - r_h\mathbf{w} , \nabla \mu_h) \quad \forall \mu_h \in \mathbb{Q}_{0h} .$$

Therefore,

(4.3) $$\| \nabla z_h \|_{0,\Omega} \leq \| \mathbf{w} - r_h\mathbf{w} \|_{0,\Omega} \leq C_4(s)\, h^{1+3/s-3/2} \| \mathbf{curl\ u}_h \|_{0,\Omega} .$$

Hence, the inverse inequality :

(4.4) $$\| \nabla z_h \|_{0,4,\Omega} \leq C_5\, h^{-3/4} \| \nabla z_h \|_{0,\Omega}$$

implies

4.5) $$\| \nabla z_h \|_{0,4,\Omega} \leq C_6(s)\, h^{1/4+3/s-3/2} \| \mathbf{curl\ u}_h \|_{0,\Omega} .$$

Collecting the above bounds, we obtain :

$$\| \mathbf{u}_h \|_{0,4,\Omega} \leq h^{1/4} [C_2(s) + C_6(s)\, h^{3/s-3/2}] \| \mathbf{curl\ u}_h \|_{0,\Omega} + C_3C_7 \| \mathbf{curl\ u}_h \|_{0,\Omega} .$$

This implies (4.1) if the overall power of h is non negative, i.e. if s lies in the range :

$$2 < s \leq 12/5 .$$

The proof for the space $\mathbb{U}_h$ is exactly alike, except that we switch to the space $\mathbb{U}$ by solving a Neumann's problem. The space $\mathbb{U}_{0h}$ is of course a subspace of $\mathbb{U}_h$ but it can also be dealt with like $\mathbb{V}_{0h}$. ∎

Remark 4.1.

With easy modifications, the above proof can be extended to $L^p(\Omega)^3$ for all real p in the range : $2 \leq p \leq 6$. The proof for the case $p = 2$ is due to Nedelec [19] . ∎

Theorem 4.2.

We retain the assumptions of Theorem 4.1. Let $(\mathbf{u}_h)$ *be a family of functions of* $\mathbb{V}_{0h}$, $\mathbb{U}_{0h}$ *or* $\mathbb{U}_h$ *respectively that satisfies :*

$$\text{weak-lim}_{h\to 0}\, \mathbf{curl}\ \mathbf{u}_h = \mathbf{w} \quad in\ L^2(\Omega)^3 .$$

Then $\mathbf{w} = \mathbf{curl}\ \mathbf{u}$ *where* $\mathbf{u}$ *belongs to* $\mathbb{V}_0$, $\mathbb{U}_0$ *or* $\mathbb{U}$ *respectively*

and

$$\lim_{h\to 0} \mathbf{u}_h = \mathbf{u} \quad in\ L^4(\Omega)^3 .$$

Proof

Let us do the proof for the space $\mathbb{V}_{0h}$, the proof for the other spaces being very similar. Since **curl** $\mathbf{u}_h$ is bounded in $L^2(\Omega)^3$, Remark 4.1 implies (in particular) that $\mathbf{u}_h$ is bounded in $H_0(\mathbf{curl};\Omega)$. Therefore there exists $\mathbf{u}$ in $H_0(\mathbf{curl};\Omega)$ such that

$$\text{weak-lim}_{h\to 0}\, \mathbf{u}_h = \mathbf{u} \quad \text{in } H(\mathbf{curl};\Omega) .$$

Furthermore, it is easy to see that $\operatorname{div} \mathbf{u} = 0$, i.e. $\mathbf{u}$ belongs to $\mathbb{V}_0$; hence $\mathbf{u}$ belongs to $H^1(\Omega)^3$.

Next, like in the proof of Theorem 4.1, we associate a divergence free function with $\mathbf{u}_h$ by introducing the solution $p(h)$ in $H_0^1(\Omega)$ of the problem :

$$(\nabla p(h), \nabla \mu) = (\mathbf{u}_h, \nabla \mu) \quad \forall \mu \in H_0^1(\Omega)$$

and the difference

$$\mathbf{u}(h) = \mathbf{u}_h - \nabla p(h) .$$

On the one hand, it stems from the weak convergence of $\mathbf{u}_h$ that

$$\text{weak-lim}_{h\to 0}\, p(h) = 0 \quad \text{in } H^1(\Omega) .$$

On the other hand, $\mathbf{curl}\ \mathbf{u}(h) = \mathbf{curl}\ \mathbf{u}_h$ and $\mathbf{u}(h)$ belongs to $\mathbb{V}_0$ (in fact, $\mathbf{u}(h)$ belongs to $W^{1,s}(\Omega)^3$ with $s > 2$). This implies that $\mathbf{u}(h)$ is bounded in $H^1(\Omega)^3$; consequently

$$\text{weak-lim}_{h\to 0}\, \mathbf{u}(h) = \mathbf{u} \quad \text{in } H^1(\Omega)^3$$

and therefore,

$$\lim_{h\to 0} \mathbf{u}(h) = \mathbf{u} \quad \text{in } L^4(\Omega)^3 .$$

Finally, let us write :

$$\| \mathbf{u} - \mathbf{u}_h \|_{0,4,\Omega} \le \| \mathbf{u} - \mathbf{u}(h) \|_{0,4,\Omega} + \| \mathbf{u}(h) - r_h\mathbf{u}(h) \|_{0,4,\Omega} + \| r_h\mathbf{u}(h) - \mathbf{u}_h \|_{0,4,\Omega} .$$

The bounds (4.2) and (4.5) give respectively :

$$\| \mathbf{u}(h) - r_h\mathbf{u}(h) \|_{0,4,\Omega} \le C_2(s)\, h^{1/4} \| \mathbf{curl}\ \mathbf{u}_h \|_{0,\Omega} ,$$

$$\| r_h\mathbf{u}(h) - \mathbf{u}_h \|_{0,4,\Omega} \le C_6(s)\, h^{1/4+3/s-3/2} \| \mathbf{curl}\ \mathbf{u}_h \|_{0,\Omega} .$$

The strong convergence in $L^4(\Omega)^3$ follows from these two inequalities provided s lies in the range $2 < s < 12/5$.

Again, this proof extends easily to $L^p(\Omega)^3$ for $2 \le p < 6$; note that the range of the real s decreases as p increases. ♦

5. Error analysis.

Let us do the analysis for the discretization (3.16) of the variational problem (2.5), (2.6), (1.2), assuming it has a unique solution. That is, we suppose that the right-hand side **f** belongs to $L^{4/3}(\Omega)^3$ and we introduce the two quantities :

(5.1)
$$N = \sup \frac{(\mathbf{curl}\ \mathbf{u} \times \mathbf{v}\ ,\ \mathbf{w})}{\|\mathbf{curl}\ \mathbf{u}\|_{0,\Omega}\ \|\mathbf{curl}\ \mathbf{v}\|_{0,\Omega}\ \|\mathbf{curl}\ \mathbf{w}\|_{0,\Omega}}$$

(5.2)
$$B = \sup \frac{\|\mathbf{u}\|_{0,4,\Omega}}{\|\mathbf{curl}\ \mathbf{u}\|_{0,\Omega}}$$

where the sup is taken over all **u** , **v** , **w** in $\mathbb{V}_0$. With a standard argument, we can show that problem (2.5) , (2.6) , (1.2) has a unique solution, provided

(5.3
$$[\,N\,B\,\|\mathbf{f}\|_{0,4/3,\Omega}\,]/\nu^2 < 1\ .$$

Likewise, we define analogous quantities for the space $\mathbb{V}_{0h}$:

(5.4)
$$N_h = \sup \frac{(\mathbf{curl}\ \mathbf{u}_h \times \mathbf{v}_h\ ,\ \mathbf{w}_h)}{\|\mathbf{curl}\ \mathbf{u}_h\|_{0,\Omega}\ \|\mathbf{curl}\ \mathbf{v}_h\|_{0,\Omega}\ \|\mathbf{curl}\ \mathbf{w}_h\|_{0,\Omega}}$$

(5.5)
$$B_h = \sup \frac{\|\mathbf{u}_h\|_{0,4,\Omega}}{\|\mathbf{curl}\ \mathbf{u}_h\|_{0,\Omega}}$$

where the sup is taken over all $\mathbf{u}_h$, $\mathbf{v}_h$, $\mathbf{w}_h$ in $\mathbb{V}_{0h}$. Theorem 4.1 implies that N_h and B_h are bounded with respect to h and a familiar argument derives from Theorem 4.2 that

$$\limsup_{h\to 0} N_h \le N \quad \text{and} \quad \limsup_{h\to 0} B_h \le B\ .$$

Therefore, if the condition (5.3) holds, say

$$[\,N\,B\,\|\mathbf{f}\|_{0,4/3,\Omega}\,]/\nu^2 \le 1 - \delta \quad \text{for some } \delta > 0\ ,$$

then for all sufficiently small h, say $h \le h_0$, we shall have :

(5.6)
$$[\,N_h\,B_h\,\|\mathbf{f}\|_{0,4/3,\Omega}\,]/\nu^2 \le 1 - \delta/2\ .$$

Now, a finite dimensional version of Brouwer's fixed point theorem establishes that the discrete problem (3.16) always has a solution. The uniqueness follows easily from condition

$$[\,N_h\,B_h\,\|\mathbf{f}\|_{0,4/3,\Omega}\,]/\nu^2 < 1\ .$$

Theorem 5.1.

Let Ω be a convex polyhedron. If the right-hand side **f** *is in* $L^{4/3}(\Omega)^3$ *and satisfies the condition:*

(5.7)
$$[\,N\,B\,\|\mathbf{f}\|_{0,4/3,\Omega}\,]/\nu^2 \le 1 - \delta \quad \textit{for some } \delta > 0\ ,$$

then the variational problem (2.5) , (2.6) , (1.2) has a unique solution :

$$\mathbf{u} \in H^1(\Omega)^3 \ \textit{with}\ \mathbf{curl}\ \mathbf{curl}\ \mathbf{u} \in L^{4/3}(\Omega)^3\ ,\quad \tilde{p} \in W^{1,4/3}(\Omega)\ .$$

If the triangulation $\mathfrak{T}_h$ *is uniformly regular and* h *sufficiently small, then the discrete problem (3.16) has a unique solution* $(\mathbf{u}_h\ ,\ \tilde{p}_h)$. *In addition, if* **curl u** *belongs to* $L^4(\Omega)^3$ *and* $\tilde{p}$ *to* $H^1(\Omega)$, *the following error estimates hold:*

$$(5.8)\qquad \| \mathbf{curl}\,(\mathbf{u}-\mathbf{u}_h)\|_{0,\Omega} \le K_1(\nu,\mathbf{u},\mathbf{f})\Big[\inf_{\mathbf{v}_h\in\mathbb{M}_{0h}} \|(\mathbf{u}-\mathbf{v}_h)\|_{H(\mathbf{curl};\Omega)} + C(s)\,h^{3/s-1/2}\inf_{q_h\in\mathbb{Q}_{0h}} |\tilde{p}-q_h|_{1,\Omega}\Big],$$

$$(5.9)\qquad \|\tilde{p}-\tilde{p}_h\|_{0,\Omega} \le K_2(\nu,\mathbf{u},\mathbf{f})\Big[\|(\mathbf{u}-\mathbf{u}_h)\|_{H(\mathbf{curl};\Omega)} + C\,h\inf_{q_h\in\mathbb{Q}_{0h}} |\tilde{p}-q_h|_{1,\Omega}\Big].$$

All constants involved are independent of h.

Proof

We fiirst establish the estimate (5.8).From the continuous and discrete formulations, we derive :

$$\nu\,(\mathbf{curl}\,(\mathbf{u}_h-\mathbf{w}_h),\mathbf{curl}\,\mathbf{v}_h) + (\mathbf{curl}\,(\mathbf{u}_h-\mathbf{w}_h)\times\mathbf{u}_h,\mathbf{v}_h) + (\mathbf{curl}\,\mathbf{u}\times(\mathbf{u}_h-\mathbf{w}_h),\mathbf{v}_h) + (\nabla(\tilde{p}_h-\tilde{p}),\mathbf{v}_h)$$

$$= \nu\,(\mathbf{curl}\,(\mathbf{u}-\mathbf{w}_h),\mathbf{curl}\,\mathbf{v}_h) + (\mathbf{curl}\,(\mathbf{u}-\mathbf{w}_h)\times\mathbf{u}_h,\mathbf{v}_h) + (\mathbf{curl}\,\mathbf{u}\times(\mathbf{u}-\mathbf{w}_h),\mathbf{v}_h)\quad \forall\,\mathbf{v}_h,\mathbf{w}_h\in\mathbb{M}_{0h}.$$

Let us restrict $\mathbf{w}_h$ to $\mathbb{V}_{0h}$ and take $\mathbf{v}_h=\mathbf{u}_h-\mathbf{w}_h$ in $\mathbb{V}_{0h}$. Then we are left with :

$$\nu\|\mathbf{curl}\,(\mathbf{u}_h-\mathbf{w}_h)\|^2_{0,\Omega} + (\mathbf{curl}\,(\mathbf{u}_h-\mathbf{w}_h)\times\mathbf{u}_h,\mathbf{u}_h-\mathbf{w}_h)$$
$$= \nu\,(\mathbf{curl}\,(\mathbf{u}-\mathbf{w}_h),\mathbf{curl}\,(\mathbf{u}_h-\mathbf{w}_h)) + (\mathbf{curl}\,(\mathbf{u}-\mathbf{w}_h)\times\mathbf{u}_h,\mathbf{u}_h-\mathbf{w}_h)$$
$$+ (\mathbf{curl}\,\mathbf{u}\times(\mathbf{u}-\mathbf{w}_h),\mathbf{u}_h-\mathbf{w}_h) + (\nabla(\tilde{p}-q_h),\mathbf{u}_h-\mathbf{w}_h)\quad \forall\,\mathbf{w}_h\in\mathbb{V}_{0h},\ \forall\,q_h\in\mathbb{Q}_{0h}.$$

Hence, repeated applications of (5.4) and (5.5) together with (5.6) and the bound :

$$\|\mathbf{curl}\,\mathbf{u}_h\|_{0,\Omega} \le (1/\nu)\,B_h\,\|\mathbf{f}\|_{0,4/3,\Omega}$$

yield :

$$(5.10)\qquad \begin{aligned}(\nu\,\delta/2)\,\|\mathbf{curl}\,(\mathbf{u}_h-\mathbf{w}_h)\|^2_{0,\Omega} &\le \nu\|\mathbf{curl}\,(\mathbf{u}-\mathbf{w}_h)\|_{0,\Omega}\|\mathbf{curl}\,(\mathbf{u}_h-\mathbf{w}_h)\|_{0,\Omega}\\ &+ (1/\nu)\,(B_h)^3\,\|\mathbf{f}\|_{0,4/3,\Omega}\|\mathbf{curl}\,(\mathbf{u}-\mathbf{w}_h)\|_{0,\Omega}\|\mathbf{curl}\,(\mathbf{u}_h-\mathbf{w}_h)\|_{0,\Omega}\\ &+ B_h\,\|\mathbf{curl}\,\mathbf{u}\|_{0,4,\Omega}\|\mathbf{u}-\mathbf{w}_h\|_{0,\Omega}\|\mathbf{curl}\,(\mathbf{u}_h-\mathbf{w}_h)\|_{0,\Omega} + |(\nabla(\tilde{p}-q_h),\mathbf{u}_h-\mathbf{w}_h)|.\end{aligned}$$

If the functions of $\mathbb{V}_{0h}$ were divergence free, the last term above would vanish. Let us show, as can reasonably be expected, that this term is "small". In the proof of Theorem 4.1, we have found that each function $\mathbf{v}_h$ of $\mathbb{M}_{0h}$ can be split into:

$$\mathbf{v}_h = r_h\mathbf{w} + \nabla z_h$$

where z_h belongs to $\mathbb{Q}_{0h}$ and $\mathbf{w}$ to $\mathbb{V}_0\cap W^{1,s}(\Omega)$ with

$$\|\mathbf{w}\|_{1,s,\Omega} \le C_1(s)\,\|\mathbf{curl}\,\mathbf{v}_h\|_{0,s,\Omega}\quad \text{for some real } s>2\,.$$

Thus, applying this result to $\mathbf{u}_h-\mathbf{w}_h$ and choosing $q_h=\mathring{P}_h\tilde{p}$, the $H_0^1(\Omega)$-projection of $\tilde{p}$ onto $\mathbb{Q}_{0h}$ we easily derive :

$$(\nabla(\tilde{p}-\mathring{P}_h\tilde{p}),\mathbf{u}_h-\mathbf{w}_h) = (\nabla(\tilde{p}-\mathring{P}_h\tilde{p}),r_h\mathbf{w}-\mathbf{w}).$$

Consequently, in view of (4.3), we get :

$$(5.11)\qquad |(\nabla(\tilde{p}-\mathring{P}_h\tilde{p}),\mathbf{u}_h-\mathbf{w}_h)| \le C_2(s)\,h^{3/s-1/2}\,\|\mathbf{curl}\,(\mathbf{u}_h-\mathbf{w}_h)\|_{0,\Omega}|\tilde{p}-\mathring{P}_h\tilde{p}|_{1,\Omega}\,.$$

Substituting into (5.10), this gives :

$$(\nu\,\delta/2)\,\|\,\mathbf{curl}\,(\mathbf{u}_h - \mathbf{w}_h)\,\|_{0,\Omega} \le \nu\,\|\,\mathbf{curl}\,(\mathbf{u} - \mathbf{w}_h)\,\|_{0,\Omega} + C_2(s)\,h^{3/s-1/2}\,|\,\tilde{p} - \mathring{P}_h\tilde{p}\,|_{1,\Omega}$$

(5.12) $$+ (1/\nu)\,(B_h)^3\,\|\,\mathbf{f}\,\|_{0,4/3,\Omega}\,\|\,\mathbf{curl}\,(\mathbf{u} - \mathbf{w}_h)\,\|_{0,\Omega} + B_h\,\|\,\mathbf{curl}\,\mathbf{u}\,\|_{0,4,\Omega}\,\|\,\mathbf{u} - \mathbf{w}_h\,\|_{0,\Omega}$$
$$\forall\ \mathbf{w}_h \in \mathbb{V}_{0h}\ .$$

Now, we extend this inequality to all functions $\mathbf{v}_h$ of $\mathbb{M}_{0h}$: define q_h in $\mathbb{Q}_{0h}$ by :

$$(\nabla q_h\,,\,\nabla \mu_h) = (\mathbf{v}_h\,,\,\nabla \mu_h) \quad \forall\ \mu_h \in \mathbb{Q}_{0h}$$

and set $\mathbf{w}_h = \mathbf{v}_h - \nabla q_h$. Then, $\mathbf{w}_h$ belongs to $\mathbb{V}_{0h}$ and $\mathbf{curl}\ \mathbf{w}_h = \mathbf{curl}\ \mathbf{v}_h$. Moreover,

$$\|\,\mathbf{u} - \mathbf{w}_h\,\|^2_{0,\Omega} = (\mathbf{u} - \mathbf{w}_h,\ \mathbf{u} - \mathbf{v}_h)$$

and therefore

(5.13) $$\|\,\mathbf{u} - \mathbf{w}_h\,\|_{0,\Omega} \le \|\,\mathbf{u} - \mathbf{v}_h\,\|_{0,\Omega}\ .$$

Hence, (5.12) holds for all $\mathbf{v}_h$ in $\mathbb{M}_{0h}$; knowing that the constant B_h is bounded with respect to h , this proves (5.8) .

To establish (5.9), we use a classical duality argument (*cf* Aubin [1] and Nitsche [20]); we write

$$\|\,\tilde{p} - \tilde{p}_h\,\|_{0,\Omega} = \sup_{g \in L2(\Omega)} \frac{(\tilde{p} - \tilde{p}_h\,,\,g)}{\|\,g\,\|_{0,\Omega}}$$

and we use the fact that : $g = -\Delta\mu\ ,\ \mu_{|\Gamma} = 0\ ,\ \mu \in H^2(\Omega)\ ,\ \|\,\mu\,\|_{2,\Omega} \le C_3\,\|\,g\,\|_{0,\Omega}$. Thus

$$(\tilde{p} - \tilde{p}_h\,,\,g) = (\nabla(\tilde{p} - \tilde{p}_h),\,\nabla\mu) = (\nabla(\tilde{p} - \tilde{p}_h),\,\nabla(\mu - \mu_h)) + (\nabla(\tilde{p} - \tilde{p}_h),\,\nabla\mu_h)$$
$$\forall\ \mu_h \in \mathbb{Q}_{0h}\ .$$

Hence, choosing $\mu_h = \mathring{P}_h\,\mu$, and taking the difference between the exact and discrete equations, we obtain :

$$(\tilde{p} - \tilde{p}_h\,,\,g) = (\nabla(\tilde{p} - q_h)\,,\,\nabla(\mu - \mathring{P}_h\mu)) + (\nabla(\tilde{p} - \tilde{p}_h)\,,\,\nabla\mathring{P}_h\mu)$$
$$= (\nabla(\tilde{p} - q_h)\,,\,\nabla(\mu - \mathring{P}\mu)) - (\mathbf{curl}\,(\mathbf{u} - \mathbf{u}_h)\times\mathbf{u}_h\,,\,\nabla\mathring{P}_h\mu) - (\mathbf{curl}\,\mathbf{u}\times(\mathbf{u} - \mathbf{u}_h)\,,\,\nabla\mathring{P}_h\mu)$$
$$\forall\ q_h \in \mathbb{Q}_{0h}\ .$$

Therefore in view of the fact that

$$|\mathring{P}_h\mu|_{1,4,\Omega} \le C_4\,\|\,\mu\,\|_{2,\Omega}\ ,\quad |\,\mu - \mathring{P}_h\mu\,|_{1,\Omega} \le C_5\,h\,\|\,\mu\,\|_{2,\Omega}\ ,$$

we derive :

(5.14) $$|\,(\tilde{p} - \tilde{p}_h\,,\,g)\,| \le C_3\,[\,C_5\,h\,|\tilde{p} - q_h|_{1,\Omega} + C_4\,B_h\,\|\,\mathbf{curl}\,\mathbf{u}_h\,\|_{0,\Omega}\,\|\,\mathbf{curl}\,(\mathbf{u} - \mathbf{u}_h)\,\|_{0,\Omega}$$
$$+ C_4\,\|\,\mathbf{curl}\,\mathbf{u}\,\|_{0,4,\Omega}\,\|\,\mathbf{u} - \mathbf{u}_h\,\|_{0,\Omega}\,]\,\|\,g\,\|_{0,\Omega} \quad \forall\ q_h \in \mathbb{Q}_{0h}\ .$$

This proves (5.9) . ■

Corollary 5.1.

Under the assumptions of Theorem 5.1, the pair $(\mathbf{u}_h\,,\,\tilde{p}_h)$ *tends to* $(\mathbf{u}\,,\,\tilde{p})$ *in* $H(\mathbf{curl};\Omega)\times L^2(\Omega)$. *Moreover, when the solution is sufficiently smooth we have the following orders of convergence:*

$$\|\,\mathbf{u} - \mathbf{u}_h\,\|_{H(\mathbf{curl};\Omega)} \le C_1\,h^k\,\{|\,\mathbf{u}\,|_{k+1,\Omega} + C_2(s)\,h^{3/s-3/2}\,|\,\tilde{p}\,|_{k,\Omega}\}\ ,$$

$$\|\,\tilde{p} - \tilde{p}_h\,\|_{0,\Omega} \le C_3\,h^k\,\{|\,\mathbf{u}\,|_{k+1,\Omega} + C_4(s)\,h^{3/s-3/2}\,|\,\tilde{p}\,|_{k,\Omega}\}\ .$$

If $\tilde{p}$ *belongs to* $H^{k+1}(\Omega)$, *the factor multiplying* $C_2(s)$ *and* $C_4(s)$ *becomes* $h^{1+3/s-3/2}$.

The error estimates of Corollary 5.1 are nearly optimal in the sense that polynomials of degree k yield an error of the order of $h^{k-\varepsilon}$ for $\mathbf{curl}\ \mathbf{u}$ and p in the L^2 norm. The next point we raise is : can we prove an h^{k+1} estimate for the error on $\mathbf{u}$ in the L^2 norm? The answer is *no*, at least not with the incomplete spaces of polynomials that we use. Indeed, a duality argument permits to establish the following bound (when the solution is sufficiently smooth) :

$$\|\mathbf{u} - \mathbf{u}_h\|_{0,\Omega} \leq C_1 h \{\|\mathbf{curl}\ (\mathbf{u} - \mathbf{u}_h)\|_{0,\Omega} + |\tilde{p} - \tilde{p}_h|_{1,\Omega}\}$$
$$+ \inf_{\mathbf{v}_h \in \mathbb{M}_{0h}} \{C_2 \|\mathbf{u} - \mathbf{v}_h\|_{0,\Omega} + C_3(s)\, h^{3/s-1/2} \|\mathbf{curl}\ (\mathbf{u} - \mathbf{v}_h)\|_{0,\Omega}\} .$$

Clearly, there is no extra accuracy to be expected from $\|\mathbf{u} - \mathbf{v}_h\|_{0,\Omega}$.

We can also ask : what about the error on $\mathbf{u}$ in the L^4 norm? At the present stage, we can only give a partial answer, namely :

$$\|\mathbf{u} - \mathbf{u}_h\|_{0,4,\Omega} \leq \|\mathbf{u} - r_h\mathbf{u}\|_{0,4,\Omega} + C_4 \|\mathbf{curl}\ (r_h\mathbf{u} - \mathbf{u}_h)\|_{0,\Omega} + |t_h|_{1,4,\Omega} ,$$

where $t_h \in \mathbb{Q}_{0h}$ is the solution of :

$$(\nabla t_h , \nabla \mu_h) = (r_h\mathbf{u} - \mathbf{u} , \nabla \mu_h) \quad \forall\, \mu_h \in \mathbb{Q}_{0h} .$$

It is most likely that t_h is stable in the L^4 norm :

(5.15) $$|t_h|_{1,4,\Omega} \leq C_4 \|r_h\mathbf{u} - \mathbf{u}\|_{0,4,\Omega} ,$$

and consequently, $\|\mathbf{u} - \mathbf{u}_h\|_{0,4,\Omega}$ is of the order of $h^{k-\varepsilon}$ when $\mathbf{u}$ is smooth enough, but although (5.15) is established in a plane, convex polygon (*cf* Rannacher & Scott [21] and Scott [23]) , it is not yet proved in a polyhedron. If this result were not true, then we should resort to the inverse inequality (4.4) and thereby lose the factor $h^{3/4}$.

The analysis for the Stokes problem (2.17) ,(2.18) is of course much simpler. The discrete problem splits into :

Find $\mathbf{u}_h$ *in* $\mathbb{V}_{0h}$ *such that*

5.16) $$\nu\, (\mathbf{curl}\ \mathbf{u}_h , \mathbf{curl}\ \mathbf{v}_h) = (\mathbf{f} , \mathbf{v}_h) \quad \forall\, \mathbf{v}_h \in \mathbb{V}_{0h} ;$$

Find p_h *in* $\mathbb{Q}_{0h}$ *such that*

(5.17) $$(\nabla p_h , \nabla q_h) = (\mathbf{f} , \nabla q_h) \quad \forall\, q_h \in \mathbb{Q}_{0h} .$$

Like above, we derive the following expressions for the error :

(5.18) $$\|\mathbf{curl}\ (\mathbf{u} - \mathbf{u}_h)\|_{0,\Omega} \leq \inf_{\mathbf{v}_h \in \mathbb{M}_{0h}} 2\|(\mathbf{u} - \mathbf{v}_h)\|_{0,\Omega} + C(s)\, h^{3/s-1/2} \inf_{q_h \in \mathbb{Q}_{0h}} |p - q_h|_{1,\Omega} ,$$

(5.19) $$|p - p_h|_{1,\Omega} = \inf_{q_h \in \mathbb{Q}_{0h}} |p - q_h|_{1,\Omega} .$$

Hence, when the solution is sufficiently smooth, this scheme has the following orders of convergence :

$$\|\mathbf{curl}\ (\mathbf{u} - \mathbf{u}_h)\|_{0,\Omega} = O(h^k) \ , \ |p - p_h|_{1,\Omega} = O(h^k) .$$

The analysis of the schemes (3.17) follows closely that of the scheme (3.16). The main difference is that the "divergence free" condition is achieved now by solving Neumann's problems instead of Dirichlet's problems for the Laplace operator. This technique appears to be necessary when we discretize the boundary condition (1.4), but not if we discretize (1.3). In this case, it can be bypassed (the proofs are a little longer) and treated as a variant of scheme (3.16). Whatever the technique used, the error estimates derived are exactly like above, with identical orders of convergence.

6. Acknowledgements.

The author is gratefully indebted to A. Bossavit, M. Dauge, P. Grisvard and O. Pironneau for many interesting and fruitful discussions.

7. References.

[1] • J. P. Aubin • Behavior of the error of the approximate solutions of boundary value problems for linear elliptic operators by Galerkin's and finite difference methods. Ann. Scuola Sup. Pisa **21**, pp. 599-637. (1967).

[2] • C. Begue, C. Conca, F. Murat, O. Pironneau • Les équations de Stokes et de Navier-Stokes avec des conditions aux limites sur la pression in Nonlinear Partial Differential Equations and their Applications. Collège de France Seminar **9**. (H. Brézis & J. L. Lions *ed*).Pitman Research Notes in Math. **181**. Longman, Harlow, pp. 179-264. (1988).

[3] • A. Bendali, J. M. Dominguez, S. Gallic • A variational approach for the vector potential formulation of the Stokes and Navier-Stokes problems in three dimensional domains . J. Math. Anal. Appl. **107**, pp. 537-560. (1985).

[4] • C. Bernardi • Méthode d'éléments finis mixtes pour les équations de Navier-Stokes. Thèse Univ. Paris VI. (1979).

[5] • C. Bernardi • Optimal finite element interpolation on curved domains. SIAM J. Num. Anal. **26**, pp. 1212-1240 (1989).

[6] • A. Bossavit • A rationale for "edge-elements" in 3-D fields computations. "COMPUMAG". IEE Trans. on Magnetism.

[7] • A. Bossavit • "Whitney Elements" : two dual mixed methods for the 3-D eddy currents problem.

[8] • F. Brezzi, J. Douglas, D. Marini • Two families of mixed finite element methods for second order elliptic problems. Numer. Math. **47**, pp. 217-235. (1985).

[9] • P. G. Ciarlet • The Finite Element Method for Elliptic Problems. North Holland. (1977).

[10] • P. Clément • Approximation by finite element functions using local regularization. R.A.I.R.O. Anal. Numer. **9**, pp. 77-84. (1975).

[11] • M. Dauge • Problèmes de Neumann et de Dirichlet sur un polyèdre dans $\Re^3$: régularité dans des espaces de Sobolev L^p . C.R.A.S. **307**, I, pp. 27-32. (1988).

[12] • F. Dubois • Discrete vector potential representation of a divergence-free vector field in three dimensional domains : numerical analysis of a model problem. Report n° 163. Centre Math. Appl. Ecole Polytechnique. (1987).(To appear in SIAM J. Num. Anal.)

[13] • G. Duvaut, J. L. Lions

- Les Inéquations en Mécanique et en Physique. Dunod. (1972).
- Inequalities in Mechanics and Physics. Series of Comp. Studies in Math. SpringerVerlag. (1976).

[14] • L. Franca, T. J. Hugues • A new finite element formulation for computational fluid dynamics: VII. The Stokes Problem with various well-posed boundary conditions: symmetric formulations that converge for all velocity/pressure spaces. Comp. Meth. in Applied Mech. and Eng. **65**, pp. 85-96. (1987).

[15] • V. Girault • Incompressible finite element methods for Navier-Stokes equations with nonstandard boundary conditions in $\Re^3$. Math. of Comp. **51**, n° 183, pp. 55-74. (1988).

[16] • V. Girault, P. A. Raviart • Finite Element Methods for Navier-Stokes Equations. Springer Series in Comp. Math. **5**. Springer-Verlag. (1986).

[17] • P. Grisvard • Behavior of the solutions of an elliptic boundary value problem in a polygonal or polyhedral domain. Numerical Solution of Partial Differential Equations. III. Synspade. (B. Hubbard *ed*). Academic Press. (1975).

[18] • J. C. Nedelec • Mixed finite elements in $\Re^3$. Numer. Math. **35**, pp. 315-341. (1980).

[19] • J. C. Nedelec • Eléments finis mixtes incompressibles pour l'équation de Stokes dans $\Re^3$. Numer. Math. **39**, pp. 97-112. (1982).

[20] • J. Nitsche • Ein Kriterium für die Quasi-Optimalität des Ritzschen Verfahrens. Numer. Math. **11**, pp. 346-348. (1968).

[21] • R. Rannacher, R. Scott • Some optimal error estimates for piecewise linear finite element approximation. Math. of Comp. **38** , pp. 437-445. (1982).

[22] • P. A. Raviart, J. M. Thomas • A mixed finite element method for 2nd order elliptic problems. Mathematical Aspects of Finite Element Methods. (A. Dold. & B. Eckmann *ed*) Lecture Notes **606**. Springer-Verlag. (1977).

[23] • R. Scott • Optimal L^∞ estimates for the finite element method on irregular meshes. Math. of Comp. **30**, pp. 681-697. (1976).

[24] • J. M. Thomas • Sur l'analyse numérique des méthodes d'éléments finis hybrides et mixtes. Thèse Univ. Paris VI. (1977).

[25] • R. Verfürth • Mixed finite element approximation of the vector potential. Numer. Math. **50**, pp. 685-695. (1987).

ON LAGRANGEAN METHODS AND VOLTERRA INTEGRAL EQUATIONS OF THE FIRST KIND FOR INCOMPRESSIBLE NAVIER STOKES PROBLEMS

F.K. Hebeker
Institute of Supercomputing and Applied Mathematics
IBM Scientific Center
Tiergartenstraße 15, D-6900 Heidelberg

INTRODUCTION

In this paper we present a mathematical analysis of an unsteady boundary element method (BEM) on Lagrangean coordinates to treat numerically the initial boundary value problem of the incompressible Navier Stokes equations governing viscous homogeneous fluid flows in a bounded cavity $\Omega \subset \mathbb{R}^3$ (with smooth boundary $\partial\Omega$):

$$v_t + v \bullet \nabla v - \mu \Delta v + \nabla p = 0 \text{ in } \Omega_T \tag{1}$$

$$\operatorname{div} v = 0 \text{ in } \Omega_T \tag{2}$$

$$v_{|\partial\Omega} = 0 \text{ for } t > 0 \tag{3}$$

$$v_{|t=0} = v_0 \text{ in } \Omega. \tag{4}$$

Here the vector field $v(t,x)$ or the scalar field $p(t,x)$ denotes the velocity (at time $t > 0$ and location $x \in \Omega$) or the pressure function of the flow, resp., and $\mu \sim \frac{1}{Re} > 0$ is the dynamic viscosity of the medium (Re the Reynolds number). For short, exterior forces have been neglected. Further, $\Omega_T = \Omega \times (0, T]$, where $(0, T]$ denotes a given finite time interval.

Our approximation procedure is as follows: on short time intervals we approximate the characteristics (of the acceleration terms of the differential equations). This "Lagrangean" approach [17] serves to handle the convective terms, so we are faced essentially to the linearized unsteady Stokes system to be solved successively on short time intervals. These Stokes systems will be reformulated in terms of an equivalent Volterra boundary integral equations system of the *first* kind. For this, we will provide for a mathematical framework that guarantees "quasioptimality" of appropriate Galerkin-type time-space boundary element schemes. Note that, in case of the unsteady Stokes system, the classical Volterra boundary integral equations system of the *second* kind contains a strongly singular integral operator [2] and so would raise some numerical problems.

The latter results have been obtained in parts jointly with G.C. Hsiao [9]. See also [8] for a model problem of viscous compressible motions. The present paper improves a previous one [6] by introducing *unsteady* boundary element methods for an *unsteady* problem: so one employs the correct singularity function!

THE LAGRANGEAN ALGORITHM

We do not dwell on the question of existence and uniqueness of problem (1)-(4) but rather assume that a unique solution of this problem exists on the time interval $[0, T]$ and is sufficiently smooth there. We generate a time grid $t_k = k \cdot \tau$ ($\tau = T/N$ time increment, $N \in \mathbb{N}, k = 0, \ldots, N$). Starting from the initial state (4) we assume that an approximate velocity field v^k has already been computed for v on the time level t_k. If v^k is "not too large" (cf. [7]), the approximate characteristic curves

$$X^k(t,x) = x + tv^k(x) \tag{5}$$

do not leave Ω for $|t| \leq \tau$ and $x \in \Omega$. Now we transform problem (1)-(4) on the time interval $[t_k, t_{k+1}]$ to Lagrangean coordinates, approximate the particles' paths appropriately by (5) and use the fact that the Lagrange transformed space derivation operators may be approximated of the second order by the original space derivation operators locally in time. This seems having rigorously been proved first by Rautmann [17], in case of vorticity transport and Fokker-Planck-Vlasov problems. For further investigations based on this matter of fact we refer to [16],[4],[21],[20],[6],[0],[18]. Consequently, we are led to this

Algorithm:

(i) Compute $\hat{v}$ on the time interval $(t_k, t_{k+1}]$ by solving

$$\begin{cases} \hat{v}_t - \mu\Delta\hat{v} + \nabla\hat{p} = 0 & \text{in } \Omega_k = \Omega \times (t_k, t_{k+1}] \\ \operatorname{div} \hat{v} = 0 & \text{in } \Omega_k \\ \hat{v}_{|\partial\Omega} = 0 & \text{for } t \in (t_k, t_{k+1}] \\ \hat{v}_{|t=t_k} = v^k & \text{in } \Omega, \end{cases}$$

(ii) Compute $X^k(-\tau, \bullet)$ on Ω by (5),

(iii) Compute new approximation v^{k+1} at time level t_{k+1} by

$$v^{k+1}(x) = \hat{v}(t_{k+1}, X^k(-\tau, x)).$$

Concerning this algorithm, by use of some methods by [17],[16], one can prove that it is consistent of the first order in the L^2- norm and, moreover, convergent as long as the numerical computations behave stable (in a rather strong sense). For more details cf. [7]. For numerical results with a related algorithm see [6].

The linear Stokes problem of step (i) is considered as the "key problem" of this algorithm and will be further investigated over the subsequent sections.

THE VOLTERRA BOUNDARY INTEGRAL EQUATION

Due to the formally simple shape of the Stokes problem of step (i) - a linear system with constant coefficients, parabolic in a generalized sense - it seems advantageous here to employ a boundary element procedure. We use Oseen's fundamental solution [19] :

$$\Gamma(t,x) = \begin{pmatrix} \delta_{ij}\gamma_\nu(t,x) + \dfrac{\partial^2}{\partial x_i \partial x_j} \dfrac{2\nu}{|x|} \displaystyle\int_0^{|x|} t\gamma_\nu(t,\rho)d\rho\,, & \dfrac{\delta(t)}{4\pi} \dfrac{x_i}{|x|^3} \\ \dfrac{\delta(t)}{4\pi} \dfrac{x_j}{|x|^3}\,, & \dfrac{\delta'(t)}{4\pi |x|} + \nu\delta(t)\delta(x) \end{pmatrix}$$

where the upper left element denotes a (3,3)-submatrix, and

$$\gamma_\nu(t,r) = \begin{cases} (4\pi\nu t)^{-3/2}\exp(-r^2/(4\nu t)) : & t > 0 \\ & 0 : \text{elsewhere} \end{cases}$$

denotes the fundamental solution of the heat operator $\dfrac{\partial}{\partial t} - \nu\Delta$. Hence, any (sufficiently regular) vector field is represented in terms of a sum of unsteady potentials based on this fundamental matrix ("Green's formula"). With the exterior normal vector n, let

$$\mathbf{t}(u,q) = -qn + \nu(\nabla u + \nabla u^T) \bullet n$$

denote the stress vector of the flow (u,q), where ∇u^T is the transposed matrix. Consequently, the vector field $\hat{v}$ appearing in step (i) is represented in operational form in terms of

$$\hat{v} = P\{v^k\} + S\{\mathbf{t}(\hat{v},\hat{p})\}\,, \quad \hat{p} \text{ analogous} \tag{6}$$

where P and S denote the unsteady hydrodynamical potentials of Poisson-Weierstraß and of the simple layer, resp., each referring to the initial time t_k. The unsteady double layer potential disappears since $\hat{v} = 0$ on $\partial\Omega$. The present approach leads to a *direct* boundary element method as nowadays employed frequently in various engineering applications.

By tending to the boundary with (6) we obtain (by use of the classical jump relations for unsteady potentials) this system of Volterra boundary integral equations of the first kind:

$$A\psi = f \quad \text{on } \partial\Omega_k = \partial\Omega \times [t_k, t_{k+1}] \tag{7}$$

for the unknown boundary stresses $\psi = \mathbf{t}(\hat{v},\hat{p})$. Here denotes A (or f) the trace of S (or $-P\{v^k\}$, resp.) on the boundary $\partial\Omega$. Hence the core of our approach to the unsteady Navier Stokes problem consists of providing for a mathematical foundation of the Volterra integral equation (7).

We introduce the anisotropic Sobolev spaces

$$\mathcal{H}_\tau^{r,s}(\partial\Omega_k) = L^2(t_k, t_{k+1}, \mathcal{H}_\tau^r(\partial\Omega)) \cap H^s(t_k, t_{k+1}, \mathcal{H}_\tau^0(\partial\Omega)),$$

where $r,s \geq 0$. Here refers $\mathcal{H}_\tau^r(\partial\Omega)$ to vector fields ϕ, living on the boundary and satisfying $\int_{\partial\Omega} \phi \bullet n do = 0$, each component of which belonging to the usual Sobolev spaces $H^r(\partial\Omega)$. These spaces form Hilbert spaces with the usual innerproducts and norms. Let us identify $\mathcal{H}_\tau^{0,0}(\partial\Omega_k)$ with its dual space. We also introduce the dual spaces

$$\mathcal{H}_\tau^{-r,-s}(\partial\Omega_k) = (\mathcal{H}_\tau^{r,s}(\partial\Omega_k))'$$

for negative exponents $(r,s > 0)$, so that we have Gelfand triplets

$$\mathcal{H}_\tau^{r,s} \subset \mathcal{H}_\tau^{0,0} \cong (\mathcal{H}_\tau^{0,0})' \subset \mathcal{H}_\tau^{-r,-s} . \tag{8}$$

These spaces are appropriate to investigate the mapping properties of the operator A .

Theorem 1: *The operator A maps*

$$A : \mathcal{H}_\tau^{-1/2,-1/4}(\partial\Omega_k) \to \mathcal{H}_\tau^{1/2,1/4}(\partial\Omega_k)$$

continuous and coercive:

$$< \phi, A\phi > \geq \beta \bullet \|\phi\|^2_{-1/2,-1/4}(\partial\Omega_k) \tag{9}$$

for all $\phi \in \mathcal{H}_\tau^{-1/2,-1/4}(\partial\Omega_k)$. *Here denotes* $<.,.>$ *the duality pairing, and* $\beta > 0$ *a fixed positive number not depending on* k, ϕ.

In particular, the Volterra integral equation (7) has a unique solution $\psi \in \mathcal{H}_\tau^{-1/2,-1/4}(\partial\Omega_k)$, *depending continuously upon the data* $f \in \mathcal{H}_\tau^{1/2,1/4}(\partial\Omega_k)$.

We will sketch the proof of the coercivity estimate (9) that is crucial for the quasioptimality of appropriate Galerkin boundary element schemes. Cf. Hebeker-Hsiao [9] for more details.

PROOF OF COERCIVITY ESTIMATE (9)

We will show at first

$$< \phi, A\phi > \;\; \geq \;\; const.\|A\phi\|^2_{1/2,1/4} \text{ for all } \phi \in \mathscr{H}_\tau^{-1/2,-1/4}(\partial\Omega_k). \tag{10}$$

Let $u = S\phi$, q the corresponding pressure potential. From the jump relation

$$[\mathbf{t}(u,q)] \;=\; \phi$$

where [.] denotes the difference between interior and exterior limits when approaching the boundary, we conclude

$$\begin{aligned} < \phi, A\phi > \;&=\; < [\mathbf{t}(u,q)], A\phi > \\ &= \frac{1}{2}\int_{\mathbb{R}^3} |\, u(t_{k+1},.)\,|^2 dx \;+\; \nu\int_{t_k}^{t_{k+1}}\int_{\mathbb{R}^3} |\,\nabla u\,|^2 dx dt \end{aligned}$$

with the help of Green's formula. Since u decays sufficiently rapidly at infinity [19], Friedrichs' inequality on $\mathbb{R}^3$ provides us with the estimate

$$< \phi, A\phi > \;\; \geq \;\; const.\|u\|^2_{1,0,\,\Omega_k}. \tag{11}$$

But, for solutions of the homogeneous Stokes equations, the norms of $H_\sigma^{1,0}(\Omega_k)$ and $H_\sigma^{1,1/2}(\Omega_k)$ are easily shown as equivalent (sub-σ denoting the subspace of divergencefree vector fields). Consequently, (11) may be strengthened up to

$$< \phi, A\phi > \;\; \geq \;\; const.\|u\|^2_{1,1/2,\,\Omega_k}.$$

This implies (10) by the trace lemma [13].

From (10), the continuity of the map

$$A \,:\, \mathscr{H}_\tau^{-1/2,-1/4}(\partial\Omega_k) \;\rightarrow\; \mathscr{H}_\tau^{1/2,1/4}(\partial\Omega_k)$$

follows easily. Also, A is one-to-one, since the fully homogeneous initial boundary value problem of the unsteady Stokes equations in $\mathbb{R}^3$ is uniquely solvable up to a pressure field depending on time only: due to the jump relations, the kernel of A consists of distributions of the shape

$$\phi_0(t,x) = a(t)n(x) \in \mathscr{H}_\tau^{-1/2,-1/4}.$$

But then $\phi_0 \in \mathscr{H}_\tau^{0,0}$ is implied by the identification (8). Hence $\phi_0 = 0$. Consequently, by Banach's theorem on the continuous inverse, (10) finally implies the desired estimate (9).

The present proof, indicated so far, extends the now standard variational techniques for steady-state strongly elliptic boundary integral equations (developed by Nedelec-Planchard [14] and Hsiao-Wendland [10]) to the unsteady Stokes system. In case of the *steady-state* Stokes system, the corresponding result has been obtained by Fischer [3] by employing the much more involved techniques of pseudodifferential operators. For an *unsteady* model problem of viscous compressible motions, the present techniques have been developed by Hebeker-Hsiao [8]. Related results for the heat equation have been recently obtained by Costabel-Wendland [1] and Arnold-Noon (see [15]).

THE GALERKIN BOUNDARY ELEMENT METHOD

With respect to numerical applications, the essence of Theorem 1 is as follows. Assume the countable set $\{\chi_1, \chi_2, \ldots\}$ total in the energy space $\mathscr{H}_\tau^{-1/2,-1/4}(\partial\Omega_k)$ of the operator A. For any $m \in \mathbb{N}$, put

$$\mathscr{H}_m = \text{span}\{\chi_1, \ldots, \chi_m\}.$$

Then the time-space Galerkin approximate

$$\psi_m(t,x) = \sum_{i=1}^{m} a_i \chi_i(t,x) \tag{12}$$

is defined by the equations

$$< \chi_j, A\psi_m - f > \; = 0 \quad \text{for all } j = 1, \dots, m \,, \tag{13}$$

which form a linear algebraic system to fix the coefficients a_i of (12). By Cea's lemma, the coercivity of A readily implies

Theorem 2: *Under the stated assumptions, Galerkin's procedure (12)-(13) is quasioptimal and convergent in the norm of* $\mathscr{H}_\tau^{-1/2,-1/4}(\partial\Omega_k)$:

$$\|\psi_m - \psi\|_{-1/2,-1/4,\,\partial\Omega_k} \leq const. \inf_{\chi \in \mathscr{H}_m} \|\chi - \psi\|_{-1/2,-1/4,\,\partial\Omega_k}$$

$$\to 0 \quad \text{when } m \to \infty .$$

The constant does not depend on m,k, and ψ.

In practice, to implement (12)-(13), one may use tensor products

$$\chi_{jl}(t,x) \; = \; ep_j(t)q_l(x)$$

of low order polynomials $p_j(t)$ and piecewise polynomials $q_l(x)$ lifted to the parameter space of the boundary, subordinate to a given decomposition of $\partial\Omega$ into portions called "boundary elements". q_l has to satisfy the constraint $\int_{\partial\Omega} q_l \bullet n do = 0$. Further, e runs over the set of unit vectors of $\mathbb{R}^3$.

Note: This paper is in final form. No similiar paper has been or is being submitted elsewhere.

REFERENCES

[0] Borchers W.: "On the characteristics method for the incompressible Navier Stokes equations". 8 pp. of E.H. Hirschel (ed.): Finite Approximations in Hydrodynamics. Braunschweig, in press.

[1] Costabel M., Wendland W.L.: Personal communication.

[2] Fabes E.B., Lewis J.E., Riviere N.M.: "Boundary value problems for the Navier Stokes equations". Amer. J. Math. 99 (1977), 626-668.

[3] Fischer T.M.: "An integral equation procedure for the exterior 3D viscous flow". Integral Equ. Oper. Th. 5 (1982), 490-505.

[4] Hebeker F.K.: "An approximation method for the Cauchy problem of the 3D equation of vorticity transport". Math. Meth. Appl. Sci. 5 (1983), 439-475.

[5] Hebeker F.K.: "Efficient boundary element methods for 3D exterior viscous flows". Numer. Meth. PDE. 2 (1986), 273-297.

[6] Hebeker F.K.: "Characteristics and boundary element methods for 3D Navier Stokes flows". pp. 305-312 of J.R. Whiteman (ed.): The Mathematics of Finite Elements and Applications. Vol. 6, London 1988.

[7] Hebeker F.K.: "Analysis of a characteristics method for some incompressible and compressible Navier Stokes problems". 22 pp., Preprint no. 1126, Fb. Mathematik, TH Darmstadt 1988.

[8] Hebeker F.K., Hsiao G.C.: "On a boundary integral equation approach to a nonstationary problem of isothermal viscous compressible flows". 20 pp., Preprint no. 1134, Fb. Mathematik, TH Darmstadt 1988, submitted for publication.

[9] Hebeker F.K., Hsiao G.C.: "On Volterra boundary integral equations of the first kind for the nonstationary Stokes equations". Manuscript to be submitted.

[10] Hsiao G.C., Wendland W.L.: "A finite element method for some integral equations of the first kind". J. Math. Anal. Appl. 58 (1977), 449-481.

[11] Hsiao G.C., Wendland W.L.: "The Aubin Nitsche lemma for integral equations". J. Integral Equ. 3 (1981), 299-315.

[12] Ladyzhenskaja O.A.: "The Mathematical Theory of Viscous Incompressible Flows". New York 1969.

[13] Lions J.L., Magenes E.: "Non-homogeneous Boundary Value Problems and Applications". Vol. 2, Berlin 1972.

[14] Nedelec J.C., Planchard J.: "Une methode variationelle d'elements finis pour la resolution numerique d'un probleme exterieur dans $\mathbb{R}^3$". RAIRO R-3, 7 (1973), 105-129.

[15] Noon P.J.: "The single layer heat potential and Galerkin boundary element methods for the heat equation". 108 pp., PhD. thesis, Univ. of Maryland 1988.

[16] Pironneau O.: "On the transport diffusion algorithm and its applications to the Navier Stokes equations". Numer. Math. 38 (1982), 309-332.

[17] Rautmann R.: "Ein Näherungsverfahren für spezielle parabolische Anfangswertaufgaben mit Operatoren". pp. 187-231 of R. Ansorge, W. Törnig (eds.): Numerische Lösung nichtlinearer partieller Differential- und Integrodifferentialgleichungen. Berlin 1972.

[18] Rautmann R.: "A convergent product formula approach to the 3D flow computations". 4 pp. of E.H. Hirschel (ed.): Finite Approximations in Hydrodynamics. Braunschweig, to appear.

[19] Solonnikov V.A.: "Estimates of the solutions of a nonstationary linearized system of Navier Stokes equations". AMS Transl., Ser. 2, Vol 75, 1968.

[20] Süli E.: "Convergence and nonlinear stability of the Lagrange-Galerkin method for the Navier Stokes equations". Numer. Math. 53 (1988), 459-483.

[21] Varnhorn W.: "Energy conserving Lagrangean approximation for the equations of Navier Stokes". J. Diff. Equ., in press.

NUMERICAL SIMULATION AND EXPERIMENTAL VERIFICATION OF CAVITY FLOWS

Bar–Yoseph, P., and Solan, A.
Department of Mechanical Engineering, TECHNION
Technion–City, IL–Haifa 32000
ISRAEL

Roesner, K.G.
Institut für Mechanik, THD
Hochschulstraße 1, D–6100 Darmstadt
WEST GERMANY

SUMMARY

Cavity flows in cylindrical and spherical geometry are investigated numerically on the basis of a finite element calculation and are compared with experimental results which were gained by a photochemical, nearly disturbance–free visualization technique. The onset of secondary flows in the axial region is analysed according to the parameters which describe the moving boundaries and the geometry of the cavity. For the cylindrical geometry the influence of an independently co– or counterrotating bottom is investigated while the lid of the cylinder is rotating at different angular velocities leading to a recirculating zone in the axial region. The aspect ratio of the cylinder and the ratio of the angular velocities of top and bottom determine the region of existence of recirculation zones in the container. The analysis of the experimental results shows the existence of stationary and even space– and time–periodic solutions of the recirculating type of axial flows.

For the spherical gap flow the numerical analysis predicts clearly for different gap sizes the existence of recirculation regions between the south poles of the spherical boundaries. In the experiments such a type of secondary flow could be observed for the concentrical and the eccentrical position of the two spherical boundaries. In the eccentrical case also a time–periodic motion in axial direction could be detected experimentally. For the spherical geometry only the inner sphere is assumed to be in motion. The outer boundary is kept at rest. The penalty finite element calculations were performed using coarse grids.

INTRODUCTION

The vortical character of fluid flows plays an important role for the overall behavior of the motion in unconfined and confined regions of the flow field. From the hudge amount of phenomena concerning vortex formation induced by different kinds of boundary conditions and flow configurations the present investigation deals with the vortical flow in confined regions where no mass flow is admitted through the boundaries of the flow field. This means that especially for the visualisation technique no injection of fluid markers into the rotating fluid is allowed. Therefore a photochromic coloring technique was developed to avoid any macroscopic disturbance of the flow field. It consists in a photochemical reaction which is known as flash photolysis. Originally it was applied by Popovich and Hummel |1| to analyse turbulent pipe flows, but because of the possibility to vary the concentration of the photochromic compound in a wide range this method can be used especially to color by UV–irradiation fluid layers which are closely adjacent to the transparent wall, when the concentration is very large. Then the absorption length of the UV–light becomes short, and the colored liquid near the wall serves as a tracer fluid if it is advected by the velocity field. A detailed description of this experimental method can be found in |2|. In the present investigation at the center of the bottom plate of a circular cylinder made of perspex UV laser light is transmitted through the walls into the organic liquid. In the axial zone of the confined vessel the colored liquid is advected towards those regions where a recirculating zone exists. Looking through the sidewall perpendicular to the axis streaksurfaces are marking those fluid domains which can be reached by the colored liquid. Fig.1 shows two recirculating zones which exist in the cylinder at high values of the aspect ratio H/R (H = cylinder height, R = radius of the cylinder) when only the lid is rotating. The region of existence of such on–axis stag–

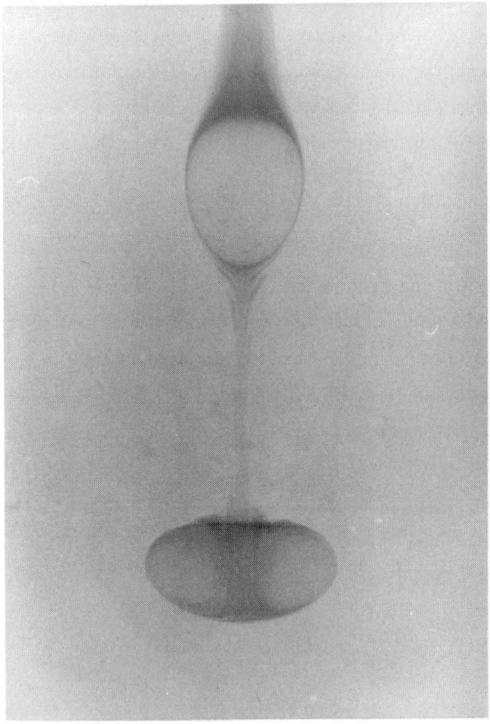

Fig. 1.

Two recirculating zones in the axial region of a circular cylinder with rotating lid

(H/R = 2.54, Re = 2000)

nation points with recirculation was analysed the first time by H.–U. Vogel experimentally and analytically |3–4|. Later a detailed experimental investigation of the velocity field of the so–called vortex breakdown phenomenon in a cylinder was given by Ronnenberg |5|. In several investigations Escudier |6–8| described all phenomena which are met in the case of a rotating lid in a cylinder at rest.

The present analysis is concerned mostly with the influence of a co– or counterrotating bottom for the cylindrical geometry. The experiments have shown that e.g. a co–rotating bottom enlarges the recirculation zones if they once have established. It can be seen that in principle a co–rotation of the bottom destabilizes the basic flow field and favors at low aspect ratios and low Reynolds numbers the phenomenon of vortex breakdown. Fig.2 shows the results of measurements which were performed for the circular cylinder.

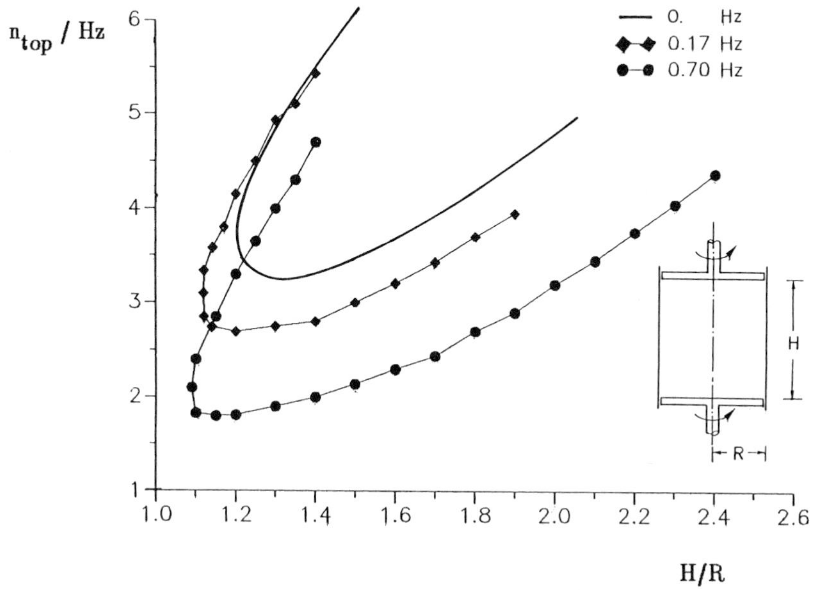

Fig. 2. Regions of existence of recirculating zones in a cylinder with rotating lid and bottom

FLOW IN A CIRCULAR CYLINDER

It is well known from |3| that below the Reynolds number of 1000 ($Re = R^2\omega/\nu$, ω = angular velocity of the lid , ν = kinematic viscosity of the liquid) and for values of H/R smaller than 1.2 no recirculation zones can be found when only the lid is rotating. As any rotating part of the boundary of the container acts like a pumping device for the adjacent liquid fluid particles are accelerated from the bottom towards the lid following a helicoidal path in the steady state situation. A stagnation point on the axis becomes more likely when the bottom is set into co–rotation with respect to the lid, because then both the rotating boundaries act on the axial region by a suction of the fluid towards the centers of the disks. Such a decelerating effect on the upwards directed stream of the liquid supports the appearance of an on–axis stagnation point with a developing recirculating zone. On the other hand the fluid particles near the rotating bottom gain a larger circumferential velocity component and consequently larger swirl angles with respect to the rotating bottom plate. The swirl angle is a very sensitive parameter describing the onset of recirculating zones as pointed out by Bossel |9|. Fig.2. shows a tendency of the liquid to form recirculating zones in the presence of a co–rotating bottom plate at Reynolds numbers which are below the known limit value for the rotating lid alone. Further experiments have indicated that the lower the aspect ratio H/R of the vessel the higher the angular velocity of the bottom plate must be to enforce a vortical motion in the axial region of recirculating type. In Fig.2. for two low frequencies of the bottom (0.17 Hz and 0.7 Hz) the onset of the recirculation is shifted to $H/R = 1.1$ and a Reynolds number of about 525. For $H/R = 0.8$ still a stagnation point on the axis exists with an extended recirculating zone. From numerical simulations due to Lopez |10| it is known that for large H/R–values time–periodic solutions exist. When analysing the experimental data one can find also a space–periodic solution which shows a three–lobed character around the axis of revolution. This mode indicates clearly that the phenomenon of vortex breakdown in a cylinder may correctly be described by a three–dimensional and time–dependent numerical simulation of the solution of the Navier–Stokes equations.

FLOW IN A SPHERICAL GAP

Calculations based on the penalty finite element method due to Bar–Yoseph et al. |11| have predicted a similar character of the flow field in the axial region between the two south poles of concentric or eccentric spheres as it was discussed for the cylindrical case. The results for different gap sizes are shown in Fig.3 and Fig.4.

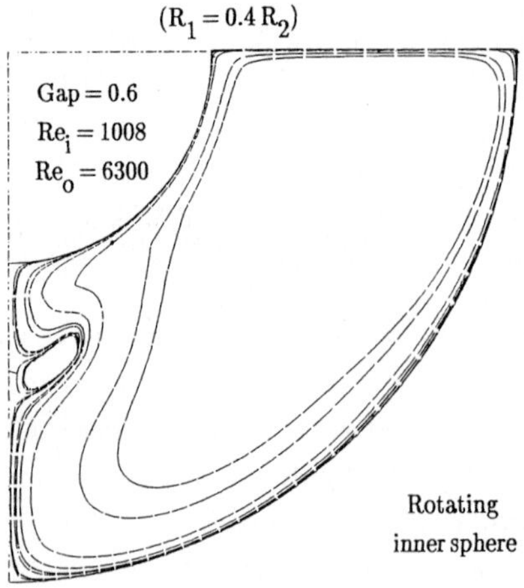

Fig. 3.

Streamlines in the concentric spherical gap with recirculation

$(Re_i = R_1^2\omega/\nu,\ Re_o = R_2^2\omega/\nu)$

For a Reynolds number of about 1000 the flow simulation in a large gap between the spherical shells shows the existence of an axially symmetric recirculation zone. In Fig.4 the same situation is met at a Reynolds number of about 1125 for the gap size 0.5. Fig.5 shows the experimental result which fits quite well to the numerical data.

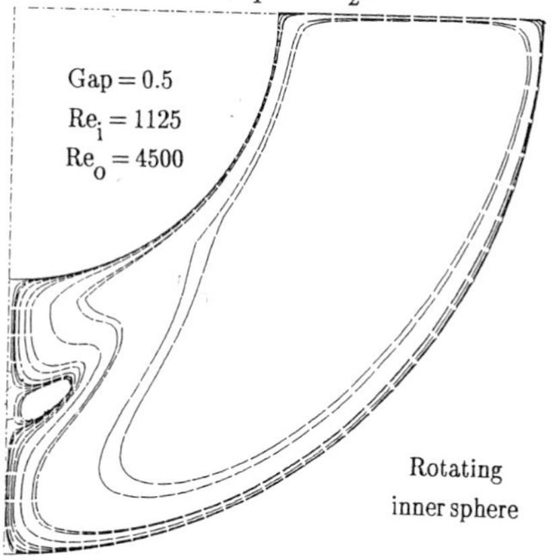

Fig. 4.

Streamlines in the concentric spherical gap with recirculation

$(Re_i = R_1^2\omega/\nu,\ Re_o = R_2^2\omega/\nu)$

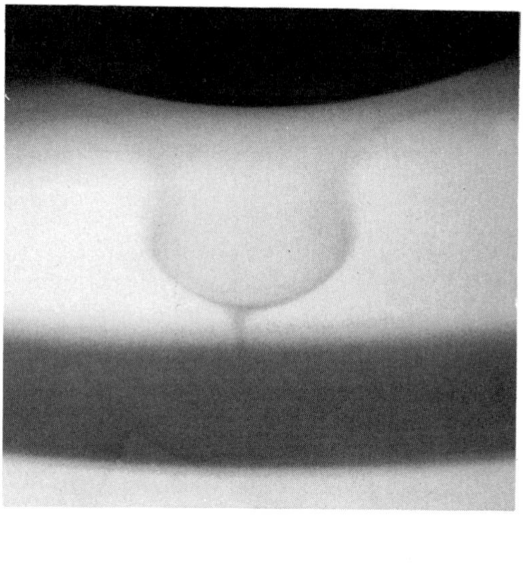

Fig. 5.

Experimental proof of on–axis stagnation points in a spherical gap

(Gap size = 0.5)

Further experimental investigations of the flow in an eccentrical spherical gap have shown that even for the eccentricity 1 when the distance between the south poles is twice the gap width a stable recirculation zone exists. This state is reached at a Reynolds number of about 600 (related to the radius of the inner sphere) and represented by the photograph in Fig.6. The experimental set–up was performed with high accuracy with respect to the rotational symmetry of the inner sphere which was done by U. Weiss from the TECHNION in Haifa in a highly sophisticated way.

Fig. 6. Stationary recirculation zone beneath a rotating sphere in an eccentrical spherical gap (Gap size = 0.5, Re_i = 601)

CONCLUSION

The numerical simulation of the time–independent Navier Stokes equations for axially symmetric flows on the basis of the penalty finite element method developed by Bar–Yoseph gives reliable results for the fine structure of the flow field in a spherical gap. On the basis of this numerical simulation the existence of vortical motions in the polar region between two spheres can be predicted accurately and was proved by physical experiments.

REFERENCES

|1| Popovich, A.T., and Hummel, R.L.: A new method for non–disturbing turbulent flow measurements very close to the wall. Chem. Eng. Sc., Vol. 22 (1967) 21–25.

|2| Larson, J., and Roesner, K.G.: Optical Flow–Velocity Measurements in Irregularly Shaped Cavities. in: Recent Contributions to Fluid Mechanics, ed. W. Haase, Springer–Verlag, 1982.

|3| Vogel, H.–U.: Experimentelle Ergebnisse über die laminare Strömung in einem zylindrischen Gehäuse mit darin rotierender Scheibe. Max–Planck–Institut für Strömungsforschung, Bericht 6 (1968).

|4| Vogel, H.–U: Rückströmungsblasen in Drallströmungen. Festschrift 50 Jahre Max–Planck–Institut für Strömungsforschung, Göttingen 1925–1975 (1975).

|5| Ronnenberg, B.: Ein selbstjustierendes 3–Komponenten–Laserdoppleranemometer nach dem Vergleichsstrahlverfahren, angewandt auf Untersuchungen in einer stationären zylindersymmetrischen Drehströmung mit einem Rückstromgebiet. Max–Planck–Institut für Strömungsforschung, Bericht 20 (1977).

|6| Escudier, M.P., and Zehnder, N.: Vortex–flow regimes. J. Fluid Mech. vol. 115 (1982) 105–121.

|7| Escudier, M.: Observations of the flow produced in a cylindrical container by a rotating endwall. Experiments in Fluids 2, (1984) 189–196.

|8| Escudier, M.: Vortex breakdown: Observations and explanations. in: Progress in Aerospace Sciences, vol. 25, No.2 (1988) 189–229.

|9| Bossel, H.H.: Vortex Breakdown Flowfield. Phys. of Fluids, vol. 12, No. 3 (1969) 498–508.

|10| Lopez, J.M.: Axisymmetric vortex breakdown in an enclosed cylinder flow. Proc. 11th Int. Conf. on Num. Meth. in Fluid Dyn., Williamsburg, VA, June 27–July 1, 1988, Springer–Verlag (1989).

|11| Bar–Yoseph, P., Seelig, S., and Roesner, K.G.: Vortex breakdown in a spherical gap, Phys. Fluids 30 (6), (1987) 1531–1583.

This paper is in final form and no similar paper is being or has been submitted elsewhere

(received May 8, 1989).

PARTICIPANTS

Amick, C.J., Prof.Dr.	Department of Mathematics, University of Chicago, 5734 University Ave., Chicago, IL 60637, USA
Asano, K., Prof.Dr.	College of General Education Kyoto, Kyoto University, 606 Japan
Beale, J.Th., Prof.Dr.	Department of Mathematics, Princeton University, New Jersey 08544, USA
Beirão da Veiga, H., Prof.Dr.	Università di Pisa, Facoltà di Ingegneria Istituto di Matematiche Applicate "U.Dini" via Bonanno, 25 B, Italy
Blum H., Dr.	Mathematisches Institut der Universität Heidelberg, D-6900 Heidelberg, Germany
Borchers, W. Dr.	Universität-GH Paderborn, FB 17, D-4790 Paderborn, Warburger Str. 100, Germany
Cottet, G.H., Dr.	Department of Mathematics, University of California, Los Angeles, USA
Duff, G., Prof. Dr.	Department of Mathematics, University of Toronto (Ontario) M5 F1 A1, Canada
Fischer, T., Dr.	Max-Planck-Institut Göttingen, D-3400 Göttingen, D-3400 Göttingen, Germany
Fursikov, A.V., Prof.Dr.	Moscow University, Department of Mechanics and Mathematics, 119899 Moscow, Lenin Hills, USSR
Giga, Y., Prof.Dr.	Department of Mathematics, Hokkaido University Sapporo 060, Japan
Girault, V., Prof.Dr.	Universite Paris VI, Analyse Numerique, Tour 55 5 E, 9, Quai Saint-Bernard, Paris 5e, France
Guillopé, C., Prof.Dr.	Université de Paris-Sud, Laboratoire d Analyse Numerique, Batiment 425, F-91405 Orsay Cedex, France
Hebeker, F.K., Doz. Dr.	IBM-Wissenschaftszentrum, Tiergartenstraße, 6900 Heidelberg
Heywood, J.G., Prof.Dr.	Department of Mathematics, University of British Columbia, # 121-1984 Mathematics Road, Vancouver B.C., V6T1Y4, Canada
Kazhikhov, A.V., Prof.Dr.	Institute of Hydrodynamics of the Sibirian Branch of Academy of Sciences, 630090 Novosibirsk, USSR
Kirchgässner, K., Prof.Dr.	Universität Stuttgart, Fachbereich Mathematik/ Informatik, Pfaffenwaldring 57, D-7000 Stuttgart 80, Germany

Kozono, H., Dr.	Department of Applied Physics, Faculty of Engeneering, Nagoya University, Chikusa-Ku, Nagoya, Japan
Kröner, D., Dr.	Mathematisches Institut der Universität Heidelberg, 6900 Heidelberg, Im Neuenheimer Feld 288
Masuda, T., Prof.Dr.	Department of Mathematics, University of Tokyo, Hongo, Tokyo, Japan 113
Miyakawa, T., Prof.Dr.	Department of Mathematics, Hiroshima University, Hiroshima, 730 Japan
Mizumachi, R., Dr.	Tohoku University, Mathematical Institute Sendai 980, Japan
Nagata, W., Prof. Dr.	Department of Mathematics, University of British Columbia, # 121-1984 Mathematics Road, Vancouver B.C., V6T1Y4, Canada
Okamoto, H., Dr.	Department of Mathematics, University of Tokyo, Hongo, Tokyo, Japan 113
Pileckas, K., Prof. Dr.	Institute of Mathematics and Cybernetics of the Academy of Sciences of the Lithuanian SSR, Akademijos str. 4, Vilnius, Lithuanian SSR,USSR
Polezhaev, V., Prof.Dr.	Institut of Problems in Mechanics the USSR Academy of Sciences, Prospect Vernadskogo, 101, 117526 Moscow, USSR
Prouse, G., Prof. Dr.	Dipartimento di Matematica, Piazza Leonardo da Vinci 32, 20133 Milano
Pukhnachov, V.V., Prof.Dr.	Institute of Hydrodynamics of the Sibirian Branch of Academy of Sciences, 630090 Novosibirsk, USSR
Rannacher, R. Prof. Dr.	Mathematisches Institut der Universität Heidelberg, D-6900 Heidelberg, Germany
Rautmann, R., Prof.Dr.	Universität-GH Paderborn, Fachbereich 17, D-4790 Paderborn, Warburger Str. 100, Germany
Rivkind, V.Ja., Prof.Dr.	Research Institute of Mathematics and Mechanics, 198904 Leningrad, Petrodvorets, Bibliotechnaya Square 2, USSR
Roesner, K.G., Prof.Dr.	Technische Hochschule Darmstadt, Institut für Mechanik, Hochschulstr. 1, 6100 Darmstadt
Rozhdestvensky, B.L., Prof.Dr.	M.V. Keldysh Institute of Applied Mathematics, 125047 Moscow, Miusskaya Square 4, USSR
Socolescu, D., Prof. Dr.	Mathematisches Institut der Universität Kaisereslautern, D-6750 Kaiserslautern, Erwin-Schrödinger Straße, Germany

Sohr, H., Prof. Dr.	Universität-GH Paderborn, Fachbereich 17, Warburger Str. 100, D-4790 Paderborn, Germany
Solonnikov, V.A., Prof. Dr.	Mathematical Institut of the Academy of Sciences USSR, Fontanka 27, Leningrad 191011, USSR
Sritharan, S.S., Prof. Dr.	USC, Department of Aerospace Engeneering, University of Southern California, Los Angeles CA USA, University park 90007
Tarunin, R., Prof. Dr.	Department of Mechanics and Mathematics of the Perm University, 614600 Perm, USSR
Titi, E.S., Prof. Dr.	Department of Cornell University, White Hall Ithaca, NY 14853, USA
Valli, A.,,Prof. Dr.	Dipartimento di Matematica e Fisica, Libera Università Trento, 38050 Povo (Trento) Italy
Varnhorn, W., Dr.	Mathematisches Institut der Universität Darmstadt, D-6100 Darmstadt, Schloßgartenstr.7, Germany
Velte, W., Prof.Dr.	Institut für Angewandte Mathematik und Statistik, D-8700 Würzburg, Am Hubland, Germany
von Wahl, W., Prof.Dr.	Fakultät für Mathematik und Physik, D-8580 Bayreuth, Postfach 101251, Germany
Wiegner, M., Prof.Dr.	Fakultät für Mathematik und Physik, D-8580 Bayreuth, Postfach 101251, Germany

Vol. 1259: F. Cano Torres, Desingularization Strategies for Three-Dimensional Vector Fields. IX, 189 pages. 1987.

Vol. 1260: N.H. Pavel, Nonlinear Evolution Operators and Semigroups. VI, 285 pages. 1987.

Vol. 1261: H. Abels, Finite Presentability of S-Arithmetic Groups. Compact Presentability of Solvable Groups. VI, 178 pages. 1987.

Vol. 1262: E. Hlawka (Hrsg.), Zahlentheoretische Analysis II. Seminar, 1984–86. V, 158 Seiten. 1987.

Vol. 1263: V.L. Hansen (Ed.), Differential Geometry. Proceedings, 1985. XI, 288 pages. 1987.

Vol. 1264: Wu Wen-tsün, Rational Homotopy Type. VIII, 219 pages. 1987.

Vol. 1265: W. Van Assche, Asymptotics for Orthogonal Polynomials. VI, 201 pages. 1987.

Vol. 1266: F. Ghione, C. Peskine, E. Sernesi (Eds.), Space Curves. Proceedings, 1985. VI, 272 pages. 1987.

Vol. 1267: J. Lindenstrauss, V.D. Milman (Eds.), Geometrical Aspects of Functional Analysis. Seminar. VII, 212 pages. 1987.

Vol. 1268: S.G. Krantz (Ed.), Complex Analysis. Seminar, 1986. VII, 195 pages. 1987.

Vol. 1269: M. Shiota, Nash Manifolds. VI, 223 pages. 1987.

Vol. 1270: C. Carasso, P.-A. Raviart, D. Serre (Eds.), Nonlinear Hyperbolic Problems. Proceedings, 1986. XV, 341 pages. 1987.

Vol. 1271: A.M. Cohen, W.H. Hesselink, W.L.J. van der Kallen, J.R. Strooker (Eds.), Algebraic Groups Utrecht 1986. Proceedings. XII, 284 pages. 1987.

Vol. 1272: M.S. Livšic, L.L. Waksman, Commuting Nonselfadjoint Operators in Hilbert Space. III, 115 pages. 1987.

Vol. 1273: G.-M. Greuel, G. Trautmann (Eds.), Singularities, Representation of Algebras, and Vector Bundles. Proceedings, 1985. XIV, 383 pages. 1987.

Vol. 1274: N. C. Phillips, Equivariant K-Theory and Freeness of Group Actions on C*-Algebras. VIII, 371 pages. 1987.

Vol. 1275: C.A. Berenstein (Ed.), Complex Analysis I. Proceedings, 1985–86. XV, 331 pages. 1987.

Vol. 1276: C.A. Berenstein (Ed.), Complex Analysis II. Proceedings, 1985–86. IX, 320 pages. 1987.

Vol. 1277: C.A. Berenstein (Ed.), Complex Analysis III. Proceedings, 1985–86. X, 350 pages. 1987.

Vol. 1278: S.S. Koh (Ed.), Invariant Theory. Proceedings, 1985. V, 102 pages. 1987.

Vol. 1279: D. Ieşan, Saint-Venant's Problem. VIII, 162 Seiten. 1987.

Vol. 1280: E. Neher, Jordan Triple Systems by the Grid Approach. XII, 193 pages. 1987.

Vol. 1281: O.H. Kegel, F. Menegazzo, G. Zacher (Eds.), Group Theory. Proceedings, 1986. VII, 179 pages. 1987.

Vol. 1282: D.E. Handelman, Positive Polynomials, Convex Integral Polytopes, and a Random Walk Problem. XI, 136 pages. 1987.

Vol. 1283: S. Mardešić, J. Segal (Eds.), Geometric Topology and Shape Theory. Proceedings, 1986. V, 261 pages. 1987.

Vol. 1284: B.H. Matzat, Konstruktive Galoistheorie. X, 286 pages. 1987.

Vol. 1285: I.W. Knowles, Y. Saitō (Eds.), Differential Equations and Mathematical Physics. Proceedings, 1986. XVI, 499 pages. 1987.

Vol. 1286: H.R. Miller, D.C. Ravenel (Eds.), Algebraic Topology. Proceedings, 1986. VII, 341 pages. 1987.

Vol. 1287: E.B. Saff (Ed.), Approximation Theory, Tampa. Proceedings, 1985–1986. V, 228 pages. 1987.

Vol. 1288: Yu. L. Rodin, Generalized Analytic Functions on Riemann Surfaces. V, 128 pages, 1987.

Vol. 1289: Yu. I. Manin (Ed.), K-Theory, Arithmetic and Geometry. Seminar, 1984–1986. V, 399 pages. 1987.

Vol. 1290: G. Wüstholz (Ed.), Diophantine Approximation and Transcendence Theory. Seminar, 1985. V, 243 pages. 1987.

Vol. 1291: C. Mœglin, M.-F. Vignéras, J.-L. Waldspurger, Correspondances de Howe sur un Corps p-adique. VII, 163 pages. 1987

Vol. 1292: J.T. Baldwin (Ed.), Classification Theory. Proceedings, 1985. VI, 500 pages. 1987.

Vol. 1293: W. Ebeling, The Monodromy Groups of Isolated Singularities of Complete Intersections. XIV, 153 pages. 1987.

Vol. 1294: M. Queffélec, Substitution Dynamical Systems – Spectral Analysis. XIII, 240 pages. 1987.

Vol. 1295: P. Lelong, P. Dolbeault, H. Skoda (Réd.), Séminaire d'Analyse P. Lelong – P. Dolbeault – H. Skoda. Seminar, 1985/1986. VII, 283 pages. 1987.

Vol. 1296: M.-P. Malliavin (Ed.), Séminaire d'Algèbre Paul Dubreil et Marie-Paule Malliavin. Proceedings, 1986. IV, 324 pages. 1987.

Vol. 1297: Zhu Y.-l., Guo B.-y. (Eds.), Numerical Methods for Partial Differential Equations. Proceedings. XI, 244 pages. 1987.

Vol. 1298: J. Aguadé, R. Kane (Eds.), Algebraic Topology, Barcelona 1986. Proceedings. X, 255 pages. 1987.

Vol. 1299: S. Watanabe, Yu.V. Prokhorov (Eds.), Probability Theory and Mathematical Statistics. Proceedings, 1986. VIII, 589 pages. 1988.

Vol. 1300: G.B. Seligman, Constructions of Lie Algebras and their Modules. VI, 190 pages. 1988.

Vol. 1301: N. Schappacher, Periods of Hecke Characters. XV, 160 pages. 1988.

Vol. 1302: M. Cwikel, J. Peetre, Y. Sagher, H. Wallin (Eds.), Function Spaces and Applications. Proceedings, 1986. VI, 445 pages. 1988.

Vol. 1303: L. Accardi, W. von Waldenfels (Eds.), Quantum Probability and Applications III. Proceedings, 1987. VI, 373 pages. 1988.

Vol. 1304: F.Q. Gouvêa, Arithmetic of p-adic Modular Forms. VIII, 121 pages. 1988.

Vol. 1305: D.S. Lubinsky, E.B. Saff, Strong Asymptotics for Extremal Polynomials Associated with Weights on $\mathbb{R}$. VII, 153 pages. 1988.

Vol. 1306: S.S. Chern (Ed.), Partial Differential Equations. Proceedings, 1986. VI, 294 pages. 1988.

Vol. 1307: T. Murai, A Real Variable Method for the Cauchy Transform, and Analytic Capacity. VIII, 133 pages. 1988.

Vol. 1308: P. Imkeller, Two-Parameter Martingales and Their Quadratic Variation. IV, 177 pages. 1988.

Vol. 1309: B. Fiedler, Global Bifurcation of Periodic Solutions with Symmetry. VIII, 144 pages. 1988.

Vol. 1310: O.A. Laudal, G. Pfister, Local Moduli and Singularities. V, 117 pages. 1988.

Vol. 1311: A. Holme, R. Speiser (Eds.), Algebraic Geometry, Sundance 1986. Proceedings. VI, 320 pages. 1988.

Vol. 1312: N.A. Shirokov, Analytic Functions Smooth up to the Boundary. III, 213 pages. 1988.

Vol. 1313: F. Colonius, Optimal Periodic Control. VI, 177 pages. 1988.

Vol. 1314: A. Futaki, Kähler-Einstein Metrics and Integral Invariants. IV, 140 pages. 1988.

Vol. 1315: R.A. McCoy, I. Ntantu, Topological Properties of Spaces of Continuous Functions. IV, 124 pages. 1988.

Vol. 1316: H. Korezlioglu, A.S. Ustunel (Eds.), Stochastic Analysis and Related Topics. Proceedings, 1986. V, 371 pages. 1988.

Vol. 1317: J. Lindenstrauss, V.D. Milman (Eds.), Geometric Aspects of Functional Analysis. Seminar, 1986–87. VII, 289 pages. 1988.

Vol. 1318: Y. Felix (Ed.), Algebraic Topology – Rational Homotopy. Proceedings, 1986. VIII, 245 pages. 1988

Vol. 1319: M. Vuorinen, Conformal Geometry and Quasiregular Mappings. XIX, 209 pages. 1988.

Vol. 1320: H. Jürgensen, G. Lallement, H.J. Weinert (Eds.), Semigroups, Theory and Applications. Proceedings, 1986. X, 416 pages. 1988.

Vol. 1321: J. Azéma, P.A. Meyer, M. Yor (Eds.), Séminaire de Probabilités XXII. Proceedings. IV, 600 pages. 1988.

Vol. 1322: M. Métivier, S. Watanabe (Eds.), Stochastic Analysis. Proceedings, 1987. VII, 197 pages. 1988.

Vol. 1323: D.R. Anderson, H.J. Munkholm, Boundedly Controlled Topology. XII, 309 pages. 1988.

Vol. 1324: F. Cardoso, D.G. de Figueiredo, R. Iório, O. Lopes (Eds.), Partial Differential Equations. Proceedings, 1986. VIII, 433 pages. 1988.

Vol. 1325: A. Truman, I.M. Davies (Eds.), Stochastic Mechanics and Stochastic Processes. Proceedings, 1986. V, 220 pages. 1988.

Vol. 1326: P.S. Landweber (Ed.), Elliptic Curves and Modular Forms in Algebraic Topology. Proceedings, 1986. V, 224 pages. 1988.

Vol. 1327: W. Bruns, U. Vetter, Determinantal Rings. VII,236 pages. 1988.

Vol. 1328: J.L. Bueso, P. Jara, B. Torrecillas (Eds.), Ring Theory. Proceedings, 1986. IX, 331 pages. 1988.

Vol. 1329: M. Alfaro, J.S. Dehesa, F.J. Marcellan, J.L. Rubio de Francia, J. Vinuesa (Eds.): Orthogonal Polynomials and their Applications. Proceedings, 1986. XV, 334 pages. 1988.

Vol. 1330: A. Ambrosetti, F. Gori, R. Lucchetti (Eds.), Mathematical Economics. Montecatini Terme 1986. Seminar. VII, 137 pages. 1988.

Vol. 1331: R. Bamón, R. Labarca, J. Palis Jr. (Eds.), Dynamical Systems, Valparaiso 1986. Proceedings. VI, 250 pages. 1988.

Vol. 1332: E. Odell, H. Rosenthal (Eds.), Functional Analysis. Proceedings, 1986–87. V, 202 pages. 1988.

Vol. 1333: A.S. Kechris, D.A. Martin, J.R. Steel (Eds.), Cabal Seminar 81–85. Proceedings, 1981–85. V, 224 pages. 1988.

Vol. 1334: Yu.G. Borisovich, Yu. E. Gliklikh (Eds.), Global Analysis – Studies and Applications III. V, 331 pages. 1988.

Vol. 1335: F. Guillén, V. Navarro Aznar, P. Pascual-Gainza, F. Puerta, Hyperrésolutions cubiques et descente cohomologique. XII, 192 pages. 1988.

Vol. 1336: B. Helffer, Semi-Classical Analysis for the Schrödinger Operator and Applications. V, 107 pages. 1988.

Vol. 1337: E. Sernesi (Ed.), Theory of Moduli. Seminar, 1985. VIII, 232 pages. 1988.

Vol. 1338: A.B. Mingarelli, S.G. Halvorsen, Non-Oscillation Domains of Differential Equations with Two Parameters. XI, 109 pages. 1988.

Vol. 1339: T. Sunada (Ed.), Geometry and Analysis of Manifolds. Procedings, 1987. IX, 277 pages. 1988.

Vol. 1340: S. Hildebrandt, D.S. Kinderlehrer, M. Miranda (Eds.), Calculus of Variations and Partial Differential Equations. Proceedings, 1986. IX, 301 pages. 1988.

Vol. 1341: M. Dauge, Elliptic Boundary Value Problems on Corner Domains. VIII, 259 pages. 1988.

Vol. 1342: J.C. Alexander (Ed.), Dynamical Systems. Proceedings, 1986–87. VIII, 726 pages. 1988.

Vol. 1343: H. Ulrich, Fixed Point Theory of Parametrized Equivariant Maps. VII, 147 pages. 1988.

Vol. 1344: J. Král, J. Lukeš, J. Netuka, J. Veselý (Eds.), Potential Theory – Surveys and Problems. Proceedings, 1987. VIII, 271 pages. 1988.

Vol. 1345: X. Gomez-Mont, J. Seade, A. Verjovski (Eds.), Holomorphic Dynamics. Proceedings, 1986. VII, 321 pages. 1988.

Vol. 1346: O. Ya. Viro (Ed.), Topology and Geometry – Rohlin Seminar. XI, 581 pages. 1988.

Vol. 1347: C. Preston, Iterates of Piecewise Monotone Mappings on an Interval. V, 166 pages. 1988.

Vol. 1348: F. Borceux (Ed.), Categorical Algebra and its Applications. Proceedings, 1987. VIII, 375 pages. 1988.

Vol. 1349: E. Novak, Deterministic and Stochastic Error Bounds in Numerical Analysis. V, 113 pages. 1988.

Vol. 1350: U. Koschorke (Ed.), Differential Topology. Proceedings, 1987. VI, 269 pages. 1988.

Vol. 1351: I. Laine, S. Rickman, T. Sorvali, (Eds.), Complex Analysis, Joensuu 1987. Proceedings. XV, 378 pages. 1988.

Vol. 1352: L.L. Avramov, K.B. Tchakerian (Eds.), Algebra – Some Current Trends. Proceedings, 1986. IX, 240 Seiten. 1988.

Vol. 1353: R.S. Palais, Ch.-l. Terng, Critical Point Theory and Submanifold Geometry. X, 272 pages. 1988.

Vol. 1354: A. Gómez, F. Guerra, M.A. Jiménez, G. López (Eds.), Approximation and Optimization. Proceedings, 1987. VI, 280 pages. 1988.

Vol. 1355: J. Bokowski, B. Sturmfels, Computational Synthetic Geometry. V, 168 pages. 1989.

Vol. 1356: H. Volkmer, Multiparameter Eigenvalue Problems and Expansion Theorems. VI, 157 pages. 1988.

Vol. 1357: S. Hildebrandt, R. Leis (Eds.), Partial Differential Equations and Calculus of Variations. VI, 423 pages. 1988.

Vol. 1358: D. Mumford, The Red Book of Varieties and Schemes. V, 309 pages. 1988.

Vol. 1359: P. Eymard, J.-P. Pier (Eds.), Harmonic Analysis. Proceedings, 1987. VIII, 287 pages. 1988.

Vol. 1360: G. Anderson, C. Greengard (Eds.), Vortex Methods. Proceedings, 1987. V, 141 pages. 1988.

Vol. 1361: T. tom Dieck (Ed.), Algebraic Topology and Transformation Groups. Proceedings, 1987. VI, 298 pages. 1988.

Vol. 1362: P. Diaconis, D. Elworthy, H. Föllmer, E. Nelson, G.C. Papanicolaou, S.R.S. Varadhan. École d'Été de Probabilités de Saint-Flour XV–XVII, 1985–87. Editor: P.L. Hennequin. V, 459 pages. 1988.

Vol. 1363: P.G. Casazza, T.J. Shura. Tsirelson's Space. VIII, 204 pages. 1988.

Vol. 1364: R.R. Phelps, Convex Functions, Monotone Operators and Differentiability. IX, 115 pages. 1989.

Vol. 1365: M. Giaquinta (Ed.), Topics in Calculus of Variations. Seminar, 1987. X, 196 pages. 1989.

Vol. 1366: N. Levitt, Grassmannians and Gauss Maps in PL-Topology. V, 203 pages. 1989.

Vol. 1367: M. Knebusch, Weakly Semialgebraic Spaces. XX, 376 pages. 1989.

Vol. 1368: R. Hübl, Traces of Differential Forms and Hochschild Homology. III, 111 pages. 1989.

Vol. 1369: B. Jiang, Ch.-K. Peng, Z. Hou (Eds.), Differential Geometry and Topology. Proceedings, 1986–87. VI, 366 pages. 1989.

Vol. 1370: G. Carlsson, R.L. Cohen, H.R. Miller, D.C. Ravenel (Eds.), Algebraic Topology. Proceedings, 1986. IX, 456 pages. 1989.

Vol. 1371: S. Glaz, Commutative Coherent Rings. XI, 347 pages. 1989.

Vol. 1372: J. Azéma, P.A. Meyer, M. Yor (Eds.), Séminaire de Probabilités XXIII. Proceedings. IV, 583 pages. 1989.

Vol. 1373: G. Benkart, J.M. Osborn (Eds.), Lie Algebras, Madison 1987. Proceedings. V, 145 pages. 1989.

Vol. 1374: R.C. Kirby, The Topology of 4-Manifolds. VI, 108 pages. 1989.

Vol. 1375: K. Kawakubo (Ed.), Transformation Groups. Proceedings, 1987. VIII, 394 pages, 1989.

Vol. 1376: J. Lindenstrauss, V.D. Milman (Eds.), Geometric Aspects of Functional Analysis. Seminar (GAFA) 1987–88. VII, 288 pages. 1989.

Vol. 1377: J.F. Pierce, Singularity Theory, Rod Theory, and Symmetry-Breaking Loads. IV, 177 pages. 1989.

Vol. 1378: R.S. Rumely, Capacity Theory on Algebraic Curves. III, 437 pages. 1989.

Vol. 1379: H. Heyer (Ed.), Probability Measures on Groups IX. Proceedings, 1988. VIII, 437 pages. 1989 ol. 1380: H.P. Schlickewei, E. Wirsing (Eds.), Number Theory, Ulm 987. Proceedings. V, 266 pages. 1989.

ol. 1381: J.-O. Strömberg, A. Torchinsky. Weighted Hardy Spaces. 193 pages. 1989.

ol. 1382: H. Reiter, Metaplectic Groups and Segal Algebras. XI, 128 ages. 1989.

ol. 1383: D.V. Chudnovsky, G.V. Chudnovsky, H. Cohn, M.B. Nathanson Eds.), Number Theory, New York 1985–88. Seminar. V, 256 pages. 1989.

ol. 1384: J. Garcia-Cuerva (Ed.), Harmonic Analysis and Partial Differential Equations. Proceedings, 1987. VII, 213 pages. 1989.

ol. 1385: A.M. Anile, Y. Choquet-Bruhat (Eds.), Relativistic Fluid Dynamics. Seminar, 1987. V, 308 pages. 1989.

Vol. 1386: A. Bellen, C.W. Gear, E. Russo (Eds.), Numerical Methods for Ordinary Differential Equations. Proceedings, 1987. VII, 136 pages. 1989.

Vol. 1387: M. Petković, Iterative Methods for Simultaneous Inclusion of Polynomial Zeros. X, 263 pages. 1989.

Vol. 1388: J. Shinoda, T.A. Slaman, T. Tugué (Eds.), Mathematical Logic and Applications. Proceedings, 1987. V, 223 pages. 1989.

Vol. 1000: Second Edition. H. Hopf, Differential Geometry in the Large. VII, 184 pages. 1989.

Vol. 1389: E. Ballico, C. Ciliberto (Eds.), Algebraic Curves and Projective Geometry. Proceedings, 1988. V, 288 pages. 1989.

Vol. 1390: G. Da Prato, L. Tubaro (Eds.), Stochastic Partial Differential Equations and Applications II. Proceedings, 1988. VI, 258 pages. 1989.

Vol. 1391: S. Cambanis, A. Weron (Eds.), Probability Theory on Vector Spaces IV. Proceedings, 1987. VIII, 424 pages. 1989.

Vol. 1392: R. Silhol, Real Algebraic Surfaces. X, 215 pages. 1989.

Vol. 1393: N. Bouleau, D. Feyel, F. Hirsch, G. Mokobodzki (Eds.), Séminaire de Théorie du Potentiel Paris, No. 9. Proceedings. VI, 265 pages. 1989.

Vol. 1394: T.L. Gill, W.W. Zachary (Eds.), Nonlinear Semigroups, Partial Differential Equations and Attractors. Proceedings, 1987. IX, 233 pages. 1989.

Vol. 1395: K. Alladi (Ed.), Number Theory, Madras 1987. Proceedings. VII, 234 pages. 1989.

Vol. 1396: L. Accardi, W. von Waldenfels (Eds.), Quantum Probability and Applications IV. Proceedings, 1987. VI, 355 pages. 1989.

Vol. 1397: P.R. Turner (Ed.), Numerical Analysis and Parallel Processing. Seminar, 1987. VI, 264 pages. 1989.

Vol. 1398: A.C. Kim, B.H. Neumann (Eds.), Groups – Korea 1988. Proceedings. V, 189 pages. 1989.

Vol. 1399: W.-P. Barth, H. Lange (Eds.), Arithmetic of Complex Manifolds. Proceedings, 1988. V, 171 pages. 1989.

Vol. 1400: U. Jannsen. Mixed Motives and Algebraic K-Theory. XIII, 246 pages. 1990.

Vol. 1401: J. Steprāns, S. Watson (Eds.), Set Theory and its Applications. Proceedings, 1987. V, 227 pages. 1989.

Vol. 1402: C. Carasso, P. Charrier, B. Hanouzet, J.-L. Joly (Eds.), Nonlinear Hyperbolic Problems. Proceedings, 1988. V, 249 pages. 1989.

Vol. 1403: B. Simeone (Ed.), Combinatorial Optimization. Seminar, 1986. V, 314 pages. 1989.

Vol. 1404: M.-P. Malliavin (Ed.), Séminaire d'Algèbre Paul Dubreil et Marie-Paul Malliavin. Proceedings, 1987 – 1988. IV, 410 pages. 1989.

Vol. 1405: S. Dolecki (Ed.), Optimization. Proceedings, 1988. V, 223 pages. 1989.

Vol. 1406: L. Jacobsen (Ed.), Analytic Theory of Continued Fractions III. Proceedings, 1988. VI, 142 pages. 1989.

Vol. 1407: W. Pohlers, Proof Theory. VI, 213 pages. 1989.

Vol. 1408: W. Lück, Transformation Groups and Algebraic K-Theory. XII, 443 pages. 1989.

Vol. 1409: E. Hairer, Ch. Lubich, M. Roche. The Numerical Solution of Differential-Algebraic Systems by Runge-Kutta Methods. VII, 139 pages. 1989.

Vol. 1410: F.J. Carreras, O. Gil-Medrano, A.M. Naveira (Eds.), Differential Geometry. Proceedings, 1988. V, 308 pages. 1989.

Vol. 1411: B. Jiang (Ed.), Topological Fixed Point Theory and Applications. Proceedings, 1988. VI, 203 pages. 1989.

Vol. 1412: V.V. Kalashnikov, V.M. Zolotarev (Eds.), Stability Problems for Stochastic Models. Proceedings, 1987. X, 380 pages. 1989.

Vol. 1413: S. Wright, Uniqueness of the Injective III₁ Factor. III, 108 pages. 1989.

Vol. 1414: E. Ramírez de Arellano (Ed.), Algebraic Geometry and Complex Analysis. Proceedings, 1987. VI, 180 pages. 1989.

Vol. 1415: M. Langevin, M. Waldschmidt (Eds.), Cinquante Ans de Polynômes. Fifty Years of Polynomials. Proceedings, 1988. IX, 235 pages. 1990.

Vol. 1416: C. Albert (Ed.), Géométrie Symplectique et Mécanique. Proceedings, 1988. V, 289 pages. 1990.

Vol. 1417: A.J. Sommese, A. Biancofiore, E.L. Livorni (Eds.). Algebraic Geometry. Proceedings, 1988. V, 320 pages. 1990.

Vol. 1418: M. Mimura, Homotopy Theory and Related Topics. Proceedings, 1988. V, 241 pages. 1990.

Vol. 1419: P.S. Bullen, P.Y. Lee, J.L. Mawhin, P. Muldowney, W.F. Pfeffer (Eds.), New Integrals. Proceedings, 1988. V, 202 pages. 1990.

Vol. 1420: M. Galbiati, A. Tognoli (Eds.), Real Analytic Geometry. Proceedings, 1988. IV, 366 pages. 1990.

Vol. 1421: H.A. Biagioni, A Nonlinear Theory of Generalized Functions. XII, 214 pages. 1990.

Vol. 1422: V. Villani (Ed.), Complex Geometry and Analysis. Proceedings, 1988. V, 109 pages. 1990.

Vol. 1423: S.O. Kochman, Stable Homotopy Groups of Spheres: A Computer-Assisted Approach. VIII, 330 pages. 1990.

Vol. 1424: F.E. Burstall, J.H. Rawnsley, Twistor Theory for Riemannian Symmetric Spaces. III, 112 pages. 1990.

Vol. 1425: R.A. Piccinini (Ed.), Groups of Self-Equivalences and Related Topics. Proceedings, 1988. V, 214 pages. 1990.

Vol. 1426: J. Azéma, P.A. Meyer, M. Yor (Eds.), Séminaire de Probabilités XXIV 1988/89. V, 490 pages. 1990.

Vol. 1428: K. Erdmann, Blocks of Tame Representation Type and Related Algebras. XV, 312 pages. 1990.

Vol. 1429: S. Homer, A. Nerode, R.A. Platek, G.E. Sacks, A. Scedrov, Logic and Computer Science. Seminar, 1988. Editor: P. Odifreddi. V, 162 pages. 1990.

Vol. 1430: W. Bruns, A. Simis (Eds.), Commutative Algebra. Proceedings, 1988. V, 160 pages. 1990.

Vol. 1431: J.G. Heywood, K. Masuda, R. Rautmann, V.A. Solonnikov (Eds.), The Navier-Stokes Equations – Theory and Numerical Methods. Proceedings, 1988. VII, 238 pages. 1990.